Visual Analysis of Behaviour

Shaogang Gong · Tao Xiang

Visual Analysis of Behaviour

From Pixels to Semantics

 Springer

Shaogang Gong
School of Electronic Engineering
and Computer Science
Queen Mary University of London
Mile End Rd.
London, E1 4NS
UK
sgg@eecs.qmul.ac.uk

Tao Xiang
School of Electronic Engineering
and Computer Science
Queen Mary University of London
Mile End Rd.
London, E1 4NS
UK
txiang@eecs.qmul.ac.uk

ISBN 978-0-85729-669-6   e-ISBN 978-0-85729-670-2
DOI 10.1007/978-0-85729-670-2
Springer London Dordrecht Heidelberg New York

British Library Cataloguing in Publication Data
A catalogue record for this book is available from the British Library

Library of Congress Control Number: 2011929785

Cover design: VTeX UAB, Lithuania

Printed on acid-free paper

Springer is part of Springer Science+Business Media (www.springer.com)

To Aleka, Philip and Alexander
Shaogang Gong

To Ning and Rachel
Tao Xiang

Preface

The human visual system is able to visually recognise and interpret object behaviours under different conditions. Yet, the goal of building computer vision based recognition systems with comparable capabilities has proven to be very difficult to achieve. Computational modelling and analysis of object behaviours through visual observation is inherently ill-posed. Many would argue that our cognitive understanding remains unclear about why we associate certain semantic meanings with specific object behaviours and activities. This is because meaningful interpretation of a behaviour is subject to the observer's a priori knowledge, which is at times rather ambiguous. The same behaviour may have different semantic meanings depending upon the context within which it is observed. This ambiguity is exacerbated when many objects are present in a scene. Can a computer based model be constructed that is able to extract all necessary information for describing a behaviour from visual observation alone? Do people behave differently in the presence of others, and if so, how can a model be built to differentiate the expected normal behaviours from those of abnormality? Actions and activities associated with the same behavioural interpretation may be performed differently according to the intended meaning, and different behaviours may be acted in a subtly similar way. The question arises as to whether these differences can be accurately measured visually and robustly computed consistently for meaningful interpretation of behaviour.

Visual analysis of behaviour requires not only to solve the problems of object detection, segmentation, tracking, motion trajectory analysis, but also the modelling of context information and utilisation of non-sensory knowledge when available, such as human annotation of input data or relevance feedback to output signals. Visual analysis of behaviour faces two fundamental challenges in computational complexity and uncertainty. Object behaviours in general exhibit complex spatio-temporal dynamics in a highly dynamical and uncertain environment, for instance, human activities in a crowded public space. Segmenting and modelling human actions and activities in a visual environment is inherently ill-posed, as information processing in visual analysis of behaviour is subject to noise, incompleteness and uncertainty in sensory data. Whilst these visual phenomena are difficult to model analytically, they can be probabilistically modelled much more effectively through statistical machine learning.

Despite these difficulties, it is compelling that one of the most significant developments in computer vision research over the last 20 years has been the rapidly growing interest in automatic visual analysis of behaviour in video data captured from closed-circuit television (CCTV) systems installed in private and public spaces. The study of visual analysis of behaviour has had an almost unique impact on computer vision and machine learning research at large. It raises many challenges and provides a testing platform for examining some difficult problems in computational modelling and algorithm design. Many of the issues raised are relevant to dynamic scene understanding in general, multivariate time series analysis and statistical learning in particular.

Much progress has been made since the early 1990s. Most noticeably, statistical machine learning has become central to computer vision in general, and to visual analysis of behaviour in particular. This is strongly reflected throughout this book as one of the underlying themes. In this book, we study plausible computational models and tractable algorithms that are capable of automatic visual analysis of behaviour in complex and uncertain visual environments, ranging from well-controlled private spaces to highly crowded public scenes. The book aims to reflect the current trends, progress and challenges on visual analysis of behaviour. We hope this book will not only serve as a sampling of recent progress but also highlight some of the challenges and open questions in automatic visual analysis of object behaviour.

There is a growing demand by both governments and commerce worldwide for advanced imaging and computer vision technologies capable of automatically selecting and identifying behaviours of objects in imagery data captured in both public and private spaces for crime prevention and detection, public transport management, personalised healthcare, information management and market studies, asset and facility management. A key question we ask throughout this book is how to design automatic visual learning systems and devices capable of extracting and mining salient information from vast quantity of data. The algorithm design characteristics of such systems aim to provide, with minimum human intervention, machine capabilities for extracting relevant and meaningful semantic descriptions of salient objects and their behaviours for aiding decision-making and situation assessment.

There have been several books on human modelling and visual surveillance over the years, including *Face Detection and Gesture Recognition for Human-Computer Interaction* by Yang and Ahuja (2001); *Analyzing Video Sequences of Multiple Humans* by Ohya, Utsumi and Yamato (2002); *A Unified Framework for Video Summarization, Browsing and Retrieval: with Applications to Consumer and Surveillance Video* by Xiong, Radhakrishnan, Divakaran, Rui and Huang (2005); *Human Identification based on Gait* by Nixon, Tan and Chellapa (2005); and *Automated Multi-Camera Surveillance* by Javed and Shah (2008). There are also a number of books and edited collections on behaviour studies from cognitive, social and psychological perspectives, including *Analysis of Visual Behaviour* edited by Ingle, Goodale and Mansfield (1982); *Hand and Mind: What Gestures Reveal about Thought* by McNeill (1992); *Measuring Behaviour* by Martin (1993); *Understanding Human Behaviour* by Mynatt and Doherty (2001); and *Understanding Human Behaviour and the Social Environment* by Zastrow and Kirst-Ashman (2003). However, there has been no book that provides a comprehensive and unified treatment of visual analysis of behaviour from a computational modelling and algorithm design perspective.

This book has been written with an emphasis on computationally viable approaches that can be readily adopted for the design and development of intelligent computer vision systems for automatic visual analysis of behaviour. We present what is fundamentally a computational algorithmic approach, founded on recent advances in visual representation and statistical machine learning theories. This approach should also be attractive to researchers and system developers who would like to both learn established techniques for visual analysis of object behaviour, and gain insight into up-to-date research focus and directions for the coming years. We hope that this book succeeds in providing such a treatment of the subject useful not only for the academic research communities, both also the commerce and industry.

Overall, the book addresses a broad range of behaviour modelling problems, from established areas of human facial expression, body gesture and action analysis to emerging new research topics in learning group activity models, unsupervised behaviour profiling, hierarchical behaviour discovery, learning behavioural context, modelling rare behaviours, 'man-in-the-loop' active learning of behaviours, multi-camera behaviour correlation, person re-identification, and 'connecting-the-dots' for global abnormal behaviour detection. The book also gives in depth treatment to some popular computer vision and statistical machine learning techniques, including the Bayesian information criterion, Bayesian networks, 'bag-of-words' representation, canonical correlation analysis, dynamic Bayesian networks, Gaussian mixtures, Gibbs sampling, hidden conditional random fields, hidden Markov models, human silhouette shapes, latent Dirichlet allocation, local binary patterns, locality preserving projection, Markov processes, probabilistic graphical models, probabilistic topic models, space-time interest points, spectral clustering, and support vector machines.

The computational framework presented in this book can also be applied to modelling behaviours exhibited by many other types of spatio-temporal dynamical systems, either in isolation or in interaction, and therefore can be beneficial to a wider range of fields of studies, including Internet network behaviour analysis and profiling, banking behaviour profiling, financial market analysis and forecasting, bioinformatics, and human cognitive behaviour studies.

We anticipate that this book will be of special interest to researchers and academics interested in computer vision, video analysis and machine learning. It should be of interest to industrial research scientists and commercial developers keen to exploit this emerging technology for commercial applications including visual surveillance for security and safety, information and asset management, public transport and traffic management, personalised healthcare in assisting elderly and disabled, video indexing and search, human computer interaction, robotics, animation and computer games. This book should also be of use to post-graduate students of computer science, mathematics, engineering, physics, behavioural science, and cognitive psychology. Finally, it may provide government policy makers and commercial managers an informed guide on the potentials and limitations in deploying intelligent video analytics systems.

The topics in this book cover a wide range of computational modelling and algorithm design issues. Some knowledge of mathematics would be useful for the reader. In particular, it would be convenient if one were familiar with vectors and matrices, eigenvectors and eigenvalues, linear algebra, optimisation, multivariate analysis, probability, statistics and calculus at the level of post-graduate mathematics. However, the non-mathematically inclined reader should be able to skip over many of the equations and still understand much of the content.

London, UK   Shaogang Gong
                                                                         Tao Xiang

Acknowledgements

We shall express our deep gratitude to the many people who have helped us in the process of writing this book. The experiments described herein would not have been possible without the work of PhD students and postdoctoral research assistants at Queen Mary University of London. In particular, we want to thank Chen Change Loy, Tim Hospedales, Jian Li, Wei-Shi Zheng, Jianguo Zhang, Caifeng Shan, Yogesh Raja, Bryan Prosser, Matteo Bregonzio, Parthipan Siva, Lukasz Zalewski, Eng-Jon Ong, Jamie Sherrah, Jeffrey Ng, and Michael Walter for their contributions to this work. We are indebted to Alexandra Psarrou who read a draft carefully and gave us many helpful comments and suggestions.

We shall thank Simon Rees and Wayne Wheeler at Springer for their kind help and patience during the preparation of this book. The book was typeset using LATEX.

We gratefully acknowledge the financial support that we have received over the years from UK EPSRC, UK DSTL, UK MOD, UK Home Office, UK TSB, US Army Labs, EU FP7, the Royal Society, BAA, and QinetiQ. Finally, we shall thank our families and friends for all their support.

Contents

Acronyms

1D	one-dimensional
2D	two-dimensional
3D	three-dimensional
BIC	Bayesian information criterion
CCA	canonical correlation analysis
CCTV	closed-circuit television
CONDENSATION	conditional density propagation
CRF	conditional random field
EM	expectation-maximisation
DBN	dynamic Bayesian network
HCI	human computer interaction
HCRF	hidden conditional random field
HOG	histogram of oriented gradients
FACS	facial action coding system
FOV	field of view
FPS	frame per second
HMM	hidden Markov model
KL	Kullback–Leibler
LBP	local binary patterns
LDA	latent Dirichlet allocation
LPP	locality preserving projection
MAP	maximum a posteriori
MCMC	Markov chain Monte Carlo
MLE	maximum likelihood estimation
MRF	Markov random field
PCA	principal component analysis
PGM	probabilistic graphical model
PTM	probabilistic topic model
PTZ	pan-tilt-zoom
SIFT	scale-invariant feature transform
SLPP	supervised locality preserving projection
SVM	support vector machine
xCCA	cross canonical correlation analysis

Part I
Introduction

Chapter 1
About Behaviour

Understanding and interpreting behaviours of objects, and in particular those of humans, is central to social interaction and communication. Commonly, one considers that behaviours are the actions and reactions of a person or animal in response to external or internal stimuli. There is, however, a plethora of wider considerations of what behaviour is, ranging from economical (Simon 1955), organisational (Rollinson 2004), social (Sherman and Sherman 1930), to sensory attentional such as *visual behaviour* (Ingle et al. 1982). Visual behaviour refers to the actions or reactions of a sensory mechanism in response to a visual stimulus, for example, the navigation mechanism of nocturnal bees in dim light (Warrant 2008), visual search by eye movement of infants (Gough 1962) or drivers in response to their surrounding environment (Harbluk and Noy 2002). If visual behaviour as a search mechanism is a perceptual function that scans actively a visual environment in order to focus attention and seek an object of interest among distracters (Ltti et al. 1998), *visual analysis of behaviour* is a perceptual task that interprets actions and reactions of objects, such as people, interacting or co-existing with other objects in a visual environment (Buxton and Gong 1995; Gong et al. 2002; Xiang and Gong 2006). The study of visual analysis of behaviour, and in particular of human behaviour, is the focus of this book.

Recognising objects visually by behaviour and activity rather than shape and size plays an important role in a primate visual system (Barbur et al. 1980; Schiller and Koerner 1971; Weiskrantz 1972). In a visual environment of multiple objects co-existing and interacting, it becomes necessary to identify objects not only by their appearance but also by what they do. The latter provides richer information about objects especially when visual data are spatially ambiguous and incomplete. For instance, some animals, such as most snakes, have a very poor visual sensing system, which is unable to capture sufficient visual appearance of objects but very sensitive to movements for detecting preys and predators. The human visual system is highly efficient for scanning through large quantity of low-level imagery data and selecting salient information for a high-level semantic interpretation and gaining situational awareness.

S. Gong, T. Xiang, *Visual Analysis of Behaviour*,
DOI 10.1007/978-0-85729-670-2_1, © Springer-Verlag London Limited 2011

1.1 Understanding Behaviour

Since 1970s, the computer vision community has endeavoured to bring about intelligent perceptual capabilities to artificial visual sensors. Computer vision aims to build artificial mechanisms and devices capable of mimicking the sensing capabilities of biological vision systems (Marr 1982). This endeavour is intensified in recent years by the need for understanding massive quantity of video data, with the aim to not only comprehend objects spatially in a snapshot but also their spatio-temporal relations over time in a sequence of images. For understanding a dynamically changing social environment, a computer vision system can be designed to interpret behaviours from object actions and interactions captured visually in that environment. A significant driver for visual analysis of behaviour is automated visual surveillance, which aims to automatically interpret human activities and detect unusual events that could pose a threat to public security and safety.

If a behaviour is considered the way how an object acts, often in relation to other objects in the same visual environment, the focus of this book is on visual analysis of human behaviour and behaviours of object that are manipulated by humans, for example, vehicles driven by people. There are many interchangeable terms used in the literature concerning behaviour, including activities, actions, events, and movements. They correspond to different spatial and temporal context within which a behaviour can be defined. One may consider a behaviour hierarchy of three layers:

1. Atomic actions correspond to instantaneous atomic entities upon which an action is formed. For example, in a running action, the atomic action could be 'left leg moving in front of the right leg'. In a returning action in tennis, it could be 'swing right hand' followed by 'rotating the upper body'.
2. Actions correspond to a sequence of atomic actions that fulfil a function or purpose. For instance, walking, running, or serving a tennis ball.
3. Activities are composed of sequences of actions over space and time. For example, 'a person walking from a living room to a kitchen to fetch a cup of water', or 'two people playing tennis'. Whilst actions are likely associated with a single object in isolation, activities are almost inevitably concerned with either interactions between objects, or an object engaging with the surrounding environment.

In general, visual analysis of behaviour is about constructing models and developing devices for automatic analysis and interpretation of object actions and activities captured in a visual environment. To that end, visual analysis of behaviour focuses on three essential functions:

1. Representation and modelling: To extract and encode visual information from imagery data in a more concise form that also captures intrinsic characteristics of objects of interest;
2. Detection and classification: To discover and search for salient, perhaps also unique, characteristics of certain object behaviour patterns from large quantity of visual observations, and to discriminate them against known categories of semantic and meaningful interpretation;

3. Prediction and association: To forecast future events based on the past and current interpretation of behaviour patterns, and to forge object identification through behavioural expectation and trend.

We consider that automated visual analysis of behaviour is information processing of visual data, capable of not only modelling previously observed object behaviours, but also detecting, recognising and predicting unseen behavioural patterns and associations.

1.1.1 Representation and Modelling

A human observer can recognise behaviours of interest directly from visual observation. This suggests that imagery data embed useful information for semantic interpretation of object behaviour. Behaviour representation addresses the question of what information must be extracted from images and in what form, so that object behaviour can be understood and recognised visually. The human visual system utilises various visual cues and contextual information for recognising objects and their behaviour (Humphreys and Bruce 1989). For instance, the specific stripe pattern and its colour is a useful cue for human to recognise a tiger and distinguish it from other cats, such as lions. Similarly, the movement as well as posture of a tiger can reveal its intended action: running, walking, or about to strike. It is clear that different sources and types of visual information need be utilised for modelling and understanding object behaviour.

A behaviour representation needs to accommodate both cumulative and temporal information about an object. In order to recognise an object and its behaviour, the human visual system relates any visual stimuli falling on to the retina to a set of knowledge and expectation about the object under observation: how it *should* look like and how it is *supposed* to behave (Gregory 1970; von Helmholtz 1962). Behaviour representation should address the need for extracting visual information that can facilitate the association of visual observation with semantic interpretation. In other words, representation of visual data is part of a computational mechanism that contributes towards constructing and accumulating knowledge about behaviour. For example, modelling the action of a person walking can be considered as to learn the prototypical and generic knowledge of walking based on limited observations of walking examples, so that when an unseen instance of walking is observed, it can be recognised by utilising the accumulated a priori knowledge. An important difference between object modelling and behaviour modelling is that a behaviour model should benefit more from capturing temporal information about behaviour. Object recognition in large only considers spatial information. For instance, a behaviour model is built based on visual observation of a person's daily routine in an office which consists of meetings, tea breaks, paper works and a lunch at certain times of every day. What has been done so far can then have a significant influence on the correct interpretation of what this person is about to do (Agre 1989).

A computational model of behaviour performs both representation and matching. For representing object behaviour, one considers a model capable of capturing distinctive characteristics of an object in action and activity. A good behaviour representation aims to describe an object sufficiently well for both generalisation and discrimination in model matching. Model matching is a computational process to either explain away new instances of observation against known object behaviours, considered as its generalisation capacity, or discriminate one type of object behaviour from the others, regarded as its discrimination ability. For effective model matching, a representation needs to separate visual observation of different object behaviour types or classes in a representational space, and to maintain such separations given noisy and incomplete visual observations.

1.1.2 Detection and Classification

Generally speaking, visual classification is a process of categorising selected visual observations of interest into known classes. Classification is based on an assumption that segmentation and selection of interesting observations have already been taken place. On the other hand, visual detection aims to discover and locate patterns of interest, regardless of class interpretation, from a vast quantify of visual observations. For instance, for action recognition, a model is required to detect and segment instances of actions from a continuous observation of a visual scene. Detection in crowded scenes, such as detecting people fighting or falling in crowd, becomes challenging as objects of interest can be swamped by distracters and background clutters. To spot and recognise actions from a sea of background activities, the task of detection often poses a greater challenge than classification.

The problem of behaviour detection is further compounded when the behaviour to be detected is unknown a priori. A common aim of visual analysis of behaviour is to learn a model that is capable of detecting unseen abnormal behaviour patterns whilst recognising novel instances of known normal behaviour patterns. To that end, an anomaly is defined as an atypical and un-random behaviour pattern not represented by sufficient observations. However, in order to differentiate anomaly from trivial unseen instances or outright statistical outliers, one should consider that an anomaly satisfies a specificity constraint to known normal behaviours, i.e. true anomalies lie in the vicinity of known normal behaviours without being recognised as any.

1.1.3 Prediction and Association

An activity is usually formed by a series of object actions executed following certain temporal order at certain durations. Moreover, the ordering and durations of constituent actions can be highly variable and complex. To model such visual observations, a behaviour can be considered as a temporal process, or a time series function. An important feature of a model of temporal processes is to make prediction. To that end, behaviour prediction is concerned with detecting a future occurrence of a known behaviour based on visual observations so far. For instance, if the daily routine of a person's activities in a kitchen during breakfast time is well understood and modelled, the model can facilitate prediction of this person's next action when certain actions have been observed: the person could be expected to make coffee after finishing frying an egg. Behaviour prediction is particularly useful for explaining away partial observations, for instance, in a crowded scene when visual observation is discontinuous and heavily polluted, or for detecting and preventing likely harmful events before they take place.

Visual analysis of behaviour can assist object identification by providing contextual knowledge on how objects of interest should behave in addition to how they look. For instance, human gait describes the way people walk and can be a useful means to identify different individuals. Similarly, the way people perform different gestures may also reveal their identities. Behaviour analysis can help to determine when and where a visual identification match is most likely to be valid and relevant. For instance, in a crowded public place such as an airport arrival hall, it is infeasible to consider facial imagery identification for all the people all the time. A key to successful visual identification in such an environment is effective visual search. Behaviour analysis can assist in determining when and where objects of interest should be sought and matched against. Moreover, behaviour analysis can provide focus of attention for visual identification. Detecting people acting out of norm can activate identification with improved effectiveness and efficiency. Conversely, in order to derive a semantic interpretation of an object's behaviour, knowing what and who the object is can help. For instance, a train station staff's behaviour can be distinctively different from that of a normal passenger. Recognising a person as a member of staff in a public space can assist in interpreting correctly the behaviour of the person in question.

1.2 Opportunities

Automated visual analysis of behaviour provides some key building blocks towards an artificial intelligent vision system. To experiment with computational models of behaviour by constructing automatic recognition devices may help us with better understanding of how the human visual system bridges sensory mechanisms and semantic understanding. Behaviour analysis offers a great deal of attractive opportunities for application, despite that deploying automated visual analysis of behaviour to a realistic environment is still at its infancy. Here we outline some of the emerging applications for automated visual analysis of behaviour.

1.2.1 Visual Surveillance

There has been an accelerated expansion of closed-circuit television (CCTV) surveillance in recent years, largely in response to rising anxieties about crime and its threat to security and safety. Visual surveillance is to monitor the behaviour of people or other objects using visual sensors, typically CCTV cameras. Substantial amount of surveillance cameras have been deployed in public spaces, ranging from transport infrastructures, such as airports and underground stations, to shopping centres, sport arenas and residential streets, serving as a tool for crime reduction and risk management. Conventional video surveillance systems rely heavily on human operators to monitor activities and determine the actions to be taken upon occurrence of an incident, for example, tracking suspicious target from one camera to another camera, or alerting relevant agencies to areas of concern. Unfortunately, many actionable incidents are simply mis-detected in such a manual system due to inherent limitations from deploying solely human operators eyeballing CCTV screens. These limitations include: (1) excessive number of video screens to monitor, (2) boredom and tiredness due to prolonged monitoring, (3) lack of a priori and readily accessible knowledge on what to look for, and (4) distraction by additional operational responsibilities. As a result, surveillance footages are often used merely as passive records for post-event investigation. Mis-detection of important events can be perilous in critical surveillance tasks such as border control or airport surveillance. It has become an operational burden to screen and search exhaustively colossal amount of video data generated from growing number of cameras in public spaces. Automated computer vision systems for visual analysis of behaviour provide the potential for deploying never-tiring computers to perform routine video analysis and screening tasks, whilst assisting human operators to focus attention on more relevant threats, thus improving the efficiency and effectiveness of a surveillance system.

1.2.2 Video Indexing and Search

We are living in a digital age with huge amount of digital media, especially videos, being generated at every single moment in the forms of surveillance videos, online news footages, home videos, mobile videos, and broadcasting videos. However, once generated, very rarely they are watched. For instance, most visual data collected by surveillance systems are never watched. The only time when they are examined is when a certain incident or crime has occurred and a law enforcement organisation needs to perform a post-event analysis. Unless specific time of the incident is known, it is extremely difficult to search for an event such as someone throws a punch in front of a nightclub. For a person with a large home video collection, it is a tedious and time consuming task to indexing the videos so that they can be searched efficiently for footages of a certain type of actions or activities from years gone by. For film or TV video archive, it is also a very challenging task to search for a specific footage without text meta information, specific knowledge about the name of a subject, or the time of an event. What is missing and increasingly desired is the ability to visually search archives by what has happened, that is, automated visual search of object behaviours with categorisation.

1.2.3 Robotics and Healthcare

A key area for robotics research in recent years is to develop autonomous robots that can see and interact with people and objects, known as social robots (Breazeal 2002). Such a robot may provide a useful device in serving an aging society, as a companion and tireless personal assistant to elderly people or people with a disability. In order to interact with people, a robot must be able to understand the behaviour of the person who is interacting with. This ability can be based on gesture recognition, such as recognising waving and initialising a hand-shake, interpreting facial expression, and inferring intent by body posture. Earlier robotics research had focused more on static object recognition, manipulation and navigation through a stationary environment. More recently, there has been a shift towards developing robots capable of mimicking human behaviour and interacting with people. To that end, enabling a robot to perform automated visual analysis of human behaviour becomes essential. Related to the development of a social robot, personalised healthcare in an aging society has gained increasing prominence in recent years. To be able to collect, disseminate and make sense of sensory information from and to an elderly person in a timely fashion is the key. To that end, automated visual analysis of human behaviour can provide quantitative and routine assessment of a person's behavioural status needed for personalised illness detection and incident detection, e.g. a fall. Such sensor-based systems can reduce the cost of providing personalised health care, enabling elderly people to lead a more healthy and socially inclusive life style (Yang 2006).

1.2.4 Interaction, Animation and Computer Games

Increasingly more intelligent and user friendly human computer interaction (HCI) are needed for applications such as a game console that can recognise a player's gesture and intention using visual sensors, and a teleconferencing system that can control cameras according to the behaviour of participants. In such automated HCI systems using sensors, effective visual analysis of human behaviour is central to meaningful interaction and communication. Not surprisingly, animation for film production and gaming industries are also relying more on automated visual analysis of human behaviour for creating special effects and building visual avatars that can interact with players. By modelling human behaviour including gesture and facial expression, animations can be generated to create virtual characters, known as avatars, in films and for computer games. With players' behaviour recognised automatically, these avatars can also interact with players in real-time gaming.

1.3 Challenges

Understanding object behaviour from visual observation alone is challenging because it is intrinsically an ill-posed problem. This is equally true for both humans and computers. Visual interpretation of behaviour can be ambiguous and is subject to changing context. Visually identical behaviours may have different meanings depending on the environment in which activities are taken place. For instance, when a person is seen waving on a beach, is he greeting somebody? swatting an insect? or calling for help as his friend is drowning? In general, visual analysis of behaviour faces two fundamental challenges.

1.3.1 Complexity

Compared to object recognition in static images, an extra dimension of time needs be considered in modelling and explaining object behaviour. This makes the problem more complex. Let us consider human behaviour as an example. Human has an articulated body and the same category of body behaviours can be acted in different ways largely due to temporal variations, for example, waving fast versus slowly. This results in behaviours of identical semantics look visually different, known as large intra-class variation. On the other hand, behaviours of different semantic classes, such as jogging versus running, can be visually similar, known as small inter-class variation. Beyond single-object behaviour, a behaviour can be of multiple interacting objects characterised by their temporal ordering. In a more extreme case, a behaviour is defined in the context of a crowd where many people co-exist both spatially and temporally. In general, behaviours are defined in different spatial and temporal context.

1.3.2 Uncertainty

Based on visual information alone to describe object behaviour is inherently partial and incomplete. Unlike a human observer, when a computer is asked to interpret behaviour without other sources of information except imagery data, the problem is compounded by visual information only available in two-dimensional images of a three-dimensional space, lack of contextual knowledge, and in the presence of imaging noise.

Two-dimensional visual data give rise to visual occlusion on objects under observation. This renders not all behavioural information can be observed visually. For instance, for two people interacting with each other, depending on the camera angle, almost inevitably part of or the full body of a person is self-occluded. As a result, semantic interpretation of behaviour is made considerably harder when only partial information is available.

Behaviour interpretation is highly context dependent. However, contextual information is not always directly observable, nor necessarily always visual. For instance, on a motorway when there is a congestion, a driver often wishes to find out the cause of the congestion in order to estimate likely time delay, whether the congestion is due to an accident or road work ahead. However, that information is often unavailable in the driver's field of view, as it is likely to be located miles away. Taking another example, on a train platform, passengers start to leave the platform due to an announcement by the station staff that the train line is closed due to signal failure. This information is in audio form therefore not captured by visual observation on passenger behaviours. To interpret behaviour by visual information alone introduces additional uncertainty due to a lack of access to non-visual contextual knowledge.

Visual data are noisy, either due to sensor limitations or because of operational constraints. This problem is particularly acute for video based behaviour analysis when video resolution is often very low both spatially and temporally. For instance, a typical 24-hours 7-days video surveillance system in use today generates video footages with a frame-rate of less than three frames per second, and with heavy compression for saving storage space. Imaging noise degrades visual details available for analysis. This can further cause visual information processing to introduce additional error. For instance, if object trajectories are used for behaviour analysis, object tracking errors can increase significantly in low frame-rate and highly compressed video data.

1.4 The Approach

We set out the scope of this book by introducing the problem of visual analysis of behaviour. We have considered the core functions of behaviour analysis from a computational perspective, and outlined the opportunities and challenges for visual analysis of behaviour. In the remaining chapters of Part I, we first give an overview on different domains of visual analysis of behaviour to highlight the importance and relevance of understanding behaviour in context. This is followed by an introduction to some of the core computational and machine learning concepts used throughout the book. Following Part I, the book is organised into further three parts according to the type of behaviour and the level of complexity involved, ranging from facial expression, human gesture, single-object action, multiple object activity, crowd behaviour analysis, to distributed behaviour analysis.

Part II describes methods for modelling single-object behaviours including facial expression, gesture, and action. Different representations and modelling tools are considered and their strengths and weaknesses are discussed.

Part III is dedicated to group behaviour understanding. We consider models for exploring context to fulfil the task of behaviour profiling and abnormal behaviour detection. Different learning strategies are investigated, including supervised learning, unsupervised learning, semi-supervised learning, incremental and adaptive learning, weakly supervised learning, and active learning. These learning strategies are designed to address different aspects of a model learning problem in different observation scenarios according to the availability of visual data and human feedback.

Whilst Parts II and III consider behaviours observed from a single camera view, Part IV addresses the problem of understanding distributed behaviours from multiple observational viewpoints. An emphasis is specially placed on non-overlapping multi-camera views. In particular, we investigate the problems of behaviour correlation across camera views for camera topology estimation and global anomaly detection, and the association of people across non-overlapping camera views, known as re-identification.

References

Agre, P.E.: The dynamic structure of everyday life. PhD thesis, Massachusetts Institute of Technology, Cambridge, MA, USA (1989)

Barbur, J.L., Ruddock, K.H., Waterfield, V.A.: Human visual responses in the absence of the geniculo-calcarine projection. Brain **103**(4), 905–928 (1980)

Breazeal, C.L. (ed.): Desiging Sociable Robots. MIT Press, Cambridge (2002)

Buxton, H., Gong, S.: Visual surveillance in a dynamic and uncertain world. Artif. Intell. **78**(1–2), 431–459 (1995)

Gong, S., Ng, J., Sherrah, J.: On the semantics of visual behaviour, structured events and trajectories of human action. Image Vis. Comput. **20**(12), 873–888 (2002)

Gough, D.: The visual behaviour of infants in the first few weeks of life. Proc. R. Soc. Med. **55**(4), 308–310 (1962)

Gregory, R.L.: The Intelligent Eye. Weidenfeld and Nicolson, London (1970)

Harbluk, J.L., Noy, Y.I.: The impact of cognitive distraction on driver visual behaviour and vehicle control. Technical Report TP 13889 E, Road Safety Directorate and Motor Vehicle Regulation Directorate, Canadian Minister of Transport (2002)

Humphreys, G.W., Bruce, V.: Visual Cognition: Computational, Experimental and Neuropsychological Perspectives. Erlbaum, Hove (1989)

Ingle, D.J., Goodale, M.A., Mansfield, R.J.W. (eds.): Analysis of Visual Behaviour. MIT Press, Cambridge (1982)

Ltti, L., Koch, C., Niebur, E.: A model of saliency-based visual attention for rapid scene analysis. IEEE Trans. Pattern Anal. Mach. Intell. **20**(11), 1254–1259 (1998)

Marr, D.: Vision: A Computational Investigation Into the Human Representation and Processing of Visual Information. Freeman, New York (1982)

Rollinson, D.: Organisational Behaviour and Analysis: An Integrated Approach. Prentice Hall, New York (2004)

Schiller, P.H., Koerner, F.: Discharge characteristics of single units in superior colliculus of the alert rhesus monkey. J. Neurophysiol. **34**(5), 920–935 (1971)

Sherman, M.D., Sherman, I.C.: The process of human behaviour. J. Ment. Sci. **76**, 337–338 (1930)

Simon, H.A.: A behavioral model of rational choice. Q. J. Econ. **69**(1), 99–118 (1955)

von Helmholtz, H.: The recent progress of the theory of vision. In: Popular Scientific Lectures. Dover, New York (1962)

Warrant, E.J.: Seeing in the dark: vision and visual behaviour in nocturnal bees and wasps. J. Exp. Biol. **211**, 1737–1746 (2008)

Weiskrantz, L.: Review lecture: behavioural analysis of the monkey's visual nervous system. Proc. R. Soc. **182**, 427–455 (1972)

Xiang, T., Gong, S.: Beyond tracking: modelling activity and understanding behaviour. Int. J. Comput. Vis. **67**(1), 21–51 (2006)

Yang, G.Z. (ed.): Body Sensor Networks. Springer, Berlin (2006)

Chapter 2
Behaviour in Context

Interpreting behaviour from object action and activity is inherently subject to the context of a visual environment within which action and activity take place. Context embodies not only the spatial and temporal setting, but also the intended functionality of object action and activity (Bar 2004; Bar and Aminoff 2003; Bar and Ullman 1993; Biederman et al. 1982; Palmer 1975; Schwartz et al. 2007). Humans employ visual context extensively for both effective object recognition and behaviour understanding. For instance, one recognises, often by inference, whether a hand-held object is a mobile phone or calculator by its relative position to other body parts such as closeness to the ears, even if they are visually similar and partially occluded by the hand. Similarly for behaviour recognition, the arrival of a bus in busy traffic is more likely to be inferred by looking at the passengers' behaviour at a bus stop. Computer vision research on visual analysis of behaviour embraces a wide range of studies on developing computational models and systems for interpreting behaviour in different context. Here we highlight some well established topics and emerging trends.

2.1 Facial Expression

Behaviour exhibited from a human face is predominantly in the form of facial expression. The ability to recognise the affective state of a person is indispensable and important for successful interpersonal social interaction. Affective arousal modulates all non-verbal communication cues such as facial expressions, body postures and movements. Facial expression is perhaps the most natural and efficient means for humans to communicate their emotions and intentions, as communication is primarily carried out face to face.

Automatic facial expression recognition has attracted much attention from behavioural scientists since the work of Darwin (1872). Suwa et al. (1978) made the first attempt to automatically analyse facial expressions from images. Much progress has been made in the last 30 years towards computer-based analysis and interpretation of facial expression (Bartlett et al. 2005; Chang et al. 2004; Cohen et al. 2003;

Donato et al. 1999; Dornaika and Davoine 2005; Essa and Pentland 1997; Fasel and Luettin 2003; Hoey and Little 2004; Jia and Gong 2006, 2008; Kaliouby and Robinson 2004; Lee and Elgammal 2005; Lyons et al. 1999; Pantic and Patras 2006; Pantic and Rothkrantz 2000, 2004; Shan et al. 2009; Yacoob and Davis 1996; Yeasin et al. 2004; Yin et al. 2004; Zalewski and Gong 2005; Zhang and Ji 2005). A recent trend in human–computer interaction design also considers the desire for affective computing in order to bring about more human-like non-verbal communication skills to computer-based systems (Pantic and Rothkrantz 2003).

Facial expressions can be described at different levels. A widely used description is the Facial Action Coding System (FACS) (Ekman and Friesen 1978), which is a human-observer-based protocol developed to represent subtle changes in facial expressions. With FACS, facial expressions are decomposed into one or more action units. The aim to develop computer-based automatic AU recognition has attracted much attention in the last decade (Donato et al. 1999; Tian et al. 2001; Valstar and Pantic 2006; Zhang and Ji 2005). Psychophysical studies also indicate that basic emotions have corresponding universal facial expressions across all cultures (Ekman and Friesen 1976). This is reflected by most facial expression recognition models attempting to recognise a set of prototypic emotional expressions, including 'disgust', 'fear', 'joy', 'surprise', 'sadness' and 'anger' (Bartlett et al. 2005; Cohen et al. 2003; Lyons et al. 1999; Yeasin et al. 2004; Yin et al. 2004). Example images of different facial expressions are shown in Fig. 2.1.

Although a facial expression is a dynamic process with important information captured in motion (Bassili 1979), imagery features extracted from static face images are often used to represent a facial expression. These features include both geometric features (Valstar and Pantic 2006; Valstar et al. 2005) and appearance features (Bartlett et al. 2005; Donato et al. 1999; Lyons et al. 1999; Tian 2004; Zhang et al. 1998). The rigidity and structural composition of facial action components make facial expression a rather constrained type of behaviour, a raison d'être for extracting information from static images to discriminate different types of facial expression. Efforts have also been made to model facial expression as a dynamical process utilising explicitly spatio-temporal or image sequence information (Cohen et al. 2003; Shan et al. 2005; Zalewski and Gong 2005; Zhang and Ji 2005). These studies aim to explore a richer description of expression dynamics so to better interpret facial emotion and expression change.

2.2 Body Gesture

The ability to interpret human gestures constitutes an essential part of our perception. It reveals the intention, emotional state or even the identity of the people surrounding us and mediates our communication (Pavlovic et al. 1997). Such human activities can be characterised by hand motion patterns and modelled as trajectories in a spatio-temporal space. Figure 2.2 shows some examples of communicative gestures. Gesture recognition usually considers the problem of measuring the similarities between hand motion pattern characteristics, such as their trajectories, and

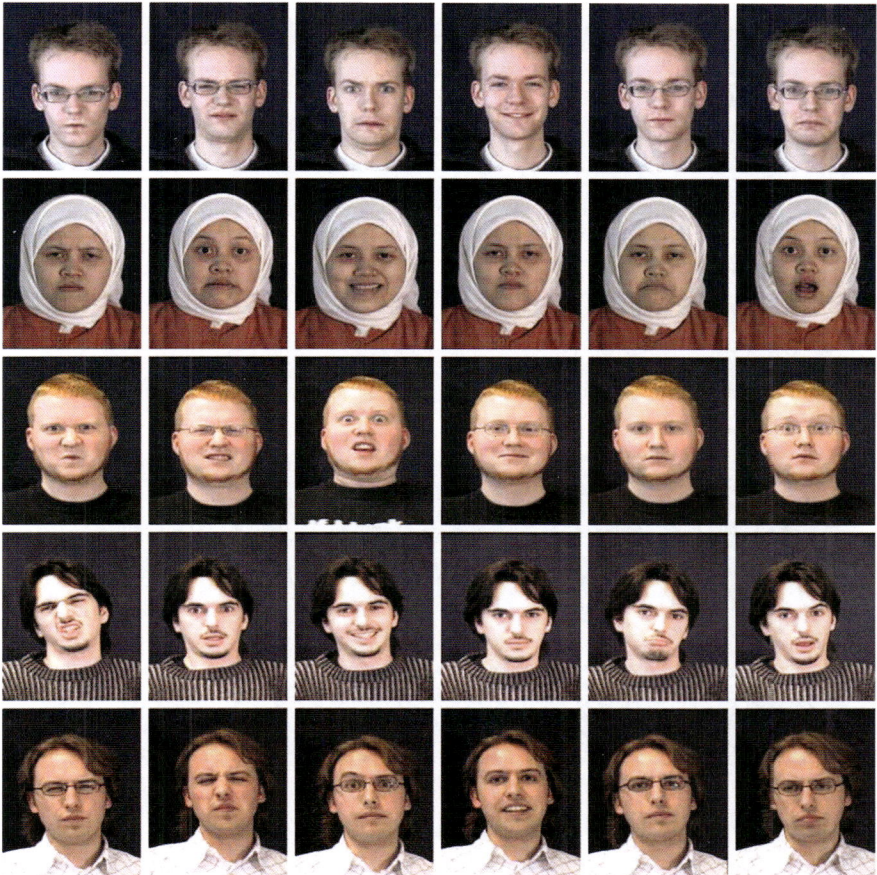

Fig. 2.1 Examples of facial expression images from the MMI database (Pantic et al. 2005)

discriminating different types of such patterns based on an assumption that each type of gesture exhibits a unique or closely similar trajectory with small variations. Based on this simple concept, well-defined gestures can be recognised by computer-based models through matching trajectory templates (Davis and Bobick 1997; McKenna and Gong 1998; Wren et al. 1997). In a well-constrained environment with little background clutter where only a single or a dominant person is assumed, for example, in computer games and close-range human–computer interaction, a holistic gesture representation can be computed without explicit body part tracking. For example, motion moment measurements projected onto the image plane (camera view-plane) axes can be used for a hand-driven interface (Freeman et al. 1996). However, most natural gestures are not well-defined in the sense that there is no clear starting or ending point for a gesture, and multiple gestures form a single continuous pattern with possible abrupt interruptions. Different gestures can also display visually similar motion patterns. To address these problems, gestures

Fig. 2.2 Examples of communicative gestures: 'waving high' (*top row*), 'waving low' (*middle row*) and 'please sit down' (*bottom row*)

Fig. 2.3 Examples of affective body gestures from the FABO database (Gunes and Piccardi 2006): 'fear' (*1st row*), 'joy' (*2nd row*), 'uncertainty' (*3rd row*), and 'surprise' (*4th row*)

can be modelled as stochastic processes under which salient phases are modelled as state transitions. Prior knowledge is built into the models statistically by estimating both state distributions and visual observation covariance at each state, and can be learned from previous observations of similar and different gesture motion patterns, known as learning from training data (Black and Jepson 1998; Bobick and Wilson 1995; Gong et al. 1999; Psarrou et al. 1995, 2002; Schlenzig et al. 1994; Starner and Fentland 1995).

Both human face and body contribute collectively in conveying the emotional state of a person (see examples in Fig. 2.3). The perception of facial expression is strongly influenced by body motion. This is probably due to a high variability in body movement when visually observed, as body language is defined collectively by both body movements including posture (static), gesture (dynamic), and facial movement including expression and gaze patterns. Studies in psychology suggest that the combined visual channels of facial expression and body gesture are more informative than each alone, and their integration is a mandatory process occurring early in the human cognitive process (Ambady and Rosenthal 1992; Meeren et al. 2005). Coulson (2004) shows experiments on attributing six universal emotions to static body postures using computer-generated mannequin figures. These experiments suggest that recognition from static body posture alone is comparable to recognition from voice, and some postures can be recognised as accurately as facial expressions.

Early studies on vision-based gesture recognition were primarily carried out on non-affective gestures, such as sign languages (Wu and Huang 2001). An affective gesture recognition system was introduced by Ravindra De Silva et al. (2006) to recognise children's emotion with a degree of intensity in the context of a game. More recent studies on affective modelling consider both facial expression and body gesture (Balomenos et al. 2005; Gunes and Piccardi 2007; Shan et al. 2007). Kapoor and Picard (2005) describe a multi-sensor affect recognition system for classifying the affective state of interest in children who are solving puzzles. This system combines the extracted sensory information from face videos (expression), a sensor chair (posture), and the state of the puzzle (context). Burgoon et al. (2005b) consider the problem of identifying emotional states from bodily cues for human behaviour understanding. Statistical techniques are often exploited to estimate a set of posture features in discriminating between emotions (Ravindra De Silva and Bianchi-Berthouze 2004). Overall, how to effectively model jointly facial and body behaviour remains an open question.

2.3 Human Action

Automatic human action recognition has become a very active research area in computer vision in recent years with applications in human–computer interaction, video indexing, visual surveillance, sport event analysis and computer games. Automatic action recognition is challenging because actions, similar to human gestures, can be performed by people of different size, appearance, and from different viewpoints. The problem is compounded by occlusion, illumination change, shadow, and possible camera movement. In general, action recognition considers a less constrained visual environment compared to that of gesture recognition. Some examples of actions can be seen in Fig. 2.4.

Action recognition requires a suitable representation. An intuitive approach to representing action can be based on either tracking a spatio-temporal shape of a person (Ali and Aggarwal 2001; Ramanan and Forsyth 2003; Rao and Shah 2001; Sheikh et al. 2005), or constructing a shape template and analysing its spatio-temporal characteristics (Gorelick et al. 2007; Ke et al. 2005; Yilmaz and Shah 2005; Zhang and Gong 2010a, 2010b). This spatio-temporal shape is often computed by extracting the silhouette of a person in motion, which may not be possible in a cluttered visual environment. To overcome this problem, less descriptive and shape-free representations can be considered. They include space-time interest points (Laptev

Fig. 2.4 Example actions in unconstrained environments. From *top* to *bottom*: 'cycling', 'diving', 'soccer juggling', and 'walking with a dog'

2005; Laptev and Lindeberg 2003; Schüldt et al. 2004), optical flow (Efros et al. 2003), motion template (Bobick and Davis 2001), space-time volumes (Yilmaz and Shah 2005), and shape context from still images (Belongie et al. 2002; Wang et al. 2006a).

Compared to other methods, a histogram representation of space-time interest point is particularly robust against noise and partially missing object features. However, a weakness of this representation is that the overall object spatio-temporal structural information is lost due to its histogram nature. A number of recent studies have tried to exploit the use of spatial or temporal structural information about interest points through a correlogram (Liu and Shah 2008), spatial location registration (Gilbert et al. 2008), motion context (Zhang et al. 2008), and spatio-temporal selection (Bregonzio et al. 2009).

Most studies on action recognition are concerned with classifying actions into different types, making an assumption that isolated visual sequences of individual actions are readily detectable and segmented. However, realistic actions are often observed continuously in the presence of other object actions. In such an environment, action detection and segmentation become difficult. Current action detection models are rather simplistic, largely based on template matching, with an action template constructed from previously seen video clips. Significant challenges remain. For instance, Ke et al. (2007) consider a semi-supervised part-based template

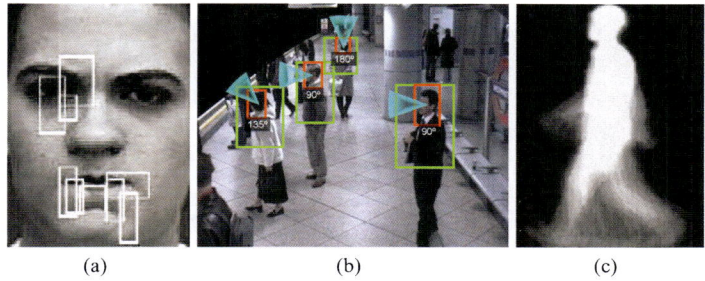

(a)   (b)   (c)

Fig. 2.5 (**a**) Facial expression, (**b**) head pose, and (**c**) gait may be used as behavioural cues to reveal human intent

for action detection. Whilst only a single training sample per action class is required, intensive user interaction including manual segmentation of body parts in each image frame is required for developing a part based shape-flow model. Alternatively, Yang et al. (2009) exploit an auxiliary annotated action dataset to compensate for the lack of detection training data. To cope with large intra-class action variations in detection, weakly supervised model learning strategies have also been studied (Hu et al. 2009; Siva and Xiang 2010; Yuan et al. 2009).

2.4 Human Intent

It could be considered that the emotional state of a person can provide non-verbal visual indicators for gauging one's intent. However, interpretation of human intent by non-verbal cues alone is challenging (Burgoon et al. 2005a; Vrij 2000). Situational body language can be visually difficult to interpret, especially if there is a deliberate attempt to deceive. This is in contrast to communicating emotional states using heightened facial expression and affective gesture. Nevertheless, there is evidence to suggest that non-verbal cues can be correlated with intent, and in particular for facilitating the detection of deceptive intent (Vrij and Mann 2005). From a visual analysis perspective, one may consider detecting intent by modelling the correlations among a range of non-verbal behavioural cues captured from facial expression, eye movement, head pose changing patterns, gesture, and gait (Fig. 2.5).

Visually gauging one's intent from a far-field distance is more difficult than in a close-range environment due to a lack of visual details. In an unconstrained open environment, analysing facial expression and eye movement patterns is unreliable due to low resolution and occlusion. In contrast, body orientation and head pose profiles over space and time can provide more reliable measurements for correlating intent. For instance, on a train platform a person facing the train with his head turning towards another direction may indicate he is checking the information board for ensuring the correct train to board (see Fig. 2.6). Two people slightly tilting their heads towards each other may suggest they are chatting. Furthermore, body orientation and head pose of a person often differ, although they are intrinsically correlated

Fig. 2.6 Visual inference of intent in an unconstrained environment is challenging

due to physiological and behavioural constraints. Both can be visually estimated and analysed collaboratively for better accuracy and robustness.

There is a large body of research on automated human detection and estimation of body and head pose. However, most existing techniques concentrate on a well-controlled environment with high-resolution imagery data (Aghajanian et al. 2009; Andriluka et al. 2009; Gong et al. 1996; Jiang 2009; Lee and Cohen 2004; Ng and Gong 2002; Ozuysal et al. 2009; Sherrah and Gong 2001; Sherrah et al. 2001; Zhang et al. 2007). A few recent studies begin to address the problem of estimating head pose given low-resolution images (Benfold and Reid 2008, 2009; Robertson and Reid 2006) and in crowded scenes (Orozco et al. 2009). For estimating articulated body orientation, one can also exploit the mapping of kinematically constrained object parts to a pictorial structure of body parts appearance (Felzenszwalb and Huttenlocher 2005). Furthermore, joint head pose and body orientation can be estimated to yield a more holistic and reliable measurement towards modelling body language (Gong et al. 2010). In addition to head pose, body posture and orientation, human gait can be considered for intent interpretation. Gait analysis typically involves the characterisation of periodic motion and spatio-temporal patterns of human silhouettes for identity recognition (Bashir et al. 2010; Han and Bhanu 2006; Nixon et al. 2005). It could be argued that analysing correlations among a person's head pose, body orientation, posture and gait, in relation to camera positions and reactions to external events, all play a role for the interpretation of intent of a person in an unconstrained environment using non-verbal cues.

2.5 Group Activity

When a sequence of different actions executed by the same object continue over time, such patterns of behaviour are considered to be activities. The activities of a person in a kitchen could be making a sandwich or washing up after lunch. Each activity consists of a series of ordered actions. For instance, making a sandwich could be composed of opening the fridge to take out bread and fillings, making the sandwich, and wrapping up. When activities involve a group of objects interacting, understanding activities is more concerned with reasoning about inter-action and

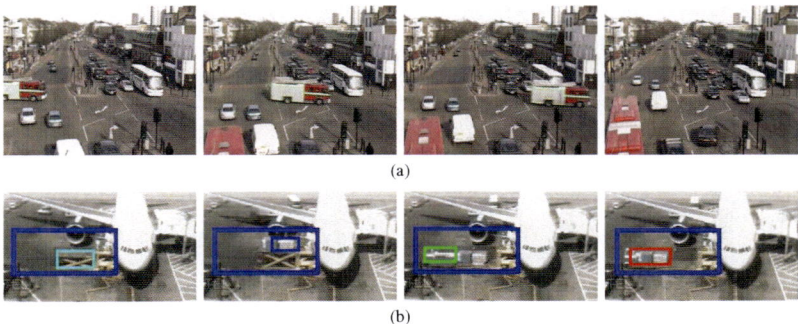

(a)

(b)

Fig. 2.7 Examples of group activities. (**a**) Different types of vehicles and pedestrians negotiate a traffic junction. (**b**) A cargo loading activity for a docked aircraft where a truck brought cargo container boxes onto a cargo lift, which then transferred the boxes into the aircraft. The cargo loading area and objects involved in the activities are highlighted by different bounding boxes

inter-person relationships including temporal order and duration among constituent actions of the individuals involved. Typical examples of group activities include passengers arriving at an airport, going through immigration desks and luggage collection; and vehicles and pedestrians negotiating a traffic junction regulated by traffic lights. Some examples of group activities can be seen in Fig. 2.7.

A fundamental element in understanding group activities is context. Visual context is the environment, background, and settings within which objects and associated behaviours are observed visually. Humans employ visual context extensively for both object recognition in a static setting and behaviour recognition in a dynamic environment. Extensive cognitive, physiological and psychophysical studies suggest that visual context plays a critical role in human visual perception (Bar and Aminoff 2003; Bar and Ullman 1993; Biederman et al. 1982; Palmer 1975). For visual analysis of behaviour, the most relevant visual context is behavioural context, as the interpretation of object behaviour is context dependent (Li et al. 2008b). In particular, a meaningful interpretation of object behaviour depends largely on knowledge of spatial and temporal context defining where and when it occurs, and correlational context specifying the expectation inferred from the correlated behaviours of other objects co-existing in the same scene. Modelling group activity can facilitate learning visual context, which in turn facilitates context-aware abnormal behaviour detection. For example, Li et al. (2008a) study the problem of detecting unusual behaviours that are visually similar to the norm when observed in isolation, but contextually incoherent when both spatial and correlational context are considered. There is a growing interest in modelling visual context for interpretation (Fernández et al. 2010; Kim and Grauman 2009; Kuettel et al. 2010; Loy et al. 2011; Streib and Davis 2010; Van Gool et al. 2009; Zaharescu and Wildes 2010).

Object tracking in a public space is a challenging problem, as object visual features are neither consistent nor unique (Gong and Buxton 1993). Understanding group activity can provide a form of dynamical context for better interpretation of individual object's whereabouts. For instance, Zheng et al. (2009) embed contextual

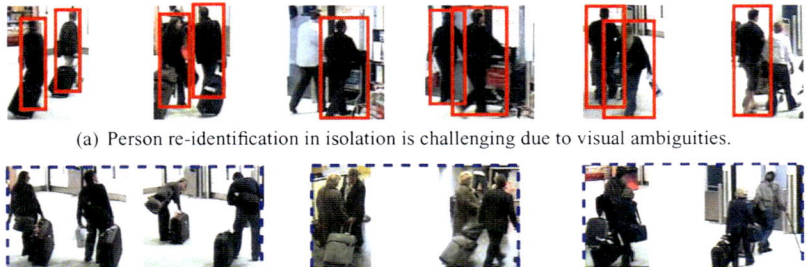

(a) Person re-identification in isolation is challenging due to visual ambiguities.

(b) Associating groups of people can reduce ambiguities in matching.

Fig. 2.8 Visual context extracted from surrounding people could be used to resolve ambiguities in person re-identification tasks

visual knowledge extracted from surrounding people into supporting descriptions for matching people across disjoint and distributed multiple views, known as person re-identification (see examples in Fig. 2.8). In another study, Ali and Shah (2008) exploit scene structure and behaviour of the crowd to assist appearance-based tracking in highly crowded scenes. More generally, tracking-by-detection models exploit either pedestrian or body part detection as categorical contextual information for more reliable tracking (Breitenstein 2009; Okuma et al. 2004; Wu and Nevatia 2007). There are also studies that exploit contextual information around a target object both spatially and temporally to facilitate more robust long-term tracking (Yang et al. 2008).

2.6 Crowd Behaviour

When the size of a group of objects increases to an extent that individual body movement cannot be distinguished, group activity becomes crowd behaviour. For instance, on a crowded train platform, people often move together in a synchronised pattern for departing or boarding a train. Visual analysis of behaviour of a single person becomes less reliable, and less relevant, whilst understanding the crowd behaviour as a whole is of more interest.

Crowd behaviour is characterised by severe occlusions due to excessive number of objects in the scene and a lack of visual details.[1] In particular, in a typical public scene as shown in Fig. 2.9, the sheer number of objects with complex activities cause severe and continuous inter-object occlusions, leading to temporal discontinuity in object movement trajectories. Tracking can be further compromised by low frame rates in image capturing, where large spatial displacement of objects occur between consecutive image frames.

---

[1]Many surveillance systems record videos at less than 5 frames per second (FPS), or compromise image resolution in order to optimise data bandwidth and storage space (Cohen et al. 2006; Kruegle 2006).

Fig. 2.9 Object tracking in crowd is challenging due to significant visual ambigu_ties. Severe inter-object occlusion and lack of visual details at times are key factors that render object tracking infeasible

Many recent studies approach the problem of modelling crowd activity using alternative representations instead of object trajectories in order to circumvent unreliable tracking. Among them, pixel-based representation that extracts pixel-level features such as colour, texture and gradient, emerges as an attractive solution since it avoids explicit object segmentation and tracking, especially applicable for representing activity in crowded scenes.

Crowd analysis can benefit from utilising region based scene decomposition, whereby a wide-area scene is segmented into local regions, followed by a global activity modelling to discover and quantify the space-time dependencies between local activities. Without explicit object segmentation, this holistic approach is applicable to dense occluded crowds in contrast to methods relying on object-based decomposition. Different techniques have been proposed following this idea. The most common strategy is to divide a video into local fixed-size spatio-temporal volumes, within which local activity patterns are extracted (Hospedales et al. 2009; Kim and Grauman 2009; Kratz and Nishino 2009; Mehran et al. 2009; Wang et al. 2009) (Fig. 2.10(a)). There are also attempts for exploiting more rigid, top-down structural knowledge in decomposing a scene into exit and entry zones, paths, and junctions (Breitenstein et al. 2008; Makris and Ellis 2005; Wang et al. 2006b) (Fig. 2.10(b)) and semantic regions (Li et al. 2008b; Loy et al. 2010) (Fig. 2.10(c)).

2.7 Distributed Behaviour

Activities can cover large spaces over different time zones. For example, a person moves from his home to work across a city; or a passenger checks in an airline desk, passes through a security check and boards a plane. To understand and interpret such activities, a single viewpoint captured from one camera is inadequate. Behaviours

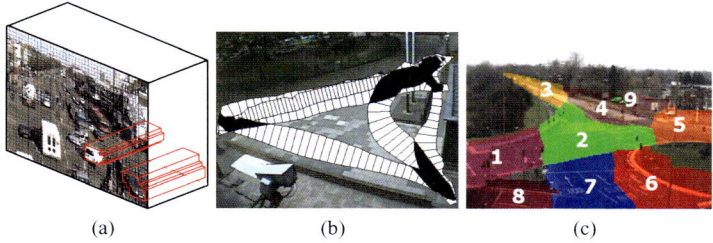

(a)   (b)   (c)

Fig. 2.10 Examples of activity representation by scene decomposition techniques: (**a**) local spatio-temporal volumes (Hospedales et al. 2009), (**b**) movement paths (*white areas*) and junctions
(*black areas*) (Makris and Ellis 2005), and (**c**) semantic regions, each of which encapsulates similar activities regardless of object class and differs from those observed in other regions (Li et al.
2008b)

need be understood in a coherent global context observed from distributed multiple
cameras. In comparison to a single camera system, the field of view of a multiple
camera system is not limited by physical layout of a scene. It offers the potential
for providing a complete record of an object's activity between spaces over time,
allowing a global interpretation of the object's behaviour. The main characteristic
of multi-camera behaviour correlation, in contrast to behavioural interpretation in
a single camera view, is that there can be uncertain spatial and temporal gaps between camera views leading to missing information for a coherent interpretation of
behaviours across all camera views.

In most distributed multiple camera systems, it is the norm to have disjoint fixed-
position cameras with non-overlapping field of views. This is designed to maximise
spatial coverage in a wide-area scene whilst minimising the deployment cost. Figure 2.11 shows an illustration of different types of overlap between multi-camera
field of views. Apart from static cameras, there exists a combination of heterogeneous sensors such as pan-tilt-zoom (PTZ) cameras, mobile wearable cameras, and
non-visual positioning sensors to jointly monitor and interpret activities in a wide-
area scene of complex structures.

Significant efforts have been made in the last few years by the computer vision community towards developing multi-camera models, focusing on learning behaviour correlations across camera views. The problem of learning inter-camera
relationships manifests itself most frequently as either the object association problem (Javed et al. 2003, 2005; Stauffer and Tieu 2003), or the topology inference
problem (Makris et al. 2004; Tieu et al. 2005). The former emphasises more on
associating cross-camera trajectory correspondence, whilst topology inference aims
to infer a scene model for the camera topology of a camera network.

A more general problem for object association across camera views is not merely
restricted to continuous spatio-temporal trajectory correspondence. This is known as
the problem of person re-identification. Solving this problem has compelling practical applications, such as monitoring long-term activity patterns of targeted individuals over large spaces and across time zones. Current models mostly achieve inter-
camera association by matching appearance and motion trends of a target across

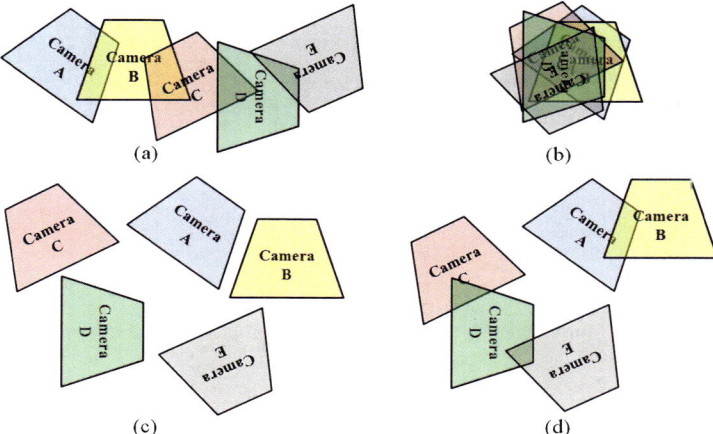

Fig. 2.11 An illustration of different degrees of overlap between the field of view of five cameras: (**a**) cameras partially overlap with adjacent cameras, (**b**) all cameras partially overlap with each other, (**c**) non-overlapping camera network, (**d**) the most common case involving different types of overlapping. The gap between two camera views is commonly known as a '*blind area*', in which a target object may be temporarily out of view of any camera

views. Appearance features are often extracted from the entire individual body since biometric features, such as iris or facial patterns, are no longer reliable in an unconstrained visual environment. Motion trends are typically represented by movement speed across views. Different strategies have been proposed to ensure those features to be robust against changes in object articulation, camera viewpoint, and illumination (Farenzena et al. 2010; Gray and Tao 2008; Hu et al. 2006; Prosser et al. 2008).

Object association based on visual appearance information alone is extremely difficult due to the visual ambiguities of people's appearance in an unconstrained public space (see Figs. 2.8 and 2.12). Apart from exploiting local context such as the group association context (Zheng et al. 2009), additional inter-camera spatio-temporal contextual information needs be explored to constrain and reduce association ambiguities across views. In general, inter-camera person re-identification cannot be achieved by matching imagery information alone. It is essential to formulate and model knowledge about inter-camera relationships as contextual constraints in assisting object re-identification over significantly different camera views (Chen et al. 2008; Loy et al. 2010).

If correspondences of many objects are established using a large pool of observations over time, a model can be built to infer the paths and transition time distributions between cameras. Makris et al. (2004) describe an unsupervised method to accumulate observations from a large set of cross-camera entrance and exit events in order to estimate an inter-camera transition time distribution. A peak in this transition time distribution implies a connection between the two camera views. Alternatively, without having to solve any object motion displacement correspondence

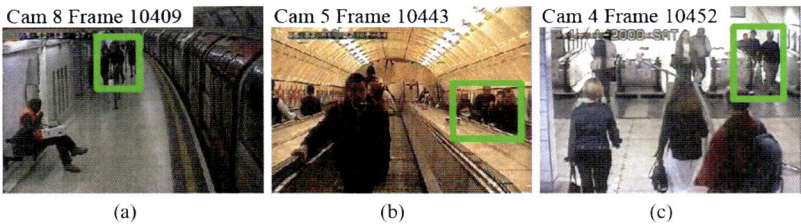

Fig. 2.12 Examples of significant changes in the visual appearance of people moving across different camera views in a public space: (**a**) A group of people (highlighted in *green boxes*) get off a train [Cam 8, frame 10409], and subsequently (**b**) take an upward escalator [Cam 5, frame 10443], which (**c**) leads them to the escalator exit view [Cam 4, frame 10452]. The same people exhibit drastic appearance variations due to changes in illumination (both intra and inter-camera), camera viewpoint, and each individual's body articulation

problem explicitly, another strategy applicable to constrain object association across disjoint cameras is to infer inter-camera relationships through modelling stochastic temporal correlation patterns across views from a population of activities instead of any individual's whereabouts (Loy et al. 2010; Marinakis and Dudek 2009; Niu and Grimson 2006; Tieu et al. 2005).

Estimation of inter-camera behaviour correlation can be made easier if the topology of a camera network can be inferred. This problem of inferring the spatial and temporal relationships among cameras is known as camera topology inference (Makris et al. 2004; Tieu et al. 2005). The aim is to estimate, given population activity patterns across camera views, camera transition probabilities, i.e. how likely an object exiting a camera view would reappear in another camera view; and inter-camera time delay distributions, which is the travel time needed to cross a blind area. The resulting camera topology can be represented by a graph, whose nodes describing population activities in camera field of views, whilst connections between nodes depict either camera transition probabilities or inter-camera time delay distributions.

2.8 Holistic Awareness: Connecting the Dots

Global activity analysis across multiple camera views and multiple information sources, as if 'connecting the dots', is an emerging problem that demands a solution. The goal is to build a global model for connecting and interpreting holistically distributed activities captured locally by different cameras and sensors, for example, detecting unusual events with a global situational awareness. In a distributed camera network, a global interpretation of localised activities is subject to uncertain and partial observations at different locations and different time, compounded by significant differences in object visual appearance (Fig. 2.12).

An intuitive approach to global activity analysis is to reconstruct a global trail of an object by merging the trajectories observed in different views (Zelniker et al. 2008). With non-overlapping camera views, one also needs to solve the camera topology inference problem (Makris et al. 2004; Tieu et al. 2005) and the object association problem (Javed et al. 2005). However, in a crowded public space, it is unrealistic to visually reconstruct exhaustively object global motion trajectories for all objects.

Wang et al. (2010) suggest a trajectory-based solution which bypasses both the topology inference and object association problems. In particular, trajectories observed in different views are first connected based on their temporal proximity, that is, two trajectories are connected if they approximately co-occur in time. A generative model based on latent Dirichlet allocation (Blei et al. 2003) is exploited to cluster object trajectories by co-occurrence. The formed clusters are used to represent activity categories for imposing expectation on inter-camera transition probabilities. A reconstructed global trajectory is considered 'unusual' if it has a low likelihood when matched against the inter-camera transition probabilities.

In contrast to a trajectory-based model, Loy et al. (2010) consider an alternative model that automatically infers the unknown time-delayed correlations among local activities across cameras without object segmentation and tracking. This model aims to explore visual context in a global sense. Human behaviours in a public space are inherently context-aware, and context-constrained by implicit or explicit physical structures and social protocols observed in a shared space. Automatically discovering global spatio-temporal correlations among different camera locations over time can help to establish more accurate interpretations of object behaviours not only visually plausible at each camera location but also globally coherent in context.

Most multi-camera systems focus on passive and stationary cameras. There is a growing interest in collaborative activity analysis using a combination of heterogeneous sensors including PTZ cameras, mobile cameras, non-visual positioning sensors, and radio tags. There are systems using different types of cameras for collaborative monitoring, for example, a static camera to monitor a wide-area scene, working with a PTZ camera to obtain closed-up shot of a facial image (Zhou et al. 2003). More recently, there are also studies that investigate cooperative object detection and tracking using multiple active PTZ cameras (Everts et al. 2007) and through wearable cameras (Alahia et al. 2010). However, the use of heterogeneous sensors remains largely unexplored, especially on the development of reliable models for harvesting and fusing multi-channel information from different types of sensors and non-sensor sources.

References

Aghajanian, J., Warrell, J., Prince, S., Li, P., Rohn, J., Baum, B.: Patch-based within-object classification. In: IEEE International Conference on Computer Vision, Kyoto, Japan, October 2009, pp. 1125–1132 (2009)

Alahia, A., Vandergheynsta, P., Bierlaireb, M., Kunt, M.: Cascade of descriptors to detect and track objects across any network of cameras. Comput. Vis. Image Underst. **114**(6), 624–640 (2010)

Ali, A., Aggarwal, J.: Segmentation and recognition of continuous human activity. In: IEEE Workshop on Detection and Recognition of Events in Video, Vancouver, Canada, July 2001, pp. 28–35 (2001)

Ali, S., Shah, M.: Floor fields for tracking in high density crowd scenes. In: European Conference on Computer Vision, Marseille, France, October 2008, pp. 1–14 (2008)

Ambady, N., Rosenthal, R.: Thin slices of expressive behaviour as predictors of interpersonal consequences: a meta-analysis. Psychol. Bull. **111**(2), 256–274 (1992)

Andriluka, M., Roth, S., Schiele, B.: Pictorial structures revisited: people detection and articulated pose estimation. In: IEEE Conference on Computer Vision and Pattern Recognition, Miami, USA, June 2009, pp. 1014–1021 (2009)

Balomenos, T., Raouzaiou, A., Ioannou, S., Drosopoulos, A., Karpouzis, K., Kollias, S.: Emotion analysis in man-machine interaction systems. In: Machine Learning for Multimodal Interaction, pp. 318–328 (2005)

Bar, M.: Visual objects in context. Nat. Rev., Neurosci. **5**, 617–629 (2004)

Bar, M., Aminoff, E.: Cortical analysis of visual context. Neuron **38**(2), 347–358 (2003)

Bar, M., Ullman, S.: Spatial context in recognition. Perception **25**, 343–352 (1993)

Bartlett, M.S., Littlewort, G., Frank, M., Lainscsek, C., Fasel, I., Movellan, J.: Recognizing facial expression: machine learning and application to spontaneous behavior. In: IEEE Conference on Computer Vision and Pattern Recognition, San Diego, USA, June 2005, pp. 568–573 (2005)

Bashir, K., Xiang, T., Gong, S.: Gait recognition without subject cooperation. Pattern Recognit. Lett. **31**(13), 2052–2060 (2010)

Bassili, J.N.: Emotion recognition: the role of facial movement and the relative importance of upper and lower area of the face. J. Pers. Soc. Psychol. **37**(11), 2049–2058 (1979)

Belongie, S., Malik, J., Puzicha, J.: Shape matching and object recognition using shape contexts. IEEE Trans. Pattern Anal. Mach. Intell. **24**(4), 509–522 (2002)

Benfold, B., Reid, I.: Colour invariant head pose classification in low resolution video. In: British Machine Vision Conference, Leeds, UK, September 2008

Benfold, B., Reid, I.: Guiding visual surveillance by tracking human attention. In: British Machine Vision Conference, London, UK, September 2009

Biederman, I., Mezzanotte, R.J., Rabinowitz, J.C.: Scene perception: detecting and judging objects undergoing relational violations. Cogn. Psychol. **14**, 143–177 (1982)

Black, M., Jepson, A.: Recognising temporal trajectories using the condensation algorithm. In: IEEE International Conference on Automatic Face and Gesture Recognition, Nara, Japan, pp. 16–21 (1998)

Blei, D.M., Ng, A.Y., Jordan, M.I.: Latent Dirichlet allocation. J. Mach. Learn. Res. **3**, 993–1022 (2003)

Bobick, A.F., Davis, J.: The recognition of human movement using temporal templates. IEEE Trans. Pattern Anal. Mach. Intell. **23**(3), 257–267 (2001)

Bobick, A.F., Wilson, A.: A state-based technique for the summarisation of recognition of gesture. In: IEEE International Conference on Computer Vision, pp. 382–388 (1995)

Bregonzio, M., Gong, S., Xiang, T.: Recognising action as clouds of space-time interest points. In: IEEE Conference on Computer Vision and Pattern Recognition, Miami, USA, June 2009, pp. 1948–1955 (2009)

Breitenstein, M.D.: Visual surveillance—dynamic behavior analysis at multiple levels. PhD thesis, ETH Zurich (2009)

Breitenstein, M.D., Sommerlade, E., Leibe, B., Van Gool, L., Reid, I.: Probabilistic parameter selection for learning scene structure from video. In: British Machine Vision Conference, Leeds, UK, September 2008

Burgoon, J., Adkins, M., Kruse, J., Jensen, M.L., Meservy, T., Twitchell, D.P., Deokar, A., Nunamaker, J.F., Lu, S., Tsechpenakis, G., Metaxas, D.N., Younger, R.E.: An approach for intent identification by building on deception detection. In: 38th Annual Hawaii International Conference on System Sciences, Big Island, Hawaii, USA, January 2005, pp. 21–28 (2005a)

Burgoon, J., Jensen, M.L., Meservy, T., Kruse, J., Nunamaker, J.F.: Augmenting human identification of emotional states in video. In: International Conference on Intelligent Data Analysis, McClean, USA (2005b)

Chang, Y., Hu, C., Turk, M.: Probabilistic expression analysis on manifolds. In: IEEE Conference on Computer Vision and Pattern Recognition, Washington DC, USA, June 2004, pp. 520–527 (2004)

Chen, K., Lai, C., Hung, Y., Chen, C.: An adaptive learning method for target tracking across multiple cameras. In: IEEE Conference on Computer Vision and Pattern Recognition, pp. 1–8 (2008)

Cohen, I., Sebe, N., Garg, A., Chen, L., Huang, T.S.: Facial expression recognition from video sequences: temporal and static modeling. Comput. Vis. Image Underst. **91**, 160–187 (2003)

Cohen, N., Gatusso, J., MacLennan-Brown, K.: CCTV Operational Requirements Manual—Is Your CCTV System Fit for Purpose? Home Office Scientific Development Branch, version 4 (55/06) edition (2006)

Coulson, M.: Attributing emotion to static body postures: recognition accuracy, confusions, and viewpoint dependence. J. Nonverbal Behav. **28**(2), 117–139 (2004)

Darwin, C.: The Expression of the Emotions in Man and Animals. Murray, London (1872)

Davis, J., Bobick, A.F.: The representation and recognition of action using temporal templates. In: IEEE Conference on Computer Vision and Pattern Recognition, Puerto Rico, June 1997, pp. 928–934 (1997)

Donato, G., Bartlett, M., Hager, J., Ekman, P., Sejnowski, T.: Classifying facial actions. IEEE Trans. Pattern Anal. Mach. Intell. **21**(10), 974–989 (1999)

Dornaika, F., Davoine, F.: Simultaneous facial action tracking and expression recognition using a particle filter. In: IEEE International Conference on Computer Vision, Beijing, China, October 2005, pp. 1733–1738 (2005)

Efros, A., Berg, A., Mori, G., Malik, J.: Recognizing action at a distance. In: IEEE International Conference on Computer Vision, Nice, France, pp. 726–733 (2003)

Ekman, P., Friesen, W.V.: Pictures of Facial Affect. Consulting Psychologists Press, Palo Alto (1976)

Ekman, P., Friesen, W.V.: Facial Action Coding System: A Technique for Measurement of Facial Movement. Consulting Psychologists Press, Palo Alto (1978)

Essa, I., Pentland, A.: Coding, analysis, interpretation, and recognition of facial expressions. IEEE Trans. Pattern Anal. Mach. Intell. **19**(7), 757–763 (1997)

Everts, I., Sebe, N., Jones, G.A.: Cooperative object tracking with multiple PTZ cameras. In: International Conference on Image Analysis and Processing, September 2007, pp. 323–330 (2007)

Farenzena, M., Bazzani, L., Perina, A., Murino, V., Cristani, M.: Person re-identification by symmetry-driven accumulation of local features. In: IEEE Conference on Computer Vision and Pattern Recognition, San Francisco, USA, June 2010, pp. 2360–2367 (2010)

Fasel, B., Luettin, J.: Automatic facial expression analysis: a survey. Pattern Recognit. **36**, 259–275 (2003)

Felzenszwalb, P., Huttenlocher, D.: Pictorial structures for object recognition. Int. J. Comput. Vis. **61**(1), 55–79 (2005)

Fernández, C., Gonzàlez, J., Roca, X.: Automatic learning of background semantics in generic surveilled scenes. In: European Conference on Computer Vision, Crete, Greece, October 2010, pp. 678–692 (2010)

Freeman, W.T., Tanaka, K., Ohta, J., Kyuma, K.: Computer vision for computer games. In: IEEE International Conference on Automatic Face and Gesture Recognition, Killington, USA, October 1996, pp. 100–105 (1996)

Gilbert, A., Illingworth, J., Bowden, R.: Scale invariant action recognition using compound features mined from dense spatio-temporal corners. In: European Conference on Computer Vision, pp. 222–233 (2008)

Gong, S., Buxton, H.: Bayesian nets for mapping contextual knowledge to computational constraints in motion segmentation and tracking. In: British Machine Vision Conference, Guildford, UK, September 1993, pp. 229–238 (1993)

Gong, S., McKenna, S., Collins, J.J.: An investigation into face pose distributions. In: IEEE International Conference on Automatic Face and Gesture Recognition, Killington, USA, October 1996, pp. 265–270 (1996)

Gong, S., Walter, M., Psarrou, A.: Recognition of temporal structures: learning prior and propagating observation augmented densities via hidden Markov states. In: IEEE International Conference on Computer Vision, Corfu, Greece, September 1999, pp. 157–162 (1999)

Gong, S., Xiang, T., Hongeng, S.: Learning human pose in crowd. In: Workshop on Multimodal Pervasive Video Analysis, ACM International Conference on Multimedia, Firenze, Italy, October 2010

Gorelick, L., Blank, M., Shechtman, E., Irani, M., Basri, R.: Actions as space-time shapes. IEEE Trans. Pattern Anal. Mach. Intell. **29**(12), 2247–2253 (2007)

Gray, D., Tao, H.: Viewpoint invariant pedestrian recognition with an ensemble of localized features. In: European Conference on Computer Vision, pp. 262–275 (2008)

Gunes, H., Piccardi, M.: A bimodal face and body gesture database for automatic analysis of human nonverbal affective behavior. In: International Conference on Pattern Recognition, pp. 1148–1153 (2006)

Gunes, H., Piccardi, M.: Bi-modal emotion recognition from expressive face and body gestures. J. Netw. Comput. Appl. **30**(4), 1334–1345 (2007)

Han, J., Bhanu, B.: Individual recognition using gait energy image. IEEE Trans. Pattern Anal. Mach. Intell. **28**(2), 316–322 (2006)

Hoey, J., Little, J.J.: Value directed learning of gestures and facial displays. In: IEEE Conference on Computer Vision and Pattern Recognition, Washington DC, USA, June 2004, pp. 1026–1033 (2004)

Hospedales, T., Gong, S., Xiang, T.: A Markov clustering topic model for mining behaviour in video. In: IEEE International Conference on Computer Vision, Kyoto, Japan, October 2009, pp. 1165–1172 (2009)

Hu, W., Hu, M., Zhou, X., Tan, T., Lou, J., Maybank, S.: Principal axis-based correspondence between multiple cameras for people tracking. IEEE Trans. Pattern Anal. Mach. Intell. **28**(4), 663–671 (2006)

Hu, Y., Cao, L., Lv, F., Yan, S., Gong, Y., Huang, T.S.: Action detection in complex scenes with spatial and temporal ambiguities. In: IEEE International Conference on Computer Vision, Kyoto, Japan, October 2009, pp. 128–135 (2009)

Javed, O., Rasheed, Z., Shafique, K., Shah, M.: Tracking across multiple cameras with disjoint views. In: IEEE International Conference on Computer Vision, pp. 952–957 (2003)

Javed, O., Shafique, K., Shah, M.: Appearance modeling for tracking in multiple non-overlapping cameras. In: IEEE Conference on Computer Vision and Pattern Recognition, pp. 26–33 (2005)

Jia, K., Gong, S.: Multi-resolution patch tensor for facial expression hallucination. In: IEEE Conference on Computer Vision and Pattern Recognition, New York, USA, June 2006, pp. 395–402 (2006)

Jia, K., Gong, S.: Generalised face super-resolution. IEEE Trans. Image Process. **17**(6), 873–886 (2008)

Jiang, H.: Human pose estimation using consistent max-covering. In: IEEE International Conference on Computer Vision, Kyoto, Japan, October 2009, pp. 1357–1364 (2009)

Kaliouby, R.E., Robinson, P.: Real-time inference of complex mental states from facial expressions and head gestures. In: IEEE Workshop on Real-Time Vision for Human-Computer Interaction, Washington DC, USA, June 2004

Kapoor, A., Picard, R.W.: Multimodal affect recognition in learning environments. In: ACM International Conference on Multimedia, pp. 677–682 (2005)

Ke, Y., Sukthankar, R., Hebert, M.: Efficient visual event detection using volumetric features. In: IEEE International Conference on Computer Vision, pp. 166–173 (2005)

Ke, Y., Sukthankar, R., Hebert, M.: Event detection in crowded videos. In: IEEE International Conference on Computer Vision, Rio de Janeiro, Brasil, October 2007, pp. 1–8 (2007)

Kim, J., Grauman, K.: Observe locally, infer globally: a space-time MRF for detecting abnormal activities with incremental updates. In: IEEE Conference on Computer Vision and Pattern Recognition, Miami, USA, June 2009, pp. 2921–2928 (2009)

Kratz, L., Nishino, K.: Anomaly detection in extremely crowded scenes using spatio-temporal motion pattern models. In: IEEE Conference on Computer Vision and Pattern Recognition, pp. 1446–1453 (2009)

Kruegle, H.: CCTV Surveillance: Video Practices and Technology. Butterworth/Heinemann, Stoneham/Berlin (2006)

Kuettel, D., Breitenstein, M.D., Van Gool, L., Ferrari, V.: What's going on? discovering spatio-temporal dependencies in dynamic scenes. In: IEEE Conference on Computer Vision and Pattern Recognition, San Francisco, USA, June 2010, pp. 1951–1958 (2010)

Laptev, I.: On space-time interest points. Int. J. Comput. Vis. **64**(2), 107–123 (2005)

Laptev, I., Lindeberg, T.: Space-time interest points. In: IEEE International Conference on Computer Vision, pp. 432–439 (2003)

Lee, C.S., Elgammal, A.: Facial expression analysis using nonlinear decomposable generative models. In: IEEE International Workshop on Analysis and Modeling of Faces and Gestures, pp. 17–31 (2005)

Lee, M.W., Cohen, I.: Human upper body pose estimation in static images. In: European Conference on Computer Vision, Cambridge, UK, pp. 126–138 (2004)

Li, J., Gong, S., Xiang, T.: Global behaviour inference using probabilistic latent semantic analysis. In: British Machine Vision Conference, Leeds, UK, pp. 193–202 (2008a)

Li, J., Gong, S., Xiang, T.: Scene segmentation for behaviour correlation. In: European Conference on Computer Vision, Marseille, France, pp. 383–395 (2008b)

Liu, J., Shah, M.: Learning human actions via information maximization. In: IEEE Conference on Computer Vision and Pattern Recognition, Anchorage, USA, June 2008, pp. 1–8 (2008)

Loy, C.C., Xiang, T., Gong, S.: Time-delayed correlation analysis for multi-camera activity understanding. Int. J. Comput. Vis. **90**(1), 106–129 (2010)

Loy, C.C., Xiang, T., Gong, S.: Detecting and discriminating behavioural anomalies. Pattern Recognit. **44**(1), 117–132 (2011)

Lyons, M.J., Budynek, J., Akamatsu, S.: Automatic classification of single facial images. IEEE Trans. Pattern Anal. Mach. Intell. **21**(12), 1357–1362 (1999)

Makris, D., Ellis, T.: Learning semantic scene models from observing activity in visual surveillance. IEEE Trans. Syst. Man Cybern. **35**(3), 397–408 (2005)

Makris, D., Ellis, T., Black, J.: Bridging the gaps between cameras. In: IEEE Conference on Computer Vision and Pattern Recognition, pp. 205–210 (2004)

Marinakis, D., Dudek, G.: Self-calibration of a vision-based sensor network. Image Vis. Comput. **27**(1–2), 116–130 (2009)

McKenna, S., Gong, S.: Gesture recognition for visually mediated interaction using probabilistic event trajectories. In: British Machine Vision Conference, Southampton, UK, pp. 498–508 (1998)

Meeren, H., Heijnsbergen, C., Gelder, B.: Rapid perceptual integration of facial expression and emotional body language. Proc. Natl. Acad. Sci. USA **102**(45), 16518–16523 (2005)

Mehran, R., Oyama, A., Shah, M.: Abnormal crowd behaviour detection using social force model. In: IEEE Conference on Computer Vision and Pattern Recognition, pp. 935–942 (2009)

Ng, J., Gong, S.: Composite support vector machines for the detection of faces across views and pose estimation. Image Vis. Comput. **20**(5–6), 359–368 (2002)

Niu, C., Grimson, W.E.L.: Recovering non-overlapping network topology using far-field vehicle tracking data. In: International Conference of Pattern Recognition, pp. 944–949 (2006)

Nixon, M.S., Tan, T., Chellappa, R.: Human Identification Based on Gait. Springer, Berlin (2005)

Okuma, K., Taleghani, A., Freitas, N.D., Little, J., Lowe, D.: A boosted particle filter: multi-target detection and tracking. In: European Conference on Computer Vision (2004)

Orozco, J., Gong, S., Xiang, T.: Head pose classification in crowded scenes. In: British Machine Vision Conference, London, UK, September 2009

Ozuysal, M., Lepetit, V., Fua, P.: Pose estimation for category specific multiview object localization. In: IEEE Conference on Computer Vision and Pattern Recognition, pp. 778–785 (2009)

Palmer, S.: The effects of contextual scenes on the identification of objects. Mem. Cogn. **3**, 519–526 (1975)

Pantic, M., Patras, I.: Dynamics of facial expression: recognition of facial actions and their temporal segments from face profile image sequences. IEEE Trans. Syst. Man Cybern. **36**(2), 433–449 (2006)

Pantic, M., Rothkrantz, L.: Automatic analysis of facial expressions: the state of the art. IEEE Trans. Pattern Anal. Mach. Intell. **22**(12), 1424–1445 (2000)

Pantic, M., Rothkrantz, L.: Toward an affect-sensitive multimodal human-computer interaction. In: Proceeding of the IEEE, vol. 91, pp. 1370–1390 (2003)

Pantic, M., Rothkrantz, L.: Facial action recognition for facial expression analysis from static face images. IEEE Trans. Syst. Man Cybern. **34**(3), 1449–1461 (2004)

Pantic, M., Valstar, M., Rademaker, R., Maat, L.: Web-based database for facial expression analysis. In: IEEE International Conference on Multimedia and Expo, pp. 317–321 (2005)

Pavlovic, V.I., Sharma, R., Huang, T.S.: Visual interpretation of hand gestures for human-computer interaction: a review. IEEE Trans. Pattern Anal. Mach. Intell. **19**(7), 677–695 (1997)

Prosser, B., Gong, S., Xiang, T.: Multi-camera matching using bi-directional cumulative brightness transfer functions. In: British Machine Vision Conference, Leeds, UK, September 2008

Psarrou, A., Gong, S., Buxton, H.: Modelling spatio-temporal trajectories and face signatures on partially recurrent neural networks. In: IEEE International Joint Conference on Neural Networks, Perth, Australia, pp. 2226–3321 (1995)

Psarrou, A., Gong, S., Walter, M.: Recognition of human gestures and behaviour based on motion trajectories. Image Vis. Comput. **20**(5–6), 349–358 (2002)

Ramanan, D., Forsyth, D.A.: Automatic annotation of everyday movements. In: Advances in Neural Information Processing Systems, Vancouver, Canada (2003)

Rao, C., Shah, M.: View-invariance in action recognition. In: IEEE Conference on Computer Vision and Pattern Recognition, pp. 316–322 (2001)

Ravindra De Silva, P., Bianchi-Berthouze, N.: Modeling human affective postures: an information theoretic characterisation of posture features. Comput. Animat. Virtual Worlds **15**, 169–276 (2004)

Ravindra De Silva, P., Osano, M., Marasinghe, A.: Towards recognizing emotion with affective dimensions through body gestures. In: IEEE International Conference on Automatic Face and Gesture Recognition, pp. 269–274 (2006)

Robertson, N., Reid, I.: Estimating gaze direction from low-resolution faces in video. In: European Conference on Computer Vision, pp. 402–415 (2006)

Schlenzig, J., Hunter, E., Jain, R.: Recursive identification of gesture inputs using hidden Markov model. In: Workshop on Applications of Computer Vision, pp. 187–194 (1994)

Schüldt, C., Laptev, I., Caputo, B.: Recognizing human actions: a local SVM approach. In: International Conference on Pattern Recognition, Cambridge, UK, pp. 32–36 (2004)

Schwartz, O., Hsu, A., Dayan, P.: Space and time in visual context. Nat. Rev., Neurosci. **8**, 522–535 (2007)

Shan, C., Gong, S., McOwan, P.: Appearance manifold of facial expression. In: Sebe, N., Lew, M.S., Huang, T.S. (eds.) Computer Vision in Human-Computer Interaction. Lecture Notes in Computer Science, vol. 3723, pp. 221–230. Springer, Berlin (2005)

Shan, C., Gong, S., McOwan, P.: Capturing correlations among facial parts for facial expression analysis. In: British Machine Vision Conference, Warwick, UK, September 2007

Shan, C., Gong, S., McOwan, P.: Facial expression recognition based on local binary patterns: a comprehensive study. Image Vis. Comput. **27**(6), 803–816 (2009)

Sheikh, Y., Sheikh, M., Shah, M.: Exploring the space of a human action. In: IEEE International Conference on Computer Vision, Beijing, China, October 2005, pp. 144–149 (2005)

Sherrah, J., Gong, S.: Fusion of perceptual cues for robust tracking of head pose and position. Pattern Recognit. **34**(8), 1565–1572 (2001)

Sherrah, J., Gong, S., Ong, E.-J.: Face distribution in similarity space under varying head pose. Image Vis. Comput. **19**(11), 807–819 (2001)

Siva, P., Xiang, T.: Action detection in crowd. In: British Machine Vision Conference, Aberystwyth, UK, September 2010

Starner, T., Pentland, A.: Visual recognition of American sign language using hidden Markov models. In: IEEE International Conference on Automatic Face and Gesture Recognition, Zurich, Switzerland, pp. 189–194 (1995)

Stauffer, C., Tieu, K.: Automated multi-camera planar tracking correspondence modeling. In: IEEE Conference on Computer Vision and Pattern Recognition, pp. 259–266 (2003)

Streib, K., Davis, J.W.: Extracting pathlets from weak tracking data. In: IEEE International Conference on Advanced Video and Signal Based Surveillance, Boston, USA, pp. 353–360 (2010)

Suwa, M., Sugie, N., Fujimora, K.: A preliminary note on pattern recognition of human emotional expression. In: International Joint Conference on Pattern Recognition, pp. 408–410 (1978)

Tian, Y.: Evaluation of face resolution for expression analysis. In: Computer Vision and Pattern Recognition Workshop, pp. 82–89 (2004)

Tian, Y., Kanade, T., Cohn, J.: Recognizing action units for facial expression analysis. IEEE Trans. Pattern Anal. Mach. Intell. **23**(2), 97–115 (2001)

Tieu, K., Dalley, G., Grimson, W.E.L.: Inference of non-overlapping camera network topology by measuring statistical dependence. In: IEEE International Conference on Computer Vision, pp. 1842–1849 (2005)

Valstar, M., Pantic, M.: Fully automatic facial action unit detection and temporal analysis. In: Computer Vision and Pattern Recognition Workshop, pp. 149–154 (2006)

Valstar, M., Patras, I., Pantic, M.: Facial action unit detection using probabilistic actively learned support vector machines on tracked facial point data. In: Computer Vision and Pattern Recognition Workshop, vol. 3, pp. 76–84 (2005)

Van Gool, L., Breitenstein, M., Gammeter, S., Grabner, H., Quack, T.: Mining from large image sets. In: ACM International Conference on Image and Video Retrieval, Santorini, Greece, pp. 10:1–10:8 (2009)

Vrij, A.: Detecting Lies and Deceit: The Psychology of Lying and the Implications for Professional Practice. Wiley, New York (2000)

Vrij, A., Mann, S.: Police use of nonverbal behaviour as indicators of deception. In: Riggio, R.E., Feldman, R.S. (eds.) Applications of Nonverbal Communication, pp. 63–94. Erlbaum, Hillsdale (2005)

Wang, S., Quattoni, A., Morency, L.-P., Demirdjian, D., Darrell, T.: Hidden conditional random fields for gesture recognition. In: IEEE Conference on Computer Vision and Pattern Recognition, pp. 1521–1527 (2006a)

Wang, X., Tieu, K., Grimson, W.E.L.: Learning semantic scene models by trajectory analysis. In: European Conference on Computer Vision, pp. 110–123 (2006b)

Wang, X., Ma, X., Grimson, W.E.L.: Unsupervised activity perception in crowded and complicated scenes using hierarchical Bayesian models. IEEE Trans. Pattern Anal. Mach. Intell. **31**(3), 539–555 (2009)

Wang, X., Tieu, K., Grimson, W.E.L.: Correspondence-free activity analysis and scene modeling in multiple camera views. IEEE Trans. Pattern Anal. Mach. Intell. **32**(1), 56–71 (2010)

Wren, C., Azarbayejani, A., Darrell, T., Pentland, A.: Pfinder: real-time tracking of the human body. IEEE Trans. Pattern Anal. Mach. Intell. **19**(7), 780–785 (1997)

Wu, B., Nevatia, R.: Detection and tracking of multiple partially occluded humans by Bayesian combination of edgelet based part detectors. Int. J. Comput. Vis. **75**(2), 247–266 (2007)

Wu, Y., Huang, T.S.: Human hand modeling, analysis and animation in the context of human computer interaction. IEEE Signal Process. Mag. **18**(3), 51–60 (2001)

Yacoob, Y., Davis, L.S.: Recognizing human facial expression from long image sequences using optical flow. IEEE Trans. Pattern Anal. Mach. Intell. **18**(6), 636–642 (1996)

Yang, M., Wu, Y., Hua, G.: Context-aware visual tracking. IEEE Trans. Pattern Anal. Mach. Intell. **31**(7), 1195–1209 (2008)

Yang, W., Wang, Y., Mori, G.: Efficient human action detection using a transferable distance function. In: Asian Conference on Computer Vision, pp. 417–426 (2009)

Yeasin, M., Bullot, B., Sharma, R.: From facial expression to level of interests: a spatiotemporal approach. In: IEEE Conference on Computer Vision and Pattern Recognition, pp. 922–927 (2004)

Yilmaz, A., Shah, M.: Actions sketch: a novel action representation. In: IEEE Conference on Computer Vision and Pattern Recognition, Washington DC, USA, pp. 984–989 (2005)

Yin, L., Loi, J., Xiong, W.: Facial expression representation and recognition based on texture augmentation and topographic masking. In: ACM Multimedia, New York, USA, pp. 236–239 (2004)

Yuan, J.S., Liu, Z.C., Wu, Y.: Discriminative subvolume search for efficient action detection. In: IEEE Conference on Computer Vision and Pattern Recognition, pp. 2442–2449 (2009)

Zaharescu, A., Wildes, R.: Anomalous behaviour detection using spatiotemporal oriented energies, subset inclusion histogram comparison and event-driven processing. In: European Conference on Computer Vision, Crete, Greece, October 2010, pp. 563–576 (2010)

Zalewski, L., Gong, S.: 2D statistical models of facial expressions for realistic 3D avatar animation. In: IEEE Conference on Computer Vision and Pattern Recognition, San Diego, USA, June 2005, pp. 217–222 (2005)

Zelniker, E.E., Gong, S., Xiang, T.: Global abnormal behaviour detection using a network of CCTV cameras. In: IEEE International Workshop on Visual Surveillance, Marseille, France, October 2008

Zhang, J., Gong, S.: Action categorisation by structural probabilistic latent semantic analysis. Comput. Vis. Image Underst. **114**(8), 857–864 (2010a)

Zhang, J., Gong, S.: Action categorisation with modified hidden conditional random field. Pattern Recognit. **43**(1), 197–203 (2010b)

Zhang, J., Zhou, S., McMillan, L., Comaniciu, D.: Joint real-time object detection and pose estimation using probabilistic boosting network. In: IEEE Conference on Computer Vision and Pattern Recognition, pp. 1–7 (2007)

Zhang, Y., Ji, Q.: Active and dynamic information fusion for facial expression understanding from image sequences. IEEE Trans. Pattern Anal. Mach. Intell. **27**(5), 1–16 (2005)

Zhang, Z., Lyons, M.J., Schuster, M., Akamatsu, S.: Comparison between geometry-based and Gabor-wavelets-based facial expression recognition using multi-layer perceptron. In: IEEE International Conference on Automatic Face and Gesture Recognition, Nara, Japan, April 1998, pp. 454–459 (1998)

Zhang, Z., Hu, Y., Chan, S., Chia, L.: Motion context: a new representation for human action recognition. In: European Conference on Computer Vision, pp. 817–829 (2008)

Zheng, W., Gong, S., Xiang, T.: Associating groups of people. In: British Machine Vision Conference, London, UK, September 2009

Zhou, X., Collins, R., Kanade, T.: A master-slave system to acquire biometric imagery of humans at distance. In: International Workshop on Video Surveillance, pp. 113–120 (2003)

Chapter 3
Towards Modelling Behaviour

Automatic interpretation of object behaviour requires constructing computational models of behaviour. In particular, one considers the problem of automatically learning behaviour models directly from visual observations. In order for a computer to learn a behaviour model from data, a number of essential issues need be addressed. These include selecting a suitable representation, developing a robust interpretation mechanism, and adopting an effective strategy for model learning.

3.1 Behaviour Representation

To validate hypotheses about object behaviours under visual observation, one first considers how to represent the characteristics of object actions and activities in space and over time. Different representational models can be considered depending not only on the nature of activity, such as activity of a single person, a group of people or a crowd, but also on the visual environment, for instance, the scene background, object distance from the camera, and relative motion between image frames. Broadly speaking, there are four different approaches to behaviour representation: object-based, part-based, pixel-based, and event-based representations.

3.1.1 Object-Based Representation

An object-based representation describes explicitly each object in space and over time based on the assumption that individual objects can be segmented reasonably well in a visual scene. Object-based representations can be further divided into two sub-categories: trajectory- and template-based representations.

Trajectory-Based Representation

A trajectory-based representation aims to construct object-centred spatio-temporal trajectories for object description centred on each object's bounding box or shape

S. Gong, T. Xiang, *Visual Analysis of Behaviour*,
DOI 10.1007/978-0-85729-670-2_3, © Springer-Verlag London Limited 2011

structure. An object descriptor consists of a set of measurable attributes describing appearance (shape, colour, texture) and motion characteristics. An object trajectory is the motion history of each object in a visual space over time. Object trajectories can be considered as a starting point for behaviour analysis (Buxton and Gong 1995; Fu et al. 2005; Gong and Buxton 1992, 1993; Johnson and Hogg 1996; Naftel and Khalid 2006; Nascimento et al. 2007; Nguyen et al. 2005; Owens and Hunter 2000; Piciarelli and Foresti 2006; Saleemi et al. 2009; Wang et al. 2006; Zhao and Nevatia 2004).

A trajectory is computed by associating a detected object in consecutive image frames using motion tracking (Yilmaz et al. 2006). Tracking methods typically utilise a set of object appearance attributes to establish inter-frame correspondence of an object over time. Since object appearance attributes can vary significantly over time, motion prediction models are commonly exploited to model expectations on speed and acceleration. Examples of such models include single-modal Kalman filtering (Kalman 1960) and multi-modal particle filtering (Isard and Blake 1996).

Ideally, a single trajectory track should correspond to each object of interest. However, this is not always guaranteed, especially in noisy and crowded environments. To alleviate this problem, techniques have been developed to cope with occlusions and lighting changes (Haritaoglu et al. 2000; McKenna et al. 2000; Raja et al. 1998), and by adopting a tracking-by-detection strategy (Breitenstein 2009; Okuma et al. 2004; Wu and Nevatia 2007). There are efforts made to improve tracking in crowded scenes by exploiting space-time proximity and trajectory coherence (Brostow and Cipolla 2006). There are also studies on exploiting contextual information about the visual environment to assist motion segmentation and object tracking (Gong and Buxton 1993), and around a target object to either facilitate more robust long-term tracking (Yang et al. 2008) or overcome occlusion by learning trajectory transition likelihood (Saleemi et al. 2009). However, problems remain unsolved on how to best combine useful information sources and filter out unreliable information sources, in order to mitigate high false positive rate in object detection, identify auxiliary contextual cues, or learn trajectory distribution in crowded and unconstrained environments.

Object-centred trajectory representation can take different forms. A simple representation of trajectory is a sequence of discrete point-based vectors. Each vector can have attributes including spatial location (Psarrou et al. 2002), velocity (Johnson and Hogg 1996), velocity variation (Owens and Hunter 2000), trajectory direction, and the temporal curvature of each trajectory (Walter et al. 2001). A trajectory can be expressed in a two-dimensional image space, on a ground plane, and in a three-dimensional space if camera models are available. A single-point trajectory only represents the centroid of an object. Given sufficient imagery resolution, and especially for modelling actions of articulated objects from a close-up view, it may have added benefits to construct a composite trajectory for a set of feature points representing an object shape structure such as the contour (Baumberg and Hogg 1996; Du et al. 2007; Galata et al. 2001; Ke et al. 2007). Early trajectory studies are centred on analysing individual trajectories, for example, learning the distribution of trajectories (Johnson and Hogg 1996) and comparing similarities at a trajectory point-level (Owens and Hunter 2000). More recent development has been shifted towards analysing trajectory trends through clustering trajectories (Zhang et al. 2006), for instance, by using hierarchical clustering (Jiang et al. 2009) and spectral clustering (Fu et al. 2005; Wang et al. 2006).

Object trajectories provide a good source of spatio-temporal information about object activities, and this approach to behaviour representation is intuitively well conceived. However, there are inherent drawbacks from using object-centred trajectory for interpreting behaviour due to a number of reasons:

1. Lack of details: Computing trajectories of object shapes assume high fidelity visual details about the objects of interest in order to extract sufficient image features. However, the assumption may not always be valid, especially in a cluttered public scene with crowded activities.
2. Severe occlusion: An object may not be tracked continuously in a large space shared by other objects, caused by frequent inter-object occlusion. This often results in brickle and discontinuous object trajectories and inconsistent labelling in object association. For example, in an overcast day with smooth and moderate traffic volume, the moving vehicles on a motorway can be tracked reasonably well (Fig. 3.1(a)). Their activities can be modelled based on trajectory analysis. In a different aircraft docking scene, aircraft arrival is followed by the activities of various ground service vehicles (Fig. 3.1(b)). The movements of different objects are heavily overlapped and visually discontinuous. As a result, a large number of fragmented trajectories, known as tracklets, are obtained. This makes trajectory-based behaviour analysis very challenging.
3. Lack of context: An object's trajectory in isolation does not always capture the most distinctive information about its behaviour. An example in Fig. 3.1(c) illustrates this problem. A laboratory shopping scene is shown where a shopper can either take a can of drink and pay for it, or just browse and leave. In this scene, very similar object trajectories can be resulted from activities with different behavioural interpretations. In an unconstrained environment, analysing trajectories of each object *in isolation* is often insufficient for interpreting behaviour.

Template-Based Representation

When a behaviour is captured in a well-controlled environment, where a single or dominant object can be assumed to be close to the camera with minimal distractions in the camera view, a template-based representation can be employed for analysing behaviour. An object's action sequence can be represented by either a single or multiple templates. For a single template representation, a straightforward approach is to accumulate all measurements of object movement over a sequence into a single descriptor, such as a motion-history image (Bobick and Davis 2001). In a motion-history image, pixel-wise differences in image intensity over time due to object movement is aligned and overlaid into a single image template in which each pixel value represents object motion history at that pixel location. That is,

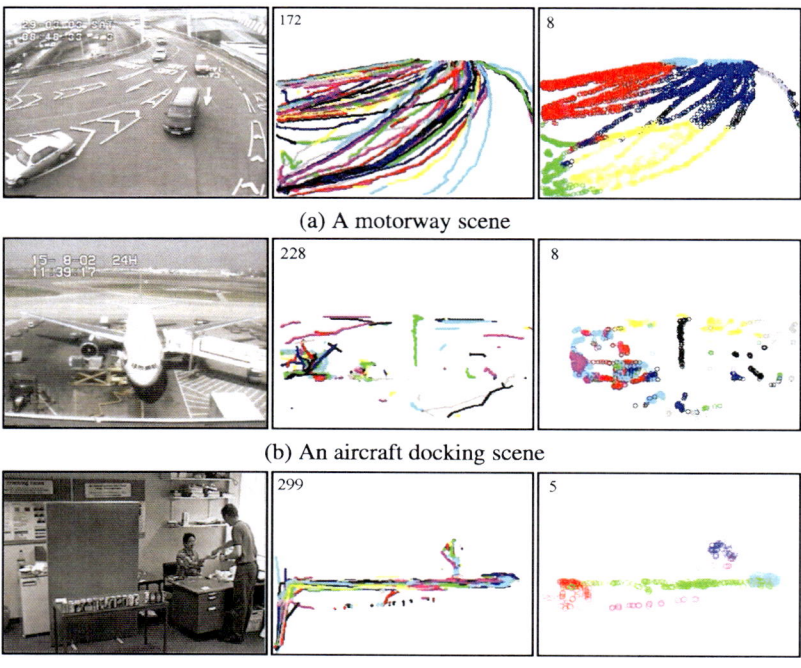

(a) A motorway scene

(b) An aircraft docking scene

(c) A laboratory shopping scene

Fig. 3.1 Comparing event-based and trajectory-based representations (Xiang and Gong 2006a). *Left column*: Example frames from three different scenes. *Middle column*: Trajectories computed over time, with different trajectories illustrated by *colour* with the accumulated total number of trajectories shown at the *top-left corner* in each diagram. *Right column*: Contextual events detected over time, with 8 class of events detected both in the motorway and aircraft docking scenes and 5 classes detected in the shopping scene. Different events are illustrated by *colour* with the number of event classes shown at the *top-left corner* in each diagram

pixel loci with more recent motion have higher intensity values. An extension of this two-dimensional motion-history-image-based representation can be considered for multiple camera views. Weinland et al. (2006) develop a motion-history volume representation for representing object actions captured by constrained multiple cameras.

In contrast to motion-history image, Boiman and Irani (2007) consider estimating pixel-wise correlation between two space-time image volumes as a template for characterising object movement patterns captured in image sequences. Efros et al. (2003) also construct object motion templates from space-time image volumes, but instead of computing exhaustively pixel-wise correlation directly, optical flow vectors are accumulated in each space-time action volume. Such a motion template descriptor aims to capture the overall spatial arrangement of local motion displacement vectors generated by an object action sequence. All single template-based motion descriptors are inherently sensitive to mis-alignment in constructing the space-time volumes from object action sequences. This is especially so when there is a large shape variation caused by object movement, or unstable camera and non-stationary background. Some efforts have been made to relax the requirement on precise image frame alignment using object segmentation and motion blurring (Efros et al. 2003).

Instead of computing a single motion template, object actions can also be represented as a sequence of shape templates. There is a good reason for constructing space-time shape templates, particularly so if an object undergoes considerable shape deformation causing difficulties to force object alignment over time. A series of two-dimensional silhouettes of a human body, extracted from a space-time volume, retain both instantaneous body shape deformation information at each time frame and body motion information over time (Brand 1999b; Gorelick et al. 2007). Using space-time shape templates shares some similarities with deriving a shape representation for constructing a composite trajectory-based representation (Baumberg and Hogg 1996). A silhouette-based representation for human action recognition has received considerable attention in recent years (Blank et al. 2005; Moeslunda et al. 2006; Sminchisescu et al. 2006; Wang et al. 2003; Zhang and Gong 2010a, 2010b).

3.1.2 Part-Based Representation

It has long been recognised that object segmentation and unit formation is fundamental to visual perception both in biological and artificial systems (Marr 1982; Nevatia and Binford 1977; Shipley et al. 2001; Spelke 1990). However, visually segmenting objects from a cluttered background with many distractors is computationally unreliable. One way to overcome the problem is to use a part-based representation of visual object. There are broadly two types of part-based object representation: constellation models (Fei-Fei et al. 2006; Fergus et al. 2007; Fischler and Elschlager 1973) and non-constellation models (Leibe et al. 2008; Yuille et al. 1992). The former represents an object by a set of parts and exploits both object visual appearance information from each part and the geometric constraints between the parts. In contrast, the latter explicitly disregards the spatial location of each part to relax the structural constraint on object formation.

A part-based representation can be built from a "bag-of-words". The bag-of-words concept was introduced originally for text document analysis (Lewis 1998), and later adopted for image representation (Fei-Fei and Perona 2005). It is a histogram representation of local image patch clusters where clustering of image patches is based on a vector descriptor of local features detected in each image patch. In principle, any part-based object representation can be extended to also represent image sequences of object action. However, spatio-temporal relationships between parts of an action are likely to be weaker than the geometric relationships between parts of an object appearance. Because of that, a part-based action representation may be more robust in a form of a non-constellation and non-structured model.

Most of part-based behaviour representations relate to the success of constructing image descriptors from sparse interesting points detected in local image patches, such as scale-invariant feature transform (SIFT) (Lowe 2004). Such a descriptor-based representation offers better invariance to affine image transforms and is more robust to background clutter. For action representation, Laptev (2005) extends this idea to model local object motion changes in a neighbourhood of space-time interest points. This idea can be further extended to a histogram-based representation using a bag of spatio-temporal words (Dollár et al. 2005; Laptev et al. 2007; Liu et al. 2009; Niebles et al. 2006, 2008; Schüldt et al. 2004; Willems et al. 2008).

Space-time interest points are feature points detected in small space-time image volumes, instead of local image patches as in the case of object representation. An action image sequence can be divided into many such small image volumes for extracting space-time interest points. Similar to an image descriptor for object representation, each small image volume is described by a vector representation of local space-time interest points. This can be called a video descriptor. Not all space-time image volumes in an action sequence are unique. To provide a more concise representation, video descriptors extracted from those space-time image volumes can be grouped into a smaller set of clusters with each cluster now expected to represent something more characteristic about the action sequence. Such a cluster is considered as a video word. By assigning video descriptors extracted from space-time image volumes to different video words during the clustering process, an action sequence can then be represented by a histogram of the accumulated numbers of video descriptors distributed over different video words. Such a bag of video words based action representation is an extension to the bag-of-words object representation.

3.1.3 Pixel-Based Representation

When the background of a visual scene becomes highly cluttered, for example, in a crowded public scene, even a part-based representation is difficult to be constructed. To relax the assumption on forming object-centred units as the basis for representation, one may consider a pixel-based behaviour representation that ignores individual object entities. This type of representation extracts all activity relevant pixel-wise space-time features directly from image colour and intensity gradient information, without object segmentation and unit association.

A basic pixel-based representation is to extract what may be considered as foreground pixels through background subtraction (Russell and Gong 2006, 2008; Stauffer and Grimson 1999). Despite its simplicity, it can be used effectively for modelling and understanding actions and activities, for instance, for unusual behaviour detection (Benezeth et al. 2009; Ng and Gong 2003). Based on foreground pixel detection, more informative pixel-based representations can be developed such as pixel-change-history (Gong and Xiang 2003b; Xiang and Gong 2006a) which models the history of foreground pixels (Fig. 3.2(a)), or average behaviour image (Jodoin et al. 2008). A foreground pixel-based activity representation is attractive not only due to its computational simplicity but also because it avoids explicit

(a)   (b)

Fig. 3.2 Examples of pixel-based activity representations: (**a**) pixel-change-history (Xiang and Gong 2006a) and (**b**) optical flow (Hospedales et al. 2009)

object segmentation, unit formation and tracking. This is especially beneficial for representing object activities in highly cluttered and crowded scenes.

For capturing spatio-temporal information about object motion, optical flow is widely used (Fig. 3.2(b)). An optical flow field is typically a dense map of imagery displacement vectors that estimate pixel-wise apparent motion between consecutive image frames over time (Beauchemin and Barron 1995; Buxton and Buxton 1983; Gong and Brady 1990; Horn and Schunck 1981; Lucas and Kanade 1981). Optical flow has been popular as a pixel-based spatio-temporal representation for activity analysis and behaviour understanding (Adam et al. 2008; Ali and Shah 2007; Andrade et al. 2006a, 2006b; Buxton and Gong 1995; Gong and Buxton 1993; Hospedales et al. 2009; Kuettel et al. 2010; Mehran et al. 2009; Wang et al. 2007, 2009; Wu et al. 2010; Yang et al. 2009). However, computing optical flow is computationally expensive. In practice, flow vectors are often computed by image grids rather than per pixel. An image is divided into cells of specific size, of which the average or median flow vectors are estimated for each image grid. Optical flow vectors at each image cell can be further quantised by discrete directions for activity modelling (Hospedales et al. 2009; Kuettel et al. 2010; Wang et al. 2009). To reduce imagery noise, flow vectors can be filtered by thresholding and a foreground mask, after which only flow vectors of sufficient magnitude caused by foreground objects are considered (Andrade et al. 2006b).

Similar to foreground pixel-based features, an optical flow-based activity representation mostly avoids tracking of individual objects, and are particularly useful for crowd flow analysis in high density crowds with severe clutter and occlusion (Andrade et al. 2006a, 2006b; Ali and Shah 2007). In contrast to foreground pixel-based features, optical flow permits straightforward extraction of motion direction and speed, which can be important for understanding certain types of activity, for example, driving behaviour at a traffic intersection. However, optical flow computation assumes relatively small inter-frame object displacement and constant brightness (Horn and Schunck 1981). Both assumptions are often invalid resulting in poor and inaccurate flow estimation.

In recent years, more sophisticated image intensity gradient-features have been exploited for activity representation (Dalal and Triggs 2005). For instance, histogram-of-oriented-gradient (HOG) features are used to detect unusual events in web-camera videos (Breitenstein et al. 2009); space-time intensity gradients are applied as salient features for learning disparities between activities (Zelnik-Manor and Irani 2006); spatio-temporal gradients are employed to characterise activities in extremely crowded scene (Kratz and Nishino 2009); and dynamic textures are used to represent activity patterns (Mahadevan et al. 2010). Whilst these gradient-feature based representations offer richer information, they are also relatively expensive to compute.

3.1.4 Event-Based Representation

Overall, direct pixel-based representations have an advantage with imagery data of low spatial and temporal resolution, and in crowded visual scenes. However, they can also be sensitive to noise due to ignoring any spatial and temporal structural constraints from object entities. To overcome some of the shortcomings of a direct pixel-based representation, a more structural flavoured pixel-level representation can be sought by constructing object-independent contextual events (Gong and Xiang 2003b; Wada and Matsuyama 2000). A contextual event is defined as a group of salient pixels sharing some common spatio-temporal attributes in a local image neighbourhood over time. They are object-independent and location specific as illustrated by the right hand side frames in Figs. 3.1(a), (b) and (c). These events are autonomous by which the number of events and their location in a scene are determined automatically bottom-up without top-down manual labelling using predefined hypotheses. Events are detected and categorised by unsupervised clustering (Xiang and Gong 2006b). An example of event clustering and discovery is shown in Fig. 3.3(a), where a laboratory shopping scene is analysed. The scene consists of a shopkeeper sitting behind a table on the right side of the view. Drink cans were laid out on a display table. Shoppers entered from the left and either browsed without paying or took a can and paid for it. Such an event-based representation does not seek to segment and form object descriptors, and it remains free from tracking and trajectory matching.

By adopting a contextual-event-based representation, an activity can be represented as a group of co-occurring events in a visual scene. Behavioural interpretation of activities can then be derived from computing temporal and causal correlations among different classes of events (Gong and Xiang 2003a; Xiang and Gong 2006b). Figure 3.1 shows some examples of object-independent contextual events, as compared to object-centred trajectories. The meaning of a contextual event differs somewhat to a more conventional definition of "atomic actions" (Babaguchi et al. 2002; Hongeng and Nevatia 2001; Medioni et al. 2001; Rao et al. 2002).

A contextual-event-based representation puts an emphasis on holistically characterising activities by temporal and causal correlations of all foreground entities in

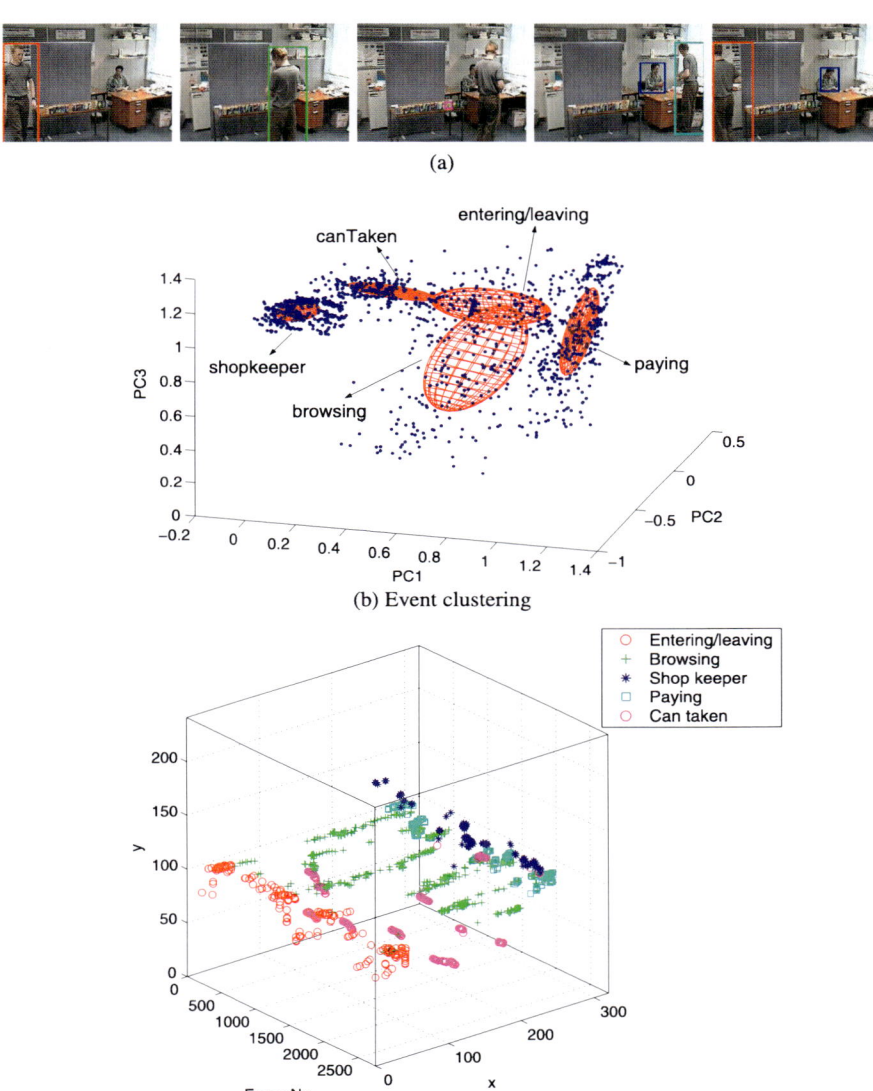

Fig. 3.3 Contextual-event discovery and interpretation in a laboratory shopping scene (Xiang and Gong 2006a). (**a**) Examples of detected and classified events in the image space. Events belonging to different classes are indicated with bounding boxes in *different colours*. (**b**) Unsupervised event clustering in a seven-dimensional feature space (showing the first three principal components for visualisation). (**c**) The location and temporal order of five classes of events being discovered. The *x* and *y* axes are the coordinates of the image plane in pixels

a visual environment, implicitly taking into account all contributing objects in the
scene. This approach to behaviour representation is arguably more robust inherently
than individual object trajectory-based representation, and more suitable for modelling complex activities of groups and interacting multiple objects. When a visual
environment becomes extremely crowded, e.g., in a busy airport, the computation
of contextual events could become unreliable. In this case, simpler local motion
events can be considered (Hospedales et al. 2009; Kuettel et al. 2010; Wang et al.
2009).

3.2 Probabilistic Graphical Models

Behavioural interpretation of activities are commonly treated as a problem of reasoning spatio-temporal correlations and causal relationships among temporal processes in a multivariate space within which activities are represented. However,
visual information about object behaviour is both subject to noise and inherently
ill-defined and incomplete. Whilst these visual phenomena are difficult to be modelled analytically, a statistical learning approach to the problem has been widely
explored. These statistical learning models are designed to not only reconstruct and
describe the characteristics of some training data, but also generalise and interpret
unseen instances of data given a learned model.

Broadly speaking, statistical behaviour models can be categorised into two
types: discriminative models and generative models (Bishop 2006). A discriminative
model provides a function for describing the dependence of an unobserved variable,
such as class label of an action, and an observed variable, such as a feature attribute of the action, by modelling the conditional probability distribution between
the two variables. Using such a model, one can infer the unobserved variable from
the observed variable, for instance, estimating the class label of an action, known as
classification. In contrast, a generative model specifies a joint probability distribution over both observed and unobserved variables. Such a model can also be used to
achieve what a discriminative model does, that is, computing the conditional probability of an unobserved variable from the value of an observed variable using the
Bayes rule. Moreover, it can do what a discriminative model cannot do, that is, to
generate plausible new values of any variable in the model, known as generalisation.
On the other hand, since a discriminative model does not need to model the distribution of the observed variables, it generally can express more complex dependences
between observed and unobserved variables given the same amount of training data.
In practice, this often leads to more accurate classification.

The vast majority of statistical behaviour models are generative models based
on probabilistic graphical models, due to the desire for a model capable of generalising to explain uncertain and complex activities. Common graphical models for
interpreting behaviour include static Bayesian networks (Buxton and Gong 1995;
Gong and Buxton 1993; Intille and Bobick 1999), dynamic Bayesian networks (Du
et al. 2006; Duong et al. 2005; Gong and Xiang 2003a; Xiang and Gong 2006a,

2008c), probabilistic topic models, which further include probabilistic latent semantic analysis (Li et al. 2008; Zhang and Gong 2010a), latent Dirichlet allocation
model (Hospedales et al. 2009; Li et al. 2009; Mehran et al. 2009), and hierarchical
Dirichlet processes model (Kuettel et al. 2010; Wang et al. 2009). Other graphical
models have also been exploited for activity modelling. For instance, propagation
net (Shi et al. 2004, 2006), a subset of dynamic Bayesian networks with the ability
to explicitly model temporal interval durations, has been studied for modelling the
duration of temporal sub-intervals in multiple events. In another study, Hamid et al.
(2007) consider a suffix tree (McCreight 1976) to extract variable length of event-
subsequence of an activity in order to improve activity class discovery accuracy.
Kim and Grauman (2009) propose to detect unusual behaviours by using a space-
time Markov random field, an undirected graphical model with nodes corresponding
to grids of local image patches, where neighbouring nodes are associated in space
and over time. Other models include Petri nets (Albanese et al. 2008), context-free
grammars (Brand 1996), and stochastic context-free grammars (Ivanov and Bobick
2000).

Besides probabilistic graphical models, there are rule-based methods for behavioural interpretation, which require manually defined behaviour rules (Dee and
Hogg 2004; Medioni et al. 2001; Shet et al. 2006; Tran and Davis 2008). There are
also model-free techniques for behavioural interpretation. For instance, behaviour
can be recognised via template matching (Boiman and Irani 2007), and unusual behaviour can be detected by clustering (Zhong et al. 2004). In the following, we pay
special attention to probabilistic graphical models due to their inherent strength for
handling uncertainty and complexity in behaviour modelling.

A probabilistic graphical model defines a graph to represent a probabilistic conditional independence structure between random variables (Jordan 1998; Koller and
Friedman 2009). It is rooted in two mathematical frameworks, graph theory and
probability theory. It is attractive for modelling behaviour because of its strength
in dealing with two fundamental challenges typically faced by a behaviour modelling problem: uncertainty and complexity (Jordan et al. 1999). The probability
theory provides ways for interfacing a model to noisy and incomplete data, as well
as enforces the consistency of a model and provides access to tractable inference
and learning algorithms. For computation, a behaviour can be considered as structured spatial and temporal patterns of correlated constituent parts, represented as
multivariate random variables in a mathematical model. By representing each constituent of a behaviour as a graph node, a graphical model is designed to capture
the underlying structure of a behaviour whilst providing an intuitively appealing
interface to data. Moreover, the semantics of behaviour, that is, a conceptually
meaningful and coherent interpretation of behaviour, can be ambiguous due to similarities among different behaviours, or a lack of details in visual observation. A
model must also cope with uncertainty in data caused by image noise, incomplete
information and a lack of contextual knowledge about behaviour. The ability to
overcome uncertainty and incompleteness by probabilistic modelling with inherent structural constraints makes graphical models appealing for behaviour modelling.

3.2.1 Static Bayesian Networks

A Bayesian network is a directed acyclic graphical model with nodes representing variables of interest such as the occurrence of a contextual event, and the links encoding causal dependencies among the variables. The strength of a dependency is parameterised by conditional probabilities that are attached to each cluster of parent-child nodes in the network (Pearl 1988).

Static Bayesian networks have been a popular tool for activity modelling due to their powerful capabilities in representing and reasoning uncertain visual observations, as well as their computational feasibility. For instance, Bayesian networks are exploited to model dependencies between a wide-area scene layout and motion patterns from vehicle behaviours in traffic (Buxton and Gong 1995; Gong and Buxton 1993). In another study on modelling multi agent interactions, Bayesian networks are applied for probabilistic representation and recognition of individual agent goals from visual observations (Intille and Bobick 1999).

Despite that static Bayesian networks are capable of reasoning about object behaviours, they are limited to modelling static causal relationships without taking into consideration temporal ordering. Therefore, they are only suitable for activities with stable causal semantics. For modelling less structured group or interactive activities involving multiple temporal processes, dynamic Bayesian networks become necessary.

3.2.2 Dynamic Bayesian Networks

A dynamic Bayesian network (DBN) extends static Bayesian networks by incorporating temporal dependencies between random variables. A DBN is described by its model structure and model parameters. The model structure is defined by the number of hidden state variables, observation variables, and the topology of the network. The model parameters quantify, at a given time instance, state transition probabilities, observation probabilities, and the initial state distributions. Hidden states are often discrete whilst observation are typically continuous random variables. Examples of DBNs are shown in Fig. 3.4.

A DBN is typically partitioned into different variable sets to represent the input, hidden and output variables of a state-space model (Murphy 2002). The simplest DBN is a hidden Markov model (HMM), with one hidden state variable and one observation variable at each time instance. HMMs have been extensively used for activity modelling and recognition. For example, Gong and Buxton (1992) exploit HMMs to model vehicle activities surrounding an aircraft docking operation. Andrade et al. (2006b) employ HMMs to analyse optical flow patterns from human crowds for anomalous flow detection.

In theory, an HMM is sufficient for modelling behaviours of any number of objects and of any complexity. However, this simple model structure is obtained at the price of having a large number of model parameters. Consequently, unless

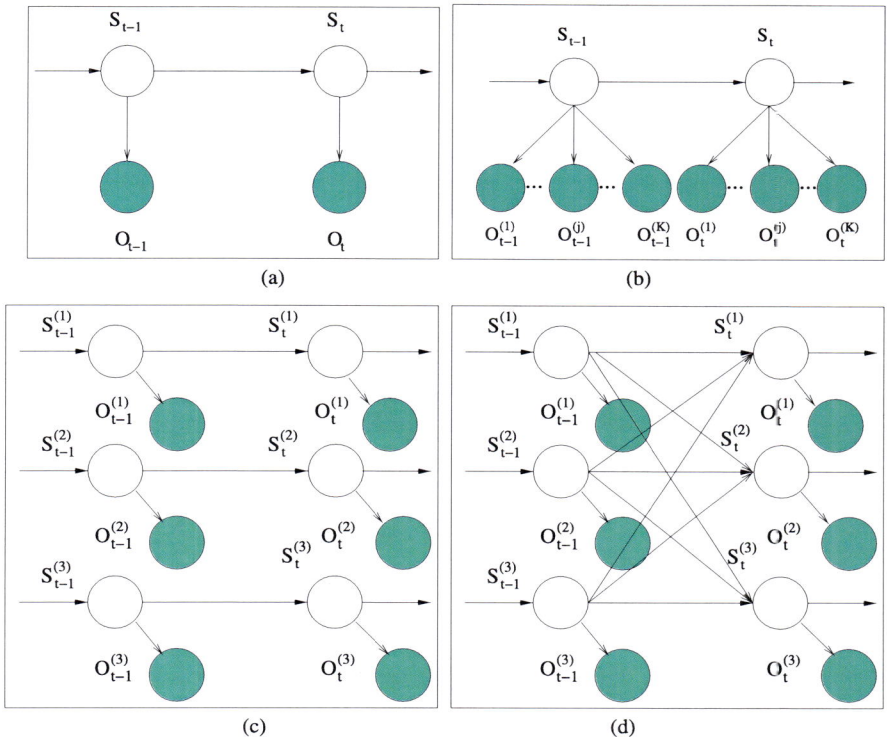

Fig. 3.4 Four different types of dynamic Bayesian networks: (**a**) A standard hidden Markov model, (**b**) a multi-observation hidden Markov model, (**c**) a parallel hidden Markov model, and (**d**) a coupled hidden Markov model. Observation nodes are shown as *shaded circles* and hidden nodes as *clear circles*

the training data size is very large and noise-free, which is not usually the case in practice, poor model learning is expected. A simple extension of an HMM, a multi-observation hidden Markov model (multi-observation HMM) is shown in Fig. 3.4(b). Compared to an HMM (Fig. 3.4(a)), the observational space of a multi-observation HMM is factorised into constituent parts and each part is assumed to be independent of each other. Consequently, the number of parameters for describing a multi-observation HMM is much lower than that for an HMM (Xiang and Gong 2008d). In reality, this independence assumption may not always be true. However, the reduced number of model parameters is beneficial for learning a better model especially given limited training data.

To model more complex activities, various topological extensions to the standard HMM have been considered. They all aim to factorise either the state space or the observation space by introducing multiple hidden state variables and observation variables. Examples of such extensions include a coupled hidden Markov model (Fig. 3.4(d)) for modelling interactions between temporal processes (Brand et al. 1997), a parallel hidden Markov model (Fig. 3.4(c)) for learning independent processes of sign language (Vogler and Metaxas 2001), and a dynamically multi-linked hidden Markov model for interpreting group activities (Gong and Xiang 2003a). There are also several attempts to embed an hierarchical behaviour structure in a model topology. Examples include hierarchical hidden Markov model (Loy et al. 2011; Nguyen et al. 2005) and switching hidden semi-Markov model (Duong et al. 2005), in which the state space is decomposed into multiple levels of states according to the hierarchical structure of a behaviour.

DBNs have in general been shown to be effective in modelling temporal and causal dependencies of complex behaviours. However, learning a DBN with multiple temporal processes is computationally expensive and often requires a large amount of training data, or extensive hand-tuning by human experts. For instance, exact inference on a coupled hidden Markov model beyond two chains, each chain corresponding to one object, is likely to be computationally intractable (Brand et al. 1997; Oliver et al. 2000). This computational complexity problem limits the applicability of DBNs in large scale settings such as global activity modelling in a multi-camera network (Loy et al. 2010b).

3.2.3 Probabilistic Topic Models

Probabilistic topic models (PTM), such as probabilistic latent semantic analysis (Hofmann 2001) and latent Dirichlet allocation (LDA) (Blei et al. 2003), have recently been applied to activity analysis and behavioural anomaly detection (Hospedales et al. 2009; Kuettel et al. 2010; Li et al. 2008, 2010; Mehran et al. 2009; Wang et al. 2009; Yang et al. 2009). Probabilistic topic models are essentially bag-of-words models that traditionally used in text mining for discovering topics from text documents according to co-occurrence of words (Blei and Lafferty 2009). A bag-of-words model can represent a text document as an unordered collection of words, discarding any grammatical structure and word order. A topic model constructs collections of words as topics by clustering co-occurring words. There may be words shared by different topics as a result. For activity modelling in video data, local motion events and segmented video clips are often treated analogously as 'words' and 'documents' as in text document analysis, and each clips may be viewed as a mixture of various 'topics' that represent events (Wang et al. 2009). Typically then, an image sequence can be divided into a sequence of short clips or video documents, each of which is represented by words that are constructed from extracted visual features or pixel spatio-temporal attributes accumulated over a temporal window, e.g. orientation value of pixel-wise optical flow vectors. Once a PTM is constructed from some training video documents, it can then be applied to explain new video documents by evaluating the expectation of detected video words in the new video data whilst considering interactions (topic) among those detected words. If certain detected events in these new data are evaluated with high expectation in the presence of other detected events, they are considered as normal. Otherwise, they are anomalies.

In general, PTMs are less demanding computationally and less sensitive to noise in comparison to DBNs due to its bag-of-words representation. This advantage, however, is based on a sacrifice of throwing away any temporal structure and ordering of activities. To address this shortcoming, Hospedales et al. (2009) formulate a Markov clustering topic model using a single Markov chain to capture explicitly temporal dependencies between activity topics. Kuettel et al. (2010) further extend this model to learn an arbitrary number of infinite hidden Markov models for multiple global temporal rules.

3.3 Learning Strategies

A statistical behaviour model is learned from some training data. The process of model learning consists of learning both a model structure and model parameters. For a graphical model, its model structure refers to (1) the number of hidden states of each hidden variable of a model and (2) the conditional independence structure of a model, that is, factorisation of the model state space in order to determine a suitable graph topology. There are extensive studies in the machine learning community on graphical model parameter learning when the structure of a model is known or assumed *a priori* (Ghahramani 1998). There are also some efforts on developing methods for learning the optimal structure of an unknown DBN, which is a more challenging problem (Beal and Ghahramani 2003; Brand 1999a; Friedman et al. 1998; Xiang and Gong 2008a, 2008b). In the following, we consider different learning strategies for building behaviour models, ranging from supervised learning, unsupervised learning, semi-supervised learning, weakly supervised learning, to active learning.

3.3.1 Supervised Learning

In a fully supervised learning paradigm, it is assumed that both positive and negative instances for each pattern class are well-defined and available during a training phase (Bishop 2006). Many earlier works on learning behaviour models take a supervised learning strategy. For example, Oliver et al. (2000) describe a coupled HMM formulation to recognise human interactions, with the assumption that there exist known *a priori* behaviour classes. In a car parking scene, Morris and Hogg (2000) detect atypical trajectories by training a statistical model with both normal and atypical trajectories. In another study, Sacchi et al. (2001) train a neural network with both normal and unusual behaviours to detect vandal acts. Gong and Xiang (2003a) recognise group activities by training a dynamically multi-linked HMM, of which the hidden states correspond to different event classes that are known beforehand. A different supervised approach can be taken whereby human operator interpretation of what is normal and unusual is hand-crafted into a model for rule-based abnormal behaviour detection (Dee and Hogg 2004; Shet et al. 2006; Tran and Davis 2008).

Supervised learning assumes that labelled examples of behaviour categories are made available. A class model is built for each target category. A supervised approach implies that the categories are known and sufficient examples for each category are available. However, both assumptions may not be valid due to: (1) Natural behaviours are usually complex; therefore in a given scene, there could be a large number of behaviour classes. To determine an exhaustive list of them is non-trivial if not impossible. Annotation of object behaviours is expensive and unreliable, much more so than labelling object image classes. (2) For visual analysis of behaviour, a primary goal is to discover automatically new and rare behaviour categories that are also genuinely interesting to a human observer. The ability for a behaviour model to discover unknown, as well as to interpret observations against pre-defined behaviour categories, is highly desirable but also considerably more challenging.

3.3.2 Unsupervised Learning

In contrast to supervised learning, unsupervised learning does not assume any prior knowledge on the pattern classes. A common strategy for unsupervised learning is clustering of unlabelled data. Clustering is the process of allocating and dividing observational data points into a number of subsets, known as clusters, so that those observations in the same cluster are similar according to a defined distance metric (Bishop 2006; Dempster et al. 1977). In general, clustering considers the problem of determining both model parameters and model order, that is, model complexity (Akaike 1973; Schwarz 1978; Xiang and Gong 2006b, 2008b).

Clustering has been exploited for modelling behaviour. For instance, Zhong et al. (2004) perform bi-partite co-clustering on video segments and identify isolated clusters as unusual behaviours. With a similar idea, Lee et al. (2006) use N-cut clustering technique to separate unusual behaviours from the normal behaviours based on a distance measurement. Hamid et al. (2005) employ graph theoretical clustering to cluster normal and abnormal behaviours. Without the need to re-cluster the entire dataset, the abnormality of an unseen behaviour pattern is determined by computing a weighted similarity between the new activity-instance and all the instances in the existing clusters. However, all previous instances have to be stored in memory for similarity comparison, which restrains the scalability of this method.

One unsupervised learning strategy that has gained growing popularity for behaviour analysis in visual surveillance is to profile vast quantity of visual observation of activities. Unusual behaviours, considered as anomalies, are subsequently detected as *outlier*, that is, those behaviour patterns that statistically deviate from the learned normal profile. In a surveillance setting, behaviours are typically characterised by vast quantity of normal activities and relatively scarce amount of unusual incidents. In addition, it is considered that normal behaviours are well-defined thus form clear clusters in a model space as compared to unusual instances with unpredictable variations. For that reason, many probabilistic graphical models for behaviour analysis are based on unsupervised learning, including DBNs (Duong et al.

2005) and PTMs (Hospedales et al. 2009; Li et al. 2008, 2009; Mehran et al. 2009; Wang et al. 2009).

Unsupervised learning can be performed in an incremental manner to make a behaviour model adaptive to changes in visual context. Visual analysis of behaviour in a realistic and unconstrained visual environment is challenging not only because the complexity and variety of visual behaviour, but also changes in visual context over time. Definitions of what constitute being normal and abnormal can also change over time (Xiang and Gong 2008b). A behaviour can be considered as either being normal or abnormal depending on when and where it takes place. This causes problems for a behaviour model learned solely from past visual observations. To overcome this problem, a behaviour model needs to be updated incrementally by new observations in order to adapt to changes in the definition and context of behaviours. Very few methods accommodate on-line model adaptation, whereby a model is capable of updating its internal parameters and structure whilst processing and interpreting live visual input. To cater for changes in visual context, Xiang and Gong (2008b) exploit unsupervised incremental learning. Specifically, a small amount of training data are used to construct an initial model. An incremental expectation-maximisation (EM) criterion is then introduced to determine whether the model parameters need be updated or new classes need be added to the model structure, when new observations are presented to the model. Alternatively, Kim and Grauman (2009) investigate incremental learning where optical flows are represented as mixture of probabilistic principal components (Tipping and Bishop 1999), and a space-time Markov random field (MRF) is used for modelling activity. Incremental updates of the mixture of probabilistic principal components and MRF parameters are carried out to adapt the model with new observations. This enables the model to adapt to changes in visual context over time. For example, earlier events detected as unusual are later detected as normal after several occurrences.

Unsupervised learning is also applicable to model-free methods. For instance, Boiman and Irani (2007) build a database of spatio-temporal patches using only regular or normal behaviour, and detect those patterns that cannot be composed from the database as being unusual. This method, however, suffers a similar scalability problem as that of Hamid et al. (2005), that is, all previous ensembles of spatio-temporal patches have to be stored. Another model-free method is introduced by Breitenstein et al. (2009), whereby typical activities are defined using nearest neighbours. Unusual event is identified by measuring its distance from the distribution of nearest neighbours. This method maintains a fixed number of clusters in the nearest neighbour model, thus avoids the memory problem encountered by Boiman and Irani (2007).

3.3.3 Semi-supervised Learning

In a visual surveillance context, unlabelled training instances is often easy to collect, whilst labelled instances, either normal or abnormal, are difficult and time-consuming to obtain. They require exhaustive annotation from human experts. In between the two extremes of supervised and unsupervised learning, there is semi-supervised learning. This approach to model learning aims to construct a predictive model by making use of typically a small amount of labelled data together with a large number of unlabelled data (Zhu 2007). Current techniques for semi-supervised learning are rather limited. A good example is proposed by Zhang et al. (2005). The key idea is to train a hidden Markov model using a large amount of labelled common behaviours. By performing iterations of likelihood test and Viterbi decoding on unlabelled video data, a number of unusual models are derived by Bayesian adaptation (Reynolds et al. 2000). A drawback of this method is that its accuracy is sensitive to the number of iterations during the semi-supervised adaptation process. It is reported that the false alarm rate increases rapidly along with the number of iterations.

A general concern with deploying semi-supervised learning strategy is that the use of unlabelled data may degrade, instead of improve, model capability. This problem may occur if the model assumption matches poorly with the problem structure for which the model aims to learn (Cozman et al. 2003).

3.3.4 Weakly Supervised Learning

For supervised learning of a behaviour model to succeed, a pool of training video data of a sufficiently large size needs be labelled with known behaviour classes. This usually requires a manual annotation process that is in practice often too expensive to undertake, and without a guarantee of consistency in labelling. The semi-supervised strategy discussed above could provide one solution to this problem, with its limitations. An alternative solution is to adopt a weakly supervised learning strategy, which requires less precise manual labelling to the training data. For instance, instead of annotating the exact location of an action and activity, only a general binary label is required which states whether an action or activity of interest is present in the video data, without having to provide detailed information on where and what. With less detailed labelling information, the learning problem becomes harder. Specifically, a model has to learn automatically parameters capable of associating where the behaviour pattern of interest is located in a scene (segmentation) as well as what that behaviour is (classification). Weakly supervised learning has been studied for both action recognition (Hu et al. 2009; Siva and Xiang 2010; Yuan et al. 2009) and rare behaviour detection (Li et al. 2010).

3.3.5 Active Learning

Existing behaviour models mostly address the unusual behaviour detection problem using an unsupervised outlier detection strategy, that is, training a model with normal patterns and flagging a newly observed pattern as an anomaly if it deviates statistically from the learned normal profile (Hospedales et al. 2009; Kim and Grauman 2009; Mehran et al. 2009; Wang et al. 2009). However, these unsupervised modelling techniques entail a number of difficulties, including detecting visually subtle and non-exaggerated anomalies, and separating true anomalies from uninteresting statistical outliers. In essence, detectable visual attributes may simply not exist in the training data and cannot be extracted by unsupervised mining. As a consequence, a model learned from data may produce a high false alarm rate due to excessive number of uninteresting outliers being detected.

Human feedback, when available, can play an important role in model learning. This is because (1) given unlimited data, it is theoretically possible to learn a model for any complex behaviour. However, the training process will be computational prohibitive. Utilising human feedback provides a solution for fast model learning. (2) Semantic interpretation of behaviour is context dependent. Context in itself can be non-visual, in which case it cannot be learned from visual data no matter how large the size of the training data pool is. In this case, integrating human feedback into model learning becomes crucial. (3) Due to large inter-class similarities, different behaviour classes often overlap in a model space. Without human supervision, distinguishing the overlapping behaviour classes is extremely difficult. (4) When an unsupervised model makes an error, for example, discovering the wrong categories, the error can be quickly recovered through learning from human feedback. Utilising human feedback is more costly than without. To learn an optimal model with minimal human feedback would be ideal.

Model learning using minimal human feedback is not the same as supervised learning by labelling a small random subset of the training data. The latter treats all examples equally therefore the selection for labelled feedback is random. Fundamentally, not all examples are equal for assisting the learning of correct decision boundaries in a model space. A solution for quantifying the relevance of different training examples in relation to their benefits to learning a particular model is known as active learning (Settles 2010). Unlike both supervised and semi-supervised learning, an active learning method is designed to identify and select automatically the most informative training examples for human labelling in order to learn a particular model most effectively.

In general, there are two different settings in active learning: pool-based setting and stream-based setting. A pool-based active learning method requires access to a fixed pool of unlabelled data for searching the most informative examples for human feedback. Stream-based setting, on the other hand, does not assume the availability of a large unlabelled training data pool. It requests labels on-the-fly based on sequential observations. This is more challenging since an immediate query decision has to be made without complete knowledge on the underlying data distribution.

There are a few studies on active learning of models for interpreting behaviour. Sillito and Fisher (2008) formulate a method to harnesses human feedback on-the-fly for improving unusual event detection performance. Using a model learned with normal instances, any instances classified as normal events in testing stage will be used together with corresponding predicted labels to re-train the model. On the other hand, human approval is sought if a newly observed instance deviates statistically

from the learned normal profile. If the suspicious instance is indeed normal, it will
be included in the re-training process, or else it will be flagged as an anomaly. This
model has its advantage over other models learned by passive unsupervised learning. Specifically it allows for incremental incorporation of normal instances in an
on-line manner to refine the model, whilst simultaneously it prevents anomalous
behaviour from being inadvertently incorporated into the model so avoiding corrup-
tion to the learned normal profile. In another study, Loy et al. (2010a) propose a
stream-based multi-criteria approach. This study shows that active learning not only
allows a model to achieve a higher classification rate with fewer training examples
compared to passive learning based on labelling random examples, it also helps in
resolving ambiguities of interest. The latter benefit leads to a more robust and accurate detection of visually subtle unusual behaviours compared to unsupervised
learning for outlier detection.

References

Adam, A., Rivlin, E., Shimshoni, I., Reinitz, D.: Robust real-time unusual event detection
    using multiple fixed-location monitors. IEEE Trans. Pattern Anal. Mach. Intell. 30(3),
    555–560 (2008)
Akaike, H.: Information theory and an extension of the maximum likelihood principle. In:
    International Symposium on Information Theory, pp. 267–281 (1973)
Albanese, M., Chellappa, R., Moscato, V., Picariello, A., Subrahmanian, V.S., Turaga, P.,
    Udrea, O.: A constrained probabilistic Petri net framework for human activity detection
    in video. IEEE Trans. Multimed. 10(6), 982–996 (2008)
Ali, S., Shah, M.: A Lagrangian particle dynamics approach for crowd flow segmentation
    and stability analysis. In: IEEE Conference on Computer Vision and Pattern Recognition,
    Minneapolis, USA, June 2007, pp. 1–6 (2007)
Andrade, E.L., Blunsden, S., Fisher, R.B.: Modelling crowd scenes for event detection. In:
    International Conference on Pattern Recognition, pp. 175–178 (2006a)
Andrade, E.L., Blunsden, S., Fisher, R.B.: Hidden Markov models for optical flow analysis
    in crowds. In: International Conference on Pattern Recognition, pp. 460–463 (2006b)
Babaguchi, N., Kawai, Y., Kitahashi, T.: Event based indexing of broadcasting sports video
    by intermodal collaboration. IEEE Trans. Multimed. 4(1), 68–75 (2002)
Baumberg, A., Hogg, D.C.: Generating spatio-temporal models from examples. Image Vis.
    Comput. 14(8), 525–532 (1996)
Beal, M., Ghahramani, Z.: The variational Bayesian EM algorithm for incomplete data: with
    application to scoring graphical model structures. Bayesian Stat. 7, 453–464 (2003)
Beauchemin, S.S., Barron, J.L.: The computation of optical flow. ACM Comput. Surv. 27(3),
    433–466 (1995)
Benezeth, Y., Jodoin, P.M., Saligrama, V., Rosenberger, C.: Abnormal events detection based
    on spatio-temporal co-occurrences. In: IEEE Conference on Computer Vision and Pattern
    Recognition, Miami, USA, June 2009, pp. 2458–2465 (2009)
Bishop, C.M.: Pattern Recognition and Machine Learning. Springer, Berlin (2006)
Blank, M., Gorelick, L., Shechtman, E., Irani, M., Basri, R.: Actions as space-time shapes.
    In: IEEE International Conference on Computer Vision, pp. 1395–1402 (2005)
Blei, D.M., Lafferty, J.: Topic Models. In Text Mining: Theory and Applications. Taylor &
    Francis, London (2009)

Blei, D.M., Ng, A.Y., Jordan, M.I.: Latent Dirichlet allocation. J. Mach. Learn. Res. **3**, 993–1022 (2003)

Bobick, A.F., Davis, J.: The recognition of human movement using temporal templates. IEEE Trans. Pattern Anal. Mach. Intell. **23**(3), 257–267 (2001)

Boiman, O., Irani, M.: Detecting irregularities in images and in video. Int. J. Comput. Vis. **74**(1), 17–31 (2007)

Brand, M.: Understanding manipulation in video. In: International Conference on Automatic Face and Gesture Recognition, Killington, USA, pp. 94–99 (1996)

Brand, M.: Structure discovery in conditional probability models via an entropic prior and parameter extinction. Neural Comput. **11**(5), 1155–1182 (1999a)

Brand, M.: Shadow puppetry. In: IEEE International Conference on Computer Vision, Corfu, Greece, September 1999, pp. 1237–1244 (1999b)

Brand, M., Oliver, N., Pentland, A.: Coupled hidden Markov models for complex action recognition. In: IEEE Conference on Computer Vision and Pattern Recognition, San Juan, Puerto Rico, pp. 994–999 (1997)

Breitenstein, M.D.: Visual surveillance—dynamic behavior analysis at multiple levels. PhD thesis, ETH Zurich (2009)

Breitenstein, M.D., Grabner, H., Van Gool, L.: Hunting Nessie—real-time abnormality detection from webcams. In: IEEE International Workshop on Visual Surveillance, Kyoto, Japan, October 2009, pp. 1243–1250 (2009)

Brostow, G.J., Cipolla, R.: Unsupervised Bayesian detection of independent motion in crowds. In: IEEE Conference on Computer Vision and Pattern Recognition, pp. 594–601 (2006)

Buxton, B.F., Buxton, H.: Monocular depth perception from optical flow by space time signal processing. Proc. R. Soc. **218**(1210), 27–47 (1983)

Buxton, H., Gong, S.: Visual surveillance in a dynamic and uncertain world. Artif. Intell. **78**(1–2), 431–459 (1995)

Cozman, F., Cohen, I., Cirelo, M.: Semi-supervised learning of mixture models. In: International Conference on Machine Learning, Washington, DC, USA, October 2003, pp. 99–106 (2003)

Dalal, N., Triggs, B.: Histograms of oriented gradients for human detection. In: IEEE Conference on Computer Vision and Pattern Recognition, San Diego, USA, June 2005, pp. 886–893 (2005)

Dee, H., Hogg, D.C.: Detecting inexplicable behaviour. In: British Machine Vision Conference, pp. 477–486 (2004)

Dempster, A., Laird, N., Rubin, D.: Maximum-likelihood from incomplete data via the EM algorithm. J. R. Stat. Soc. B **39**, 1–38 (1977)

Dollár, P., Rabaud, V., Cottrell, G., Belongie, S.: Behavior recognition via sparse spatio-temporal features. In: IEEE International Workshop on Visual Surveillance and Performance Evaluation of Tracking and Surveillance, pp. 65–72 (2005)

Du, Y., Chen, F., Xu, W., Li, Y.: Recognizing interaction activities using dynamic Bayesian network. In: International Conference on Pattern Recognition, Hong Kong, China, pp. 618–621 (2006)

Du, Y., Chen, F., Xu, W.: Human interaction representation and recognition through motion decomposition. IEEE Signal Process. Lett. **14**(12), 952–955 (2007)

Duong, T., Bui, H., Phung, D., Venkatesh, S.: Activity recognition and abnormality detection with the switching hidden semi-Markov model. In: IEEE Conference on Computer Vision and Pattern Recognition, San Diego, USA, June 2005, pp. 838–845 (2005)

Efros, A., Berg, A., Mori, G., Malik, J.: Recognizing action at a distance. In: IEEE International Conference on Computer Vision, Nice, France, pp. 726–733 (2003)

Fei-Fei, L., Perona, P.: A Bayesian hierarchical model for learning natural scene categories. In: IEEE Conference on Computer Vision and Pattern Recognition, San Diego, USA, June 2005, pp. 524–531 (2005)

Fei-Fei, L., Fergus, R., Perona, P.: One-shot learning of object categories. IEEE Trans. Pattern Anal. Mach. Intell. **28**(4), 594–611 (2006)

Fergus, R., Perona, P., Zisserman, A.: Weakly supervised scale-invariant learning of models for visual recognition. Int. J. Comput. Vis. **71**(3), 273–303 (2007)

Fischler, M.A., Elschlager, R.A.: The representation and matching of pictorial structures. IEEE Trans. Comput. **2**(1), 67–92 (1973)

Friedman, N., Murphy, K., Russell, S.: Learning the structure of dynamic probabilistic networks. In: Uncertainty in Artificial Intelligence, pp. 139–147 (1998)

Fu, Z., Hu, W., Tan, T.: Similarity based vehicle trajectory clustering and anomaly detection. In: International Conference on Image Processing, pp. 602–605 (2005)

Galata, A., Johnson, N., Hogg, D.C.: Learning variable length Markov models of behaviour. Comput. Vis. Image Underst. **81**(3), 398–413 (2001)

Ghahramani, Z.: Learning dynamic Bayesian networks. In: Adaptive Processing of Sequences and Data Structures. Lecture Notes in AI, pp. 168–197 (1998)

Gong, S., Brady, M.: Parallel computation of optic flow. In: European Conference on Computer Vision, Antibes, France, pp. 124–134 (1990)

Gong, S., Buxton, H.: On the visual expectation of moving objects: A probabilistic approach with augmented hidden Markov models. In: European Conference on Artificial Intelligence, Vienna, Austria, August 1992, pp. 781–786 (1992)

Gong, S., Buxton, H.: Bayesian nets for mapping contextual knowledge to computational constraints in motion segmentation and tracking. In: British Machine Vision Conference, Guildford, UK, September 1993, pp. 229–238 (1993)

Gong, S., Xiang, T.: Recognition of group activities using dynamic probabilistic networks. In: IEEE International Conference on Computer Vision, Nice, France, October 2003, pp. 742–749 (2003a)

Gong, S., Xiang, T.: Scene event recognition without tracking. Acta Autom. Sin. **29**(3), 321–331 (2003b)

Gorelick, L., Blank, M., Shechtman, E., Irani, M., Basri, R.: Actions as space-time shapes. IEEE Trans. Pattern Anal. Mach. Intell. **29**(12), 2247–2253 (2007)

Hamid, R., Johnson, A., Batta, S., Bobick, A.F., Isbell, C., Coleman, G.: Detection and explanation of anomalous activities—representing activities as bags of event n-grams. In: IEEE Conference on Computer Vision and Pattern Recognition, pp. 1031–1038 (2005)

Hamid, R., Maddi, S., Bobick, A.F., Essa, M.: Structure from statistics—unsupervised activity analysis using suffix trees. In: IEEE International Conference on Computer Vision, Rio de Janeiro, Brasil, October 2007, pp. 1–8 (2007)

Haritaoglu, I., Harwood, D., Davis, L.S.: W^4: Real-time surveillance of people and their activities. IEEE Trans. Pattern Anal. Mach. Intell. **22**(8), 809–830 (2000)

Hofmann, T.: Unsupervised learning by probabilistic latent semantic analysis. Mach. Learn. **42**(1/2), 177–196 (2001)

Hongeng, S., Nevatia, R.: Multi-agent event recognition. In: IEEE International Conference on Computer Vision, pp. 80–86 (2001)

Horn, B.K.P., Schunck, B.G.: Determining optical flow. Artif. Intell. **17**, 185–203 (1981)

Hospedales, T., Gong, S., Xiang, T.: A Markov clustering topic model for mining behaviour in video. In: IEEE International Conference on Computer Vision, Kyoto, Japan, October 2009, pp. 1165–1172 (2009)

Hu, Y., Cao, L., Lv, F., Yan, S., Gong, Y., Huang, T.S.: Action detection in complex scenes with spatial and temporal ambiguities. In: IEEE International Conference on Computer Vision, Kyoto, Japan, October 2009, pp. 128–135 (2009)

Intille, S.S., Bobick, A.F.: A framework for recognizing multi-agent action from visual evidence. In: National Conference on Artificial Intelligence, Menlo Park, USA, pp. 518–525 (1999)

Isard, M., Blake, A.: Contour tracking by stochastic propagation of conditional density. In: European Conference on Computer Vision, pp. 343–356 (1996)

Ivanov, Y.A., Bobick, A.F.: Recognition of visual activities and interactions by stochastic parsing. IEEE Trans. Pattern Anal. Mach. Intell. **22**(8), 852–872 (2000)

Jiang, F., Wu, Y., Katsaggelos, A.K.: A dynamic hierarchical clustering method for trajectory-based unusual video event detection. IEEE Trans. Image Process. **18**(4), 907–913 (2009)

Jodoin, P.M., Konrad, J., Saligrama, V.: Modeling background activity for behavior subtraction. In: International Conference on Distributed Smart Cameras, pp. 1–10 (2008)

Johnson, N., Hogg, D.C.: Learning the distribution of object trajectories for event recognition. Image Vis. Comput. **14**(8), 609–615 (1996)

Jordan, M.I., Ghahramani, Z., Jaakkola, T., Saul, L.: An introduction to variational methods for graphical models. Mach. Learn. **37**, 183–233 (1999)

Jordan, M.I.: Learning in Graphical Models. MIT Press, Cambridge (1998)

Kalman, R.E.: A new approach to linear filtering and prediction problems. Trans. ASME, J. Basic Eng., Ser. D **82**, 35–45 (1960)

Ke, Y., Sukthankar, R., Hebert, M.: Event detection in crowded videos. In: IEEE International Conference on Computer Vision, Rio de Janeiro, Brasil, October 2007, pp. 1–8 (2007)

Kim, J., Grauman, K.: Observe locally, infer globally: a space-time MRF for detecting abnormal activities with incremental updates. In: IEEE Conference on Computer Vision and Pattern Recognition, pp. 2921–2928 (2009)

Koller, D., Friedman, N.: Probabilistic Graphical Models: Principles and Techniques. MIT Press, Cambridge (2009)

Kratz, L., Nishino, K.: Anomaly detection in extremely crowded scenes using spatio-temporal motion pattern models. In: IEEE Conference on Computer Vision and Pattern Recognition, pp. 1446–1453 (2009)

Kuettel, D., Breitenstein, M.D., Van Gool, L., Ferrari, V.: What's going on? discovering spatio-temporal dependencies in dynamic scenes. In: IEEE Conference on Computer Vision and Pattern Recognition, San Francisco, USA, June 2010, pp. 1951–1958 (2010)

Laptev, I.: On space-time interest points. Int. J. Comput. Vis. **64**(2), 107–123 (2005)

Laptev, I., Caputo, B., Schüldt, C., Lindeberg, T.: Local velocity-adapted motion events for spatio-temporal recognition. Comput. Vis. Image Underst. **108**(3), 207–229 (2007)

Lee, C.K., Ho, M.F., Wen, W.S., Huang, C.L.: Abnormal event detection in video using N-cut clustering. In: International Conference on Intelligent Information Hiding and Multimedia Signal Processing, pp. 407–410 (2006)

Leibe, B., Leonardis, A., Schiele, B.: Robust object detection with interleaved categorization and segmentation. Int. J. Comput. Vis. **77**(3), 259–289 (2008)

Lewis, D.: Naive Bayes at forty: the independence assumption in information retrieval. In: European Conference on Machine Learning, Chemnitz, Germany, April 1998, pp. 4–15 (1998)

Li, J., Gong, S., Xiang, T.: Global behaviour inference using probabilistic latent semantic analysis. In: British Machine Vision Conference, Leeds, UK, pp. 193–202 (2008)

Li, J., Gong, S., Xiang, T.: Discovering multi-camera behaviour correlations for on-the-fly global activity prediction and anomaly detection. In: IEEE International Workshop on Visual Surveillance, Kyoto, Japan, October 2009

Li, J., Hospedales, T., Gong, S., Xiang, T.: Learning rare behaviours. In: Asian Conference on Computer Vision, Queenstown, New Zealand, November 2010

Liu, J., Yang, Y., Shah, M.: Learning semantic visual vocabularies using diffusion distance. In: IEEE Conference on Computer Vision and Pattern Recognition, Miami, USA, June 2009, pp. 461–468 (2009)

Lowe, D.: Distinctive image features from scale-invariant keypoints. Int. J. Comput. Vis. **60**(2), 91–110 (2004)

Loy, C.C., Xiang, T., Gong, S.: Stream-based active unusual event detection. In: Asian Conference on Computer Vision, Queenstown, New Zealand, November 2010a

Loy, C.C., Xiang, T., Gong, S.: Time-delayed correlation analysis for multi-camera activity understanding. Int. J. Comput. Vis. **90**(1), 106–129 (2010b)

Loy, C.C., Xiang, T., Gong, S.: Detecting and discriminating behavioural anomalies. Pattern Recognit. **44**(1), 117–132 (2011)

Lucas, B.D., Kanade, T.: An iterative image registration technique with an application to stereo vision. In: DARPA Image Understanding Workshop, pp. 121–130 (1981)

Mahadevan, V., Li, W., Bhalodia, V., Vasconcelos, N.: Anomaly detection in crowded scenes. In: IEEE Conference on Computer Vision and Pattern Recognition, San Francisco, USA, June 2010

Marr, D.: Vision: A Computational Investigation Into the Human Representation and Processing of Visual Information. Freeman, New York (1982)

McCreight, E.M.: A space-economical suffix tree construction algorithm. J. ACM **23**(2), 262–272 (1976)

McKenna, S., Jabri, S., Duric, Z., Rosenfeld, A., Wechsler, H.: Tracking group of people. Comput. Vis. Image Underst. **80**, 42–56 (2000)

Medioni, G., Cohen, I., Bremond, F., Hongeng, S., Nevatia, R.: Event detection and analysis from video streams. IEEE Trans. Pattern Anal. Mach. Intell. **23**(8), 873–889 (2001)

Mehran, R., Oyama, A., Shah, M.: Abnormal crowd behaviour detection using social force model. In: IEEE Conference on Computer Vision and Pattern Recognition, pp. 935–942 (2009)

Moeslunda, T.B., Hilton, A., Krügerc, V.: A survey of advances in vision-based human motion capture and analysis. Comput. Vis. Image Underst. **104**(2–3), 90–126 (2006)

Morris, R., Hogg, D.C.: Statistical models of object interaction. Int. J. Comput. Vis. **37**(2), 209–215 (2000)

Murphy, K.P.: Hidden Semi-Markov Models HSMMs. Unpublished notes (2002)

Naftel, A., Khalid, S.: Classifying spatiotemporal object trajectories using unsupervised learning in the coefficient feature space. Multimed. Syst. 227–238 (2006)

Nascimento, J.C., Figueiredo, M.A.T., Marques, J.S.: Semi-supervised learning of switched dynamical models for classification of human activities in surveillance applications. In: IEEE International Conference on Image Processing, pp. 197–200 (2007)

Nevatia, R., Binford, T.O.: Description and recognition of curved objects. Artif. Intell. **8**(1), 77–98 (1977)

Ng, J., Gong, S.: Learning pixel-wise signal energy for understanding semantics. Image Vis. Comput. **21**(12–13), 1183–1189 (2003)

Nguyen, N.T., Phung, D.Q., Venkatesh, S., Bui, H.H.: Learning and detecting activities from movement trajectories using the hierarchical hidden Markov model. In: IEEE Conference on Computer Vision and Pattern Recognition, San Diego, USA, pp. 955–960 (2005)

Niebles, J.C., Wang, H., Fei-Fei, L.: Unsupervised learning of human action categories using spatial-temporal words. In: British Machine Vision Conference, Edinburgh, UK (2006)

Niebles, J.C., Wang, H., Fei-Fei, L.: Unsupervised learning of human action categories using spatial-temporal words. Int. J. Comput. Vis. **79**(3), 299–318 (2008)

Okuma, K., Taleghani, A., de Freitas, N., Little, J.J., Lowe, D.: A boosted particle filter: multitarget detection and tracking. In: European Conference on Computer Vision, Prague, Czech Republic, May 2004, pp. 28–29 (2004)

Oliver, N., Rosario, B., Pentland, A.: A Bayesian computer vision system for modeling human interactions. IEEE Trans. Pattern Anal. Mach. Intell. **22**(8), 831–843 (2000)

Owens, J., Hunter, A.: Application of the self-organizing map to trajectory classification. In: IEEE International Workshop on Visual Surveillance, pp. 77–83 (2000)

Pearl, J.: Probabilistic Reasoning in Intelligent Systems: Networks of Plausible Inference. Morgan Kaufmann, San Mateo (1988)

Piciarelli, C., Foresti, G.L.: On-line trajectory clustering for anomalous events detection. Pattern Recognit. Lett. **27**, 1835–1842 (2006)

Psarrou, A., Gong, S., Walter, M.: Recognition of human gestures and behaviour based on motion trajectories. Image Vis. Comput. **20**(5–6), 349–358 (2002)

Raja, Y., McKenna, S., Gong, S.: Tracking and segmenting people in varying lighting conditions using colour. In: IEEE International Conference on Automatic Face & Gesture Recognition, Nara, Japan, pp. 228–233 (1998)

Rao, C., Yilmaz, A., Shah, M.: View-invariant representation and recognition of actions. Int. J. Comput. Vis. **50**, 203–226 (2002)

Reynolds, D.A., Quatieri, T.F., Dunn, R.B.: Speaker verification using adapted Gaussian mixture models. Digit. Signal Process. **10**(1–3), 19–41 (2000)

Russell, D., Gong, S.: Minimum cuts of a time-varying background. In: British Machine Vision Conference, Edinburgh, UK, September 2006, pp. 809–818 (2006)

Russell, D., Gong, S.: Multi-layered decomposition of recurrent scene. In: European Conference on Computer Vision, Marseille, France, October 2008, pp. 574–587 (2008)

Sacchi, C., Regazzoni, C., Gera, G., Foresti, G.: A neural network-based image processing system for detection of vandal acts in unmanned railway environments. In: International Conference on Image Analysis and Processing, pp. 529–534 (2001)

Saleemi, I., Shafique, K., Shah, M.: Probabilistic modeling of scene dynamics for applications in visual surveillance. IEEE Trans. Pattern Anal. Mach. Intell. **31**(8), 1472–1485 (2009)

Schüldt, C., Laptev, I., Caputo, B.: Recognizing human actions: A local SVM approach. In: International Conference on Pattern Recognition, Cambridge, UK, pp. 32–36 (2004)

Schwarz, G.: Estimating the dimension of a model. Ann. Math. Stat. **6**(2), 461–464 (1978)

Settles, B.: Active learning literature survey. Technical report, University of Wisconsin-Madison (2010)

Shet, V., Harwood, D., Davis, L.S.: Multivalued default logic for identity maintenance in visual surveillance. In: European Conference on Computer Vision, pp. 119–132 (2006)

Shi, Y., Huang, Y., Minnen, D., Bobick, A.F., Essa, I.: Propagation networks for recognition of partially ordered sequential action. In: IEEE Conference on Computer Vision and Pattern Recognition, pp. 862–869 (2004)

Shi, Y., Bobick, A.F., Essa, I.: Learning temporal sequence model from partially labeled data. In: IEEE Conference on Computer Vision and Pattern Recognition, New York, USA, pp. 1631–1638 (2006)

Shipley, T.F., Kellman, P.J., Shipley, T.F.: From Fragments to Objects: Segmentation and Grouping in Vision. North-Holland, Amsterdam (2001)

Sillito, R.R., Fisher, R.B.: Semi-supervised learning for anomalous trajectory detection. In: British Machine Vision Conference, Leeds, UK, September 2008

Siva, P., Xiang, T.: Action detection in crowd. In: British Machine Vision Conference, Aberystwyth, UK, September 2010

Sminchisescu, C., Kanaujia, A., Metaxas, D.: Conditional models for contextual human motion recognition. Comput. Vis. Image Underst. **104**(2–3), 210–220 (2006)

Spelke, E.S.: Principles of object perception. Cogn. Sci. **14**, 29–56 (1990)

Stauffer, C., Grimson, W.E.L.: Adaptive background mixture models for real time tracking. In: IEEE Conference on Computer Vision and Pattern Recognition, vol. 2, pp. 246–252 (1999)

Tipping, M.E., Bishop, C.M.: Mixtures of probabilistic principal component analyzers. Neural Comput. **11**(2), 443–482 (1999)

Tran, S., Davis, L.S.: Event modeling and recognition using Markov logic networks. In: European Conference on Computer Vision, Marseille, France, pp. 610–623 (2008)

Vogler, C., Metaxas, D.: A framework for recognizing the simultaneous aspects of American sign language. Comput. Vis. Image Underst. **81**(3), 358–384 (2001)

Wada, T., Matsuyama, T.: Multiobject behavior recognition by event driven selective attention method. IEEE Trans. Pattern Anal. Mach. Intell. **22**(8), 873–887 (2000)

Walter, M., Psarrou, A., Gong, S.: Data driven gesture model acquisition using minimum description length. In: British Machine Vision Conference, pp. 673–683 (2001)

Wang, L., Tan, T., Ning, H., Hu, W.: Silhouette analysis-based gait recognition for human identification. IEEE Trans. Pattern Anal. Mach. Intell. **25**(12), 1505–1518 (2003)

Wang, X., Tieu, K., Grimson, W.E.L.: Learning semantic scene models by trajectory analysis. In: European Conference on Computer Vision, pp. 110–123 (2006)

Wang, X., Ma, X., Grimson, W.E.L.: Unsupervised activity perception by hierarchical Bayesian models. In: IEEE Conference on Computer Vision and Pattern Recognition, pp. 1–8 (2007)

Wang, X., Ma, X., Grimson, W.E.L.: Unsupervised activity perception in crowded and complicated scenes using hierarchical Bayesian models. IEEE Trans. Pattern Anal. Mach. Intell. **31**(3), 539–555 (2009)

Weinland, D., Ronfard, R., Boyer, E.: Free viewpoint action recognition using motion history volumes. Comput. Vis. Image Underst. **104**(2–3), 249–257 (2006)

Willems, G., Tuytelaars, T., Van Gool, L.: An efficient dense and scale-invariant spatio-temporal interest point detector. In: European Conference on Computer Vision, pp. 650–663 (2008)

Wu, B., Nevatia, R.: Detection and tracking of multiple partially occluded humans by Bayesian combination of edgelet based part detectors. Int. J. Comput. Vis. **75**(2), 247–266 (2007)

Wu, S., Moore, B.E., Shah, M.: Chaotic invariants of Lagrangian particle trajectories for anomaly detection in crowded scenes. In: IEEE Conference on Computer Vision and Pattern Recognition, San Francisco, USA, pp. 2054–2060 (2010)

Xiang, T., Gong, S.: Beyond tracking: modelling activity and understanding behaviour. Int. J. Comput. Vis. **67**(1), 21–51 (2006a)

Xiang, T., Gong, S.: Model selection for unsupervised learning of visual context. Int. J. Comput. Vis. **69**(2), 181–201 (2006b)

Xiang, T., Gong, S.: Optimising dynamic graphical models for video content analysis. Comput. Vis. Image Underst. **112**(3), 310–323 (2008a)

Xiang, T., Gong, S.: Incremental and adaptive abnormal behaviour detection. Comput. Vis. Image Underst. **111**(1), 59–73 (2008b)

Xiang, T., Gong, S.: Video behaviour profiling for anomaly detection. IEEE Trans. Pattern Anal. Mach. Intell. **30**(5), 893–908 (2008c)

Xiang, T., Gong, S.: Activity based surveillance video content modelling. Pattern Recognit. **41**(7), 2309–2326 (2008d)

Yang, M., Wu, Y., Hua, G.: Context-aware visual tracking. IEEE Trans. Pattern Anal. Mach. Intell. **31**(7), 1195–1209 (2008)

Yang, Y., Liu, J., Shah, M.: Video scene understanding using multi-scale analysis. In: IEEE International Conference on Computer Vision, pp. 1669–1676 (2009)

Yilmaz, A., Javed, O., Shah, M.: Object tracking: a survey. ACM J. Comput. Surv. **38**(4), 1–45 (2006)

Yuan, J.S., Liu, Z.C., Wu, Y.: Discriminative subvolume search for efficient action detection. In: IEEE Conference on Computer Vision and Pattern Recognition, pp. 2442–2449 (2009)

Yuille, A., Hallinan, P., Cohen, D.: Feature extraction from faces using deformable templates. Int. J. Comput. Vis. **8**(2), 99–111 (1992)

Zelnik-Manor, L., Irani, M.: Statistical analysis of dynamic actions. IEEE Trans. Pattern Anal. Mach. Intell. **28**(9), 1530–1535 (2006)

Zhang, D., Gatica-Perez, D., Bengio, S., McCowan, I.: Semi-supervised adapted HMMs for unusual event detection. In: IEEE Conference on Computer Vision and Pattern Recognition, pp. 611–618 (2005)

Zhang, J., Gong, S.: Action categorisation by structural probabilistic latent semantic analysis. Comput. Vis. Image Underst. **114**(8), 857–864 (2010a)

Zhang, J., Gong, S.: Action categorisation with modified hidden conditional random field. Pattern Recognit. **43**(1), 197–203 (2010b)

Zhang, Z., Huang, K., Tan, T.: Comparison of similarity measures for trajectory clustering in outdoor surveillance scenes. In: International Conference on Pattern Recognition, pp. 1135–1138 (2006)

Zhao, T., Nevatia, R.: Tracking multiple humans in complex situations. IEEE Trans. Pattern Anal. Mach. Intell. **26**, 1208–1221 (2004)

Zhong, H., Shi, J., Visontai, M.: Detecting unusual activity in video. In: IEEE Conference on Computer Vision and Pattern Recognition, Washington DC, USA, pp. 819–826 (2004)

Zhu, X.: Semi-supervised learning literature survey. Technical Report 1530, Computer Sciences, University of Wisconsin-Madison (2007)

Part II
Single-Object Behaviour

Chapter 4
Understanding Facial Expression

Facial expression is a natural and efficient means for humans to communicate their emotions and intentions, as communication is primarily carried out face to face. Expression can be recognised by either static face images in isolation, or sequences of face images. For the former, it is assumed that the static visual appearance of a face contains enough information for conveying an expression. The latter exploits information from facial movement generated by expressions (Cohen et al. 2003; Essa and Pentland 1997).

Computer-based automatic facial expression recognition considers two problems: face image representation and expression classification. A good representational scheme aims to derive a set of features from face images that can most effectively capture the characteristics of facial expression. The optimal features should not only minimise visual appearance differences from the same type of expression, known as within-class variations, but also maximise differences between two different types of expressions, known as between-class variations. If indiscriminative image features are selected for a representation, it is difficult to achieve good recognition regardless of the choice of a classification mechanism. In the following, we consider the problems of how to construct a suitable representation and design an effective classification model for both static image-based and dynamic sequence-based automatic facial expression recognition.

4.1 Classification of Images

There are geometric and appearance facial features that can be extracted from face images. Geometric features capture explicitly the shape and locations of facial parts, and can be extracted to form a feature vector that represents the face geometry. In general, models based on geometric features provide similar or better expression recognition compared to those based on appearance features (Valstar et al. 2005). However, computing reliable geometric features is difficult, as it requires accurate and consistent facial feature detection and tracking (McKenna et al. 1997). Appearance features are commonly extracted by filtering images of either a whole face or

Fig. 4.1 Illustration of the basic LBP operator

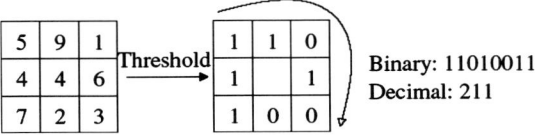

specific face regions. This approach aims to capture facial appearance changes without any explicit registration of face shape information, therefore avoiding the need for precise feature tracking and alignment. Many appearance feature-based models exploit Gabor wavelets[1] (Würtz 1995) due to its compact representation of rich orientational information at multiple image scales (Bartlett et al. 2005; Donato et al. 1999; Lyons et al. 1999; Tian 2004; Zhang et al. 1998), although it is both computation and memory intensive to extract multi-scale and multi-orientational Gabor coefficients.

Local binary patterns (LBP) are introduced as computationally inexpensive appearance features for facial image analysis (Ahonen et al. 2004; Feng et al. 2004; Hadid et al. 2004). Two important properties of LBP features are their tolerance against illumination changes and computational simplicity. Shan et al. (2005b) compare LBP features with Gabor features for facial expression recognition over a range of image resolutions, and show the effectiveness of LBP against Gabor features. Liao et al. (2006) further suggest an enhanced LBP operator to extract features in both image intensity and gradient maps for facial expression recognition, and study its effectiveness on low-resolution face images. These studies reinforce the view that, compared to Gabor wavelets, LBP features can be computed with much lower cost. Another attribute of LBP features is that they lie in a low-dimensional feature space, retaining discriminative facial information with a more compact representation.

4.1.1 Local Binary Patterns

The original LBP operator was introduced by Ojala et al. (1996) for image texture description. The operator labels each pixel of an image with a string of binary numbers computed in a local neighbourhood of the pixel. This can be done by thresholding a 3×3 neighbourhood of a pixel with the centre value and encoding the results as a 8-digit binary number (Fig. 4.1) which can assume 256 different values. A 256-bin histogram of pixel-wise LBP labels is then computed over an image region and used as a texture descriptor. The derived binary numbers in each pixel neighbourhood, known as local binary patterns or LBP codes, codify local image gradient primitives including different types of curved edges, spots, and flat areas (Fig. 4.2). Each LBP code can be regarded as a micro-texton (Hadid et al. 2004).

A limitation of this basic LBP operator can be caused its small 3×3 neighbourhood which cannot capture larger scale dominant features. This can be addressed by

---

[1]In this book, we alternate the use of terms 'Gabor wavelets' and 'Gabor filters'.

Fig. 4.2 Examples of texture primitives which can be detected by LBP (*white circles* represent binary value 1, *black circles* are for 0)

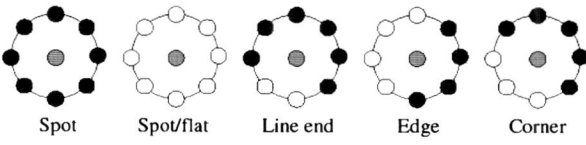

Spot   Spot/flat   Line end   Edge   Corner

Fig. 4.3 Three examples of the extended LBP (Ojala et al. 2002): a circular (8, 1) neighbourhood, a circular (12, 1.5) neighbourhood, and a circular (16, 2) neighbourhood, respectively

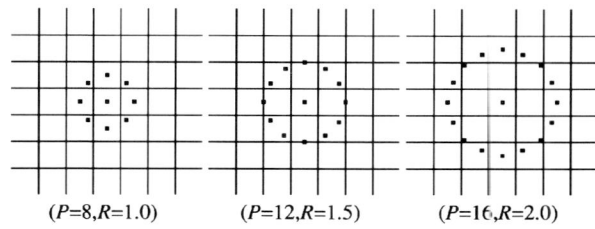

$(P=8,R=1.0)$ $(P=12,R=1.5)$ $(P=16,R=2.0)$

extending the neighbourhood size (Ojala et al. 2002). By using circular neighbourhoods and bilinearly interpolating the pixel values, the process allows any radius and number of pixels in the neighbourhood. Figure 4.3 shows three examples of the extended LBP operator, where (P, R) denotes a neighbourhood of P equally spaced sampling points. These points lie on a circle with radius of R and form a circularly symmetric neighbour set.

The extended LBP operator, denoted as $\text{LBP}_{P,R}$, produces 2^P different output values, corresponding to the 2^P different binary patterns that can be formed by the P pixels in a neighbour set. It is shown that certain bins contain more information than others (Ojala et al. 2002). Therefore, it is possible to use only a subset of the 2^P local binary patterns to describe texture of images, known as fundamental patterns, or uniform patterns. A local binary pattern is called uniform if it contains at most two bit-wise transitions from 0 to 1, or 1 to 0, when the binary string is considered circular. For example, 00000000, 00111000 and 11100001 are uniform patterns. It is observed that uniform patterns account for nearly 90% of all patterns in a (8, 1) neighbourhood and for about 70% in a (16, 2) neighbourhood in texture images (Ojala et al. 2002). Accumulating patterns with more than two transitions into a single bin yields a new LBP operator, denoted as $\text{LBP}_{P,R}^{u2}$, with less than 2^P bins. For example, the number of labels for a neighbourhood of 8 pixels is 256 for the standard LBP, but 59 for $\text{LBP}_{P,R}^{u2}$.

After labelling an image with a LBP operator, a histogram of the labelled image $f_l(x, y)$ can be defined as

$$H_i = \sum_{x,y} I\big(f_l(x, y) = i\big), \quad i = 0, \ldots, n - 1 \tag{4.1}$$

where n is the number of different labels produced by the LBP operator and

$$I(A) = \begin{cases} 1, & A \text{ is true} \\ 0, & A \text{ is false} \end{cases} \tag{4.2}$$

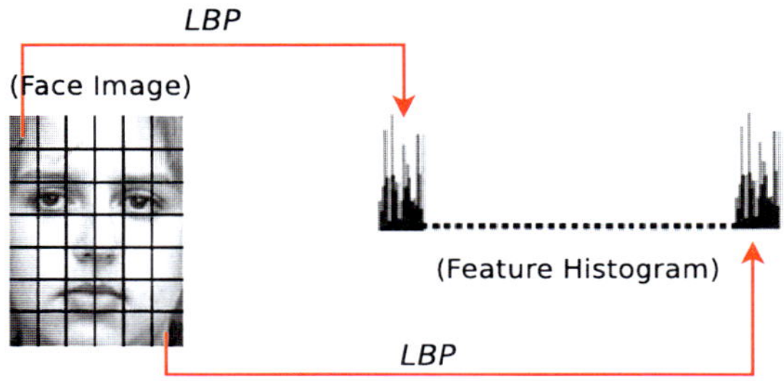

Fig. 4.4 A face image is divided into small patches from which LBP histograms are extracted and concatenated into a single, spatially enhanced LBP histogram

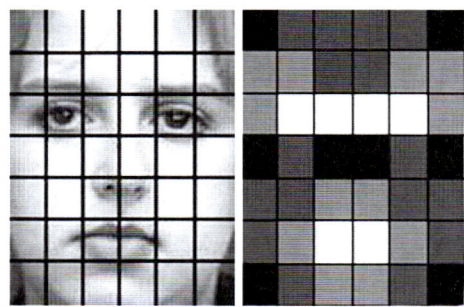

Fig. 4.5 *Left*: A face image divided into 6 × 7 patches. *Right*: The weights set for weighted dissimilarity measures. *Black squares* indicate weight 0.0, *dark gray* 1.0, *light gray* 2.0 and *white* 4.0

This LBP histogram contains information about the distribution of the local micro-patterns, such as edges, spots and flat areas, over the whole image, so it can be used to statistically describe image characteristics.

A face image can be considered as a composition of micro-patterns described by a LBP histogram (Ahonen et al. 2004; Feng et al. 2004; Hadid et al. 2004). A LBP histogram computed over the whole-face image encodes only the occurrences of the micro-patterns without any indication about their spatial locations. To also consider spatial information about facial micro-patterns, face images can be equally divided into small patches $R_0, R_1, \ldots, R_m$ where local LBP histograms are extracted (Fig. 4.4). These local LBP histograms can then be concatenated into a single spatially enhanced global feature histogram defined as

$$H_{i,j} = \sum_{x,y} I\{f_l(x, y) = i\} I\{(x, y) \in R_j\} \tag{4.3}$$

where $i = 0, \ldots, n - 1$, $j = 0, \ldots, m - 1$.

Such a spatially enhanced LBP histogram captures both the local image texture and the structure about a face. Two factors contribute towards the effectiveness of LBP feature extraction. One is the choice of LBP operator, and the other is the size of local image patches. For example, Ahonen et al. (2004) suggest that a 59-bin $\text{LBP}^{u2}_{8,2}$ operator is optimal for face images of 110×150 pixels in size, where a face image is divided into 42 (6×7) local patches of 18×21 pixels, and represented by a LBP histogram of 2,478 (59×42) bins, as shown in Fig 4.5.

4.1.2 Designing Classifiers

Different techniques can be employed to classify facial expressions into categories, such as neural networks (Padgett and Cottrell 1997; Tian 2004; Zhang et al. 1998), support vector machines (Bartlett et al. 2005), Bayesian networks (Cohen et al. 2003), and rule-based classifiers (Pantic and Patras 2006; Pantic and Rothkrantz 2000, 2004). There are a number of comparative studies on the effect of different classification models. Cohen et al. (2003) study the effect of different Bayes classifiers, and suggest that the Gaussian tree-augmented naive Bayes classifier is most effective. Bartlett et al. (2005) compare AdaBoost, support vector machines, and linear discriminant analysis for facial expression recognition. They suggest that selecting Gabor filtered features by AdaBoost, and then training a support vector machine based on those selected features, is a good model for expression classification.

In the following, we discuss in more details the use of different classification techniques with a LBP feature-based representation for expression classification, including template matching, support vector machines, linear discriminant analysis, and linear programming.

Template Matching

Histogram-based template matching is shown to be effective for facial identity recognition using a LBP-based representation (Ahonen et al. 2004). A similar scheme can also be readily applied to facial expression recognition. More precisely, LBP histograms of expression images for a particular expression class can be averaged to construct a single histogram template for this expression. To classify a target histogram extracted from a probe expression image against known templates, the chi square statistic (χ^2) can be employed as a histogram dissimilarity measure:

$$\chi^2(\mathbf{S}, \mathbf{M}) = \sum_i \frac{(S_i - M_i)^2}{S_i + M_i} \tag{4.4}$$

where $\mathbf{S}$ and $\mathbf{M}$ are two LBP histograms, and S_i and M_i are their ith elements, respectively.

Certain facial regions contain more useful information about expression than others. For example, facial features contributing to expressions mainly lie near the eye and mouth regions. Different weights can thus be set for different facial regions. An example is shown in Fig. 4.5. A weighted χ^2 statistic can be defined as

$$\chi_w^2(\mathbf{S}, \mathbf{M}) = \sum_{i,j} w_j \frac{(S_{i,j} - M_{i,j})^2}{S_{i,j} + M_{i,j}} \tag{4.5}$$

where $\mathbf{S}$ and $\mathbf{M}$ are two LBP histograms, and w_j is the weight for facial region j.

Support Vector Machine

As a machine learning technique for data classification, a support vector machine (Vapnik 1995, 1998) performs a mapping of feature vectors as data points from a low-dimensional feature space to a much higher dimensional kernel space where linear hyperplanes can be found with maximum margins for separating those data points of different classes that are linearly non-separable in the original low-dimensional feature space. For this reason, a support vector machine is an effective technique for modelling linearly non-separable facial appearance changes (Gong et al. 2002; Li et al. 2003; Romdhani et al. 1999). It is also effective for separating visually similar facial expressions and can be used for expression classification (Bartlett et al. 2003, 2005; Valstar and Pantic 2006; Valstar et al. 2005).

More precisely, given a set of labelled examples as training data $\{(x_i, y_i), i = 1, \ldots, l\}$, where $x_i \in R^n$ and $y_i \in \{1, -1\}$, a new test example x is classified by the following function:

$$f(x) = \mathrm{sgn}\left(\sum_{i=1}^{l} \alpha_i y_i K(x_i, x) + b\right) \qquad (4.6)$$

where α_i are Lagrange multipliers for a constrained optimisation of a cost function that finds a separating hyperplane, $K(\cdot, \cdot)$ is a kernel function, $\mathrm{sgn}(\cdot)$ is a function that returns the sign of its input, and b is a threshold parameter of the hyperplane. Those data points in the training samples x_i that satisfy $\alpha_i > 0$ are known as support vectors. A support vector machine finds a set of hyperplanes that maximise the distances between the support vectors and those hyperplanes. A support vector machine allows domain-specific selection of the kernel function. Though new kernels are being proposed, the most frequently used kernel functions are the polynomial and radial basis functions.

A native support vector machine makes binary decisions. A multi-class discrimination problem as in the case of expression classification is accomplished by using recursively an one-versus-rest strategy. This process trains a set of multiple binary classifiers to discriminate one expression from all others, and outputs the class with the largest output from the multiple binary classifiers.[2]

Linear Discriminant Analysis

By using support vector machines to classify facial expressions, one considers face images as data points in a multi-dimensional feature space. Depending on the number of features used to represent each image, the dimensionality of this feature space

---

[2]There are practical considerations for selecting support vector machine parameters (Hsu et al. 2003).

can be high. However, the parameters that describe facial deformations due to expression changes intrinsically lie in a much lower dimensional subspace. To discover the true nature of variations in facial expression, subspace analysis is exploited (Belhumeur et al. 1997; Lyons et al. 1999; Padgett and Cottrell 1997; Turk and Pentland 1991). Linear discriminant analysis (Belhumeur et al. 1997) is a supervised subspace learning technique, and has been applied to facial expression recognition (Lyons et al. 1999). Linear discriminant analysis searches for the projection axes on which data points of different expression classes are far from each other whilst data points of the same class are closer to each other.

Given m multi-dimensional data samples $x_1, x_2, \ldots, x_m$ in R^n that belong to c classes, linear discriminant analysis finds a transformation matrix W that maps these m points to $y_1, y_2, \ldots, y_m$ in R^l ($l \leq c$), where $y_i = W^T x_i$. The objective function of linear discriminant analysis is

$$\max_{\mathbf{w}} \frac{\mathbf{w}^T S_B \mathbf{w}}{\mathbf{w}^T S_W \mathbf{w}} \tag{4.7}$$

$$S_B = \sum_{i=1}^{c} n_i \left(\mathbf{m}^{(i)} - \mathbf{m}\right)\left(\mathbf{m}^{(i)} - \mathbf{m}\right)^T \tag{4.8}$$

$$S_W = \sum_{i=1}^{c} \left(\sum_{j=1}^{n_i} \left(x_j^{(i)} - \mathbf{m}^{(i)}\right)\left(x_j^{(i)} - \mathbf{m}^{(i)}\right)^T\right) \tag{4.9}$$

where $\mathbf{m}$ is the mean of all the samples, n_i is the number of samples in the ith class, $\mathbf{m}^{(i)}$ is the average vector of the ith class, $x_j^{(i)}$ is the jth sample in the ith class, S_B is known as a between-class scatter matrix, and S_W is known as a within-class scatter matrix. After projecting facial expression images as feature vectors onto a linear discriminant space, a nearest-neighbour classifier (Bishop 2006) can be adopted for classification.

In practice, the dimension of the feature space (n) is often much larger than the number of samples in a training set (m). So the within-class scatter matrix S_W is likely to be singular. To overcome this problem, the data set is commonly first projected into a lower dimensional principal component space (Bishop 2006).

Linear Programming

Feng et al. (2005) consider the multi-class facial expression classification problem as a set of one-to-one pairs of binary classifications, where each binary classifier is produced by linear programming. These binary classifiers are combined with a voting scheme to output a final classification score.

Given two sets of data examples $\mathcal{A}$ and $\mathcal{B}$ in R^n, linear programming seeks a linear function such that $f(x) > 0$ if $x \in \mathcal{A}$, and $f(x) \leq 0$ if $x \in \mathcal{B}$. This function is given by $f(x) = w^T x - \gamma$, and determines a plane $w^T x = \gamma$ with normal $w \in R^n$ that separates $\mathcal{A}$ from $\mathcal{B}$. Let the set of m examples in $\mathcal{A}$ be represented by a matrix

$A \in R^{m \times n}$ and the set of k examples in $\mathcal{B}$ be represented by a matrix $B \in R^{k \times n}$. One wishes to satisfy

$$Aw \geq e\gamma + e, \qquad Bw \leq e\gamma - e \qquad (4.10)$$

where e is a vector of all 1s with an appropriate dimension. Practically, because of the overlap between the two classes, one has to minimise the norm of the average error in (4.10):

$$\min_{w,\gamma} f(w,\gamma) = \min_{w,\gamma} \frac{1}{m} \left\| (-Aw + e\gamma + e)_+ \right\|_1 + \frac{1}{k} \left\| (Bw - e\gamma + e)_+ \right\|_1 \qquad (4.11)$$

where x_+ denotes the vector with components satisfying $(x_+)_i = \max\{x_i, 0\}, i = 1, \ldots, n$, and $\| \cdot \|_1$ denotes the 1-norm. Equation (4.11) can be modelled as a robust linear programming problem (Guo and Dyer 2003):

$$\min_{w,\gamma,y,z} \frac{e^T y}{m} + \frac{e^T z}{k}$$

$$\text{subject to} \quad \begin{cases} -Aw + e\gamma + e \leq y, \\ Bw - e\gamma + e \leq z, \\ y \geq 0, z \geq 0 \end{cases} \qquad (4.12)$$

which minimises the average sum of misclassification errors. Equation (4.12) is used to solve the classification problem.

Shan et al. (2009) carried out a comparative study on different classifiers for expression classification, all using a LBP feature-based representation. The study shows that a linear classifier, such as linear discriminant analysis, performs very poorly for facial expression classification, worse than that of template matching. A linear programming-based model improves classification accuracy. Overall, a support vector machine shows most potential for classifying facial expressions in static images. This is due to both the high dimensionality of the feature space and linearly non-separable facial expression images.

4.1.3 Feature Selection by Boosting

For constructing LBP histograms, face images are commonly divided into small image patches of equal size. By doing so, LBP features can only be extracted from rigid patches at fixed positions. By shifting and scaling a sub-window over face images, more local image patches can be extracted to generate overlapped local LBP histograms that give the effect of oversampling the image space, yielding a richer and more robust description of expression images. However, this process can potentially generate a large pool of local LBP histograms, resulting in a very high-dimensional space for the concatenated single joint histogram representation of an image. To reduce this dimensionality, boosting learning (Schapire and Singer 1999)

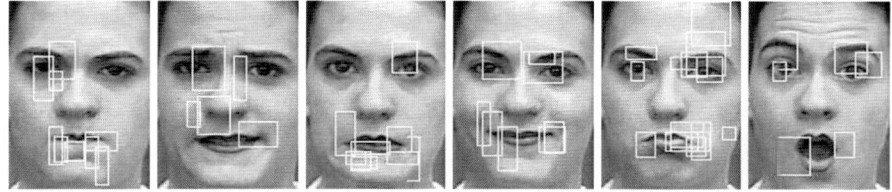

Fig. 4.6 Local image patches for different expressions selected by an AdaBoost learning process. Expressions shown from *left* to *right* are: 'anger', 'disgust', 'fear', 'joy', 'sadness', 'surprise'

can be exploited effectively for learning to select those local LBP histograms that contain the most discriminative information.

AdaBoost (Freund and Schapire 1997; Schapire and Singer 1999) is a supervised method for learning a classifier from training data. AdaBoost first learns a small number of weak classifiers whose performances are only just little better than random guessing. These weak classifiers are then boosted iteratively to form a strong classifier. The process of AdaBoost maintains a distribution on the training examples. At each iteration, a weak classifier giving the minimum weighted error rate is selected, and the distribution is updated to increase the weights of the misclassified examples and reduce the importance of the others. This learning process of AdaBoost is both simple and effective for cascaded learning of a nonlinear classification function. AdaBoost has been successfully used for addressing many problems, such as face detection (Viola and Jones 2001) and pedestrian detection (Viola et al. 2003).

For selecting local LBP histograms where each histogram is calculated from a local image patch, AdaBoost can be deployed to find those local image patches that are most representative of each expression class. A histogram-based template matching model can be considered as a weak classifier for AdaBoost. More precisely, for a given set of training images, local LBP histograms computed from images of a particular expression class are averaged to produce local class templates. Different local class templates are deployed to classify local image patches using the chi square statistic χ^2 (4.4). Given a set of labelled positive and negative training examples for each expression class, AdaBoost is used to boost those local template classifiers through re-classifying image patches.[3] AdaBoost continues until the patch classification distribution for the positive and negative examples are well separated. From this patch selection by re-classification process, the number of local LBP histograms selected for each expression is automatically determined. Figure 4.6 shows some examples of AdaBoost selected local image patches for different expressions. It is evident that different expressions are represented by different local image patches

---

[3] A native AdaBoost model works only for a two-class problem. The multi-class problem considered here is solved by using the one-versus-rest strategy, which trains AdaBoost between one expression in one class and all others in another class. For each AdaBoost learner, the images of one expression class are used as positive examples, while the images of all other expressions are used as negative examples.

with variable sizes at non-fixed positions. Shan et al. (2009) show some benefits from building a support vector machine classifier using LBP patch features selected by AdaBoost.

4.2 Manifold and Temporal Modelling

Until now, we only considered the problem of interpreting facial expressions by static visual appearances. However, a facial expression is inherently dynamic, more naturally conveyed by spatio-temporal information. Psychological experiments by Bassili (1979) suggest that facial expressions are more accurately recognised from a 'dynamic image', that is, an image sequence, than from a static image. A study by Zhao and Pietikainen (2007) considers a spatio-temporal extension of LBP for modelling dynamic textures, with an application to facial expression recognition in image sequences.

4.2.1 Locality Preserving Projections

Instead of representing individual face images as static feature patterns, a face image with N pixels can also be considered as a point in a N-dimensional image space. Image capturing variations in facial appearances can be represented as low-dimensional manifolds embedded in this N-dimensional image space, considered as the ambient space of a manifold (Fidaleo and Trivedi 2003; Gong et al. 1996; He and Niyogi 2003; He et al. 2005; Li et al. 2003; Saul and Roweis 2003; Tenenbaum et al. 2000). Similarly, it would be desirable to analyse a facial expression in a low-dimensional subspace capturing its space-time characteristics. This is also computationally attractive because a model learned from a lower dimensional space is less likely to suffer from the "curse of dimensionality" (Bishop 2006).

A number of nonlinear techniques are studied to learn the structures of a manifold. Early examples include Isomap (Tenenbaum et al. 2000), locally linear-embedding (Saul and Roweis 2003), and Laplacian eigenmap (Belkin and Niyogi 2001). However, these techniques all construct maps that are defined only on the training data. It is unclear how well these maps generalise to unseen data. The ability to generalise can be critical for modelling facial expressions. Alternatively, He and Niyogi (2003) describe a manifold learning method, known as locality preserving projection (LPP), for finding the optimal linear approximations to the eigenfunctions of a Laplace–Beltrami operator on a manifold. Different from principal component analysis, which implicitly assumes that the data space is only Euclidean, LPP assumes that the data space can be nonlinear. LPP shares some properties with locally linear-embedding and Laplacian eigenmap, such as locality preserving. But importantly, LPP is defined everywhere in the ambient space rather than just on the training data. This gives it significant advantages over locally linear-embedding and Laplacian eigenmap in explaining unseen data. It is shown that LPP has a superior discriminating power than both principal component analysis and linear discriminant analysis for face recognition (He et al. 2005). In the following, we consider the problem of modelling manifolds of facial expressions. In particular, we focus on the question on how LPP can be exploited to learn manifolds of different expressions. We also address the problem of manifold alignment for expressions from different people in order to compute a generalised manifold that is both person-independent and expression discriminative.

The generic problem of linear dimensionality reduction is the following. Given a set $x_1, x_2, \ldots, x_m$ in R^n, find a transformation matrix W that maps these m points to $y_1, y_2, \ldots, y_m$ in R^l ($l \ll n$), such that y_i represents x_i, where $y_i = W^T x_i$. Let $\mathbf{w}$ denote a transformation vector, the optimal projections preserving locality can be found by solving the following minimisation problem (He and Niyogi 2003):

$$\min_{\mathbf{w}} \sum_{i,j} \left(\mathbf{w}^T x_i - \mathbf{w}^T x_j \right)^2 S_{ij} \tag{4.13}$$

where S_{ij} evaluates the local structure of the data space. It can be defined as follows:

$$S_{ij} = \begin{cases} e^{-\frac{\|x_i - x_j\|^2}{t}} & \text{if } x_i \text{ and } x_j \text{ are close} \\ 0 & \text{otherwise} \end{cases} \tag{4.14}$$

or in a simpler form as

$$S_{ij} = \begin{cases} 1 & \text{if } x_i \text{ and } x_j \text{ are close} \\ 0 & \text{otherwise} \end{cases} \tag{4.15}$$

where 'close' can be defined by $\|x_i - x_j\|^2 < \epsilon$, x_i is among k nearest neighbours of x_j, or x_j is among k nearest neighbours of x_i. The objective function with symmetric weights S_{ij} ($S_{ij} = S_{ji}$) incurs a heavy penalty if neighbouring points x_i and x_j are mapped far apart. Therefore, minimising this objective function is to ensure that if x_i and x_j are close, $y_i (= \mathbf{w}^T x_i)$ and $y_j (= \mathbf{w}^T x_j)$ are close as well. S_{ij} can be seen as a similarity measure.   The objective function can be reduced to:

$$\frac{1}{2} \sum_{ij} \left(\mathbf{w}^T x_i - \mathbf{w}^T x_j \right)^2 S_{ij} = \sum_i \mathbf{w}^T x_i D_{ii} x_i^T \mathbf{w} - \sum_{ij} \mathbf{w}^T x_i S_{ij} x_j^T \mathbf{w}$$

$$= \mathbf{w}^T X (D - S) X^T \mathbf{w} = \mathbf{w}^T X L X^T \mathbf{w} \tag{4.16}$$

where $X = [x_1, x_2, \ldots, x_m]$, D is a diagonal matrix whose entries are column sums of S, and $D_{ii} = \sum_j S_{ji}$. $L = D - S$ is the Laplacian matrix. The bigger the value D_{ii} is, which corresponds to y_i, the more important y_i is. Therefore, a constraint is imposed as follows:

$$\mathbf{y}^T D \mathbf{y} = 1 \quad \Rightarrow \quad \mathbf{w}^T X D X^T \mathbf{w} = 1 \tag{4.17}$$

Fig. 4.7   Example face expressions from the Cohn–Kanade database (Kanade et al. 2000)

The transformation vector **w** that minimises the objective function is given by the minimum eigenvalue solution to the following generalised eigenvalue problem:

$$XLX^T \mathbf{w} = \lambda XDX^T \mathbf{w} \tag{4.18}$$

where λ is an eigenvalue. The obtained projections are the optimal linear approximation to the eigenfunctions of the Laplace–Beltrami operator on the manifold (He and Niyogi 2003). Therefore, despite that LPP is a linear technique, it is able to recover important aspects of a nonlinear manifold by preserving local structure.

Expression Manifolds

Shan et al. (2005a) study LPP-based learning of facial expression manifolds using the Cohn–Kanade database (Kanade et al. 2000). This database consists of facial expression image sequences from 100 university students between 18 to 30 years old. Figure 4.7 shows some examples. In this study, six subjects are selected from the database, each of which has six image sequences corresponding to six expressions. LPP is applied to all expression image sequences of each person to find the expression manifold for that person. A visualisation of one person's expression manifold learned in a LBP feature space is shown in Fig. 4.8. Similar, but different, expression manifolds are learned for different people (Fig. 4.9). It is found that facial expression image sequences of every person are embedded on a smooth manifold. It is also found that each expression sequence is mapped to a continuous curve on a manifold that begins from the neutral expression and extends in well-defined distinctive directions of different expressions.

The study also considers the problem of whether a single manifold constructed from different people is able to separate people regardless of their expressions, or

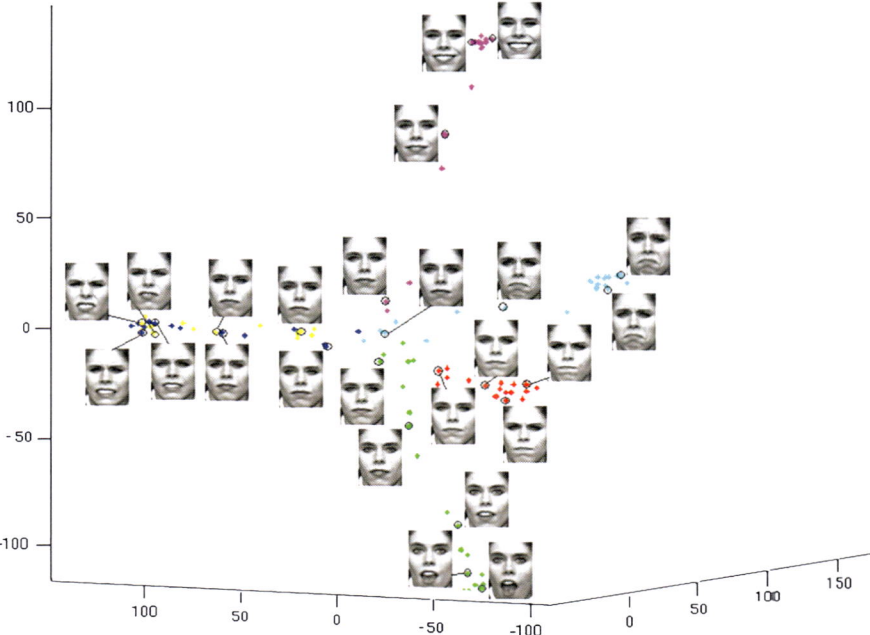

Fig. 4.8 When facial expression image sequences of a person are mapped into an embedding space described by the first 3 coordinates of LPP, they lie in separable structures (Shan et al. 2005a, 2006b). Different expressions are *colour-coded* as follows: *red*—'anger', *yellow*—'disgust', *blue*—'fear', *magenta*—'joy', *cyan*—'sadness', *green*—'surprise'. Note: the same *colour codes* of expressions are used in Figs. 4.9 to 4.11

to separate expressions regardless of a person's identity. It is found that when LPP is applied to different people, there are separate clusters in an embedded space with different people mapped into different clusters (Fig. 4.10). However, Fig. 4.11 indicates that LPP mapping is unable to project images of similar expressions from a large number of different people into separable clusters on a manifold.

Alignment of Expression Manifolds

Although image sequences of facial expressions from a single person display a well structured continuous manifold (Fig. 4.9), the scope of manifolds from different people varies significantly in terms of spread, location and shape of structure. This is evident from Fig. 4.10 where different people display different clusters without overlapping. To place them in the same space in order to compare person-independent expressions, manifolds of different people need be aligned so that similar expressions from different people are mapped to nearby locations in that space. Chang et al. (2003) propose one solution by suggesting a model to align the manifolds of different subjects in a Lipschitz embedding space.

Fig. 4.9 Expression
manifolds of three different
people in three-dimensional
LPP spaces

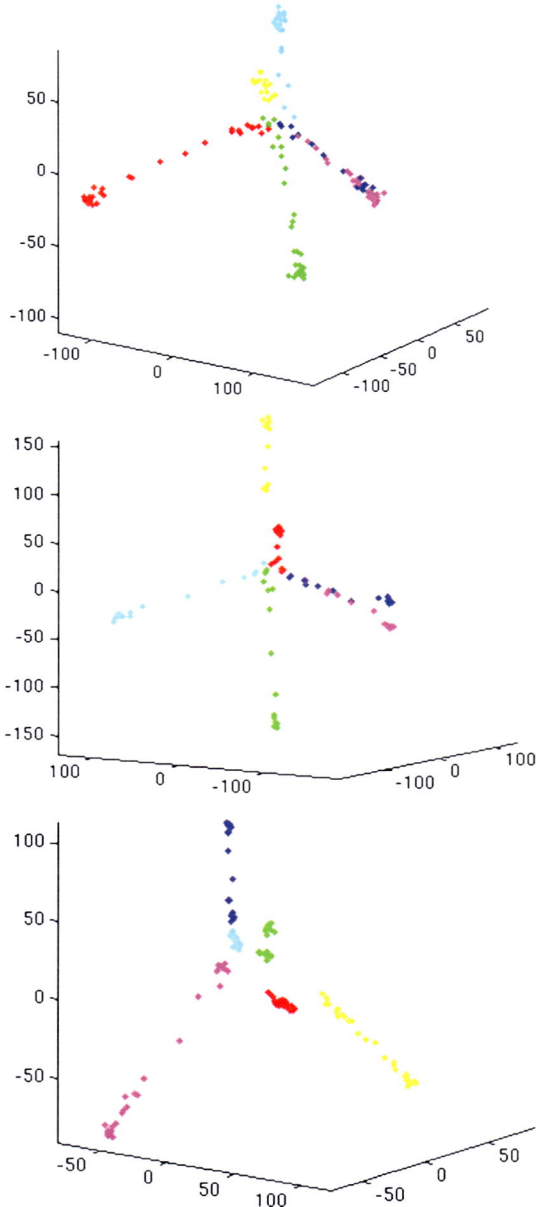

Using LPP embedding, an expression image sequence is embedded as a well
formed structure on a LPP manifold, centred at the neutral expression and expanding
towards different expressions with increasing intensiveness in the expression. This
structural characteristics is shared by manifolds from different people, as shown
in Fig. 4.9. The problem of aligning expression manifolds of different people is

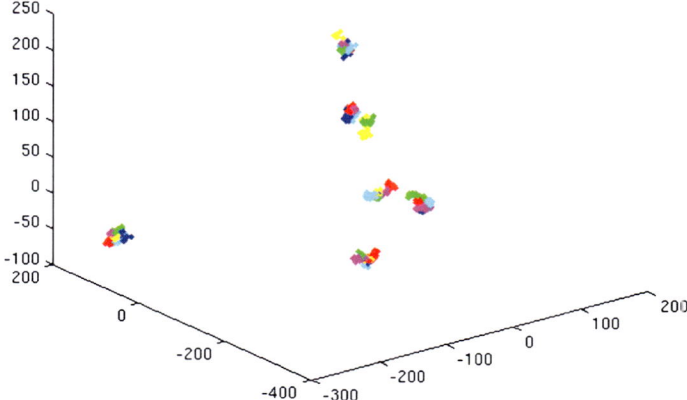

Fig. 4.10 Different expression image sequences from six different people are mapped into a single embedded space described by the first three coordinates of LPP

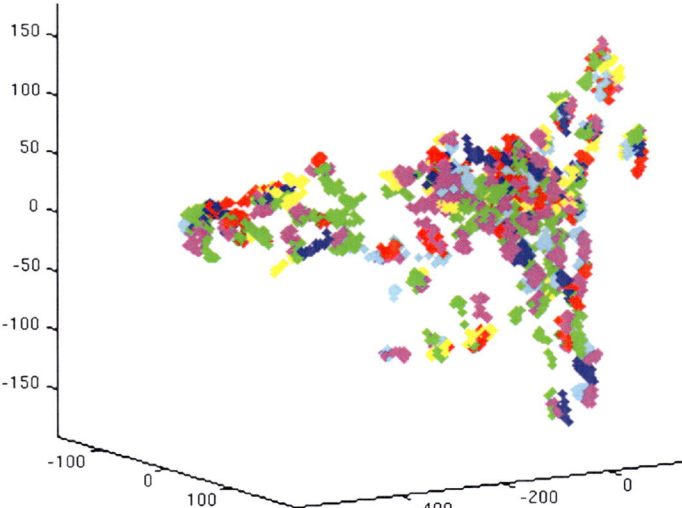

Fig. 4.11 Different expression image sequences from 96 people are projected into a single embedded space described by the first three coordinates of LPP

then reduced to the problem of aligning such structures with a common neutral expression location. To that end, a global coordinate space is defined based on image sequences of typical expressions. There are usually some similar expression images from different expressions, as certain phases of different expressions look alike. As a standard LPP is computed based on unsupervised learning, it is not well suited to embed different expressions of different people in a single space. The problem can be addressed using a supervised locality preserving projection (SLPP) algorithm

Fig. 4.12 Image sequences of seven typical facial expressions are mapped into an embedding space described by the first three coordinates of SLPP

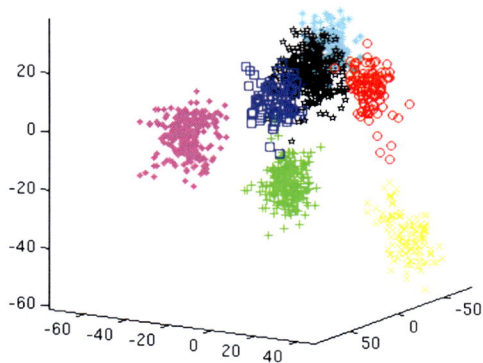

(Shan et al. 2005a), which not only preserves local structures, but also encodes class information during embedding.

More precisely, SLPP preserves class information when constructing a neighbourhood graph. The local neighbourhood of an example x_i from class c is composed of examples belonging to class c only. This can be achieved by increasing the distances between examples belonging to different classes, but leaving them unchanged if they are from the same class. Let $\mathrm{Dis}(i, j)$ denote the distance between x_i and x_j, then the distance after incorporating class information is defined as

$$\mathrm{SupDis}(i, j) = \mathrm{Dis}(i, j) + \alpha M \delta(i, j) \quad \alpha \in [0, 1] \tag{4.19}$$

where $M = \max_{i,j} \mathrm{Dis}(i, j)$; $\delta(i, j) = 1$ if x_i and x_j belong to the same class, and 0 otherwise. SLPP introduces an additional parameter α to quantify the degree of supervised learning. When $\alpha = 0$, the model converges to unsupervised LPP. When $\alpha = 1$, the model is fully supervised. For a fully supervised LPP, distances between examples in different classes are greater than the maximum distance in the entire data set. This ensures that neighbours of an example will always be picked from the class it belongs to. Varying α between 0 and 1 gives a partially supervised LPP, where an embedding is found by introducing some separation between classes.

By preserving local structure of data belonging to the same class, SLPP yields a subspace in which different expression classes are well separated regardless of whether they are from the same person. The subspace provides global coordinates for the manifolds of different subjects, which are aligned on one generalised manifold. An image sequence of a facial expression from beginning to apex is mapped onto this generalised manifold as a curve from a neutral face to a cluster for that expression. Figure 4.12 illustrates the effect of SLPP mapping of expressions. One neutral face and three peak frames at the apex of an expression from every expression sequence in the Cohn–Kanade database are used to build an SLPP embedding of seven different classes of expressions including neutral and six typical expressions as defined by the Cohn–Kanade database. It is evident that different expressions are well clustered and separated in this SLPP subspace. Furthermore, these distributions somewhat reflect a commonly perceived conception that 'joy' and 'sur-

Fig. 4.13 The manifolds of the six subjects shown in Fig. 4.10 are aligned in a single space

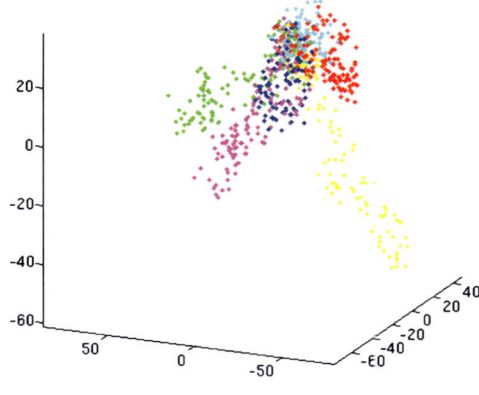

Fig. 4.14 The manifolds of the 96 subjects shown in Fig. 4.11 are aligned in a single space

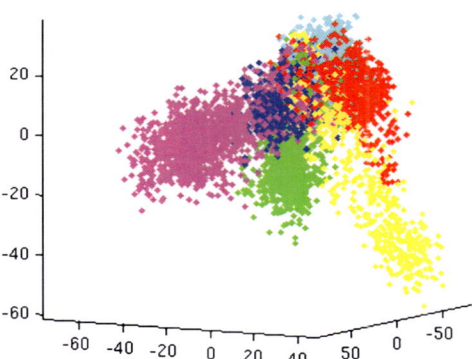

prise' is clearly separable, whilst 'anger', 'disgust', 'fear' and 'sadness' are easily confused.

An expression subspace derived by SLPP provides aligned manifolds of different people in a single space. Figure 4.13 shows aligned expression manifolds of the six people shown in Fig. 4.10. Expression manifolds from the 96 people shown in Fig. 4.11 are aligned in a single manifold shown in Fig. 4.14. It is evident that images from different people but of the same expression can now be embedded in the same locality on this SLPP manifold.

4.2.2 Bayesian Temporal Models

Once facial expression image sequences are represented by a manifold, we can consider modelling explicitly expression temporal dynamics on the manifold. Figure 4.15 depicts a schema for constructing a Bayesian temporal model of expression manifold in an SLPP embedded expression subspace. Given a generalised expression manifold learned using the SLPP from some expression gallery image sequences, a Bayesian temporal model of the manifold is formulated to represent facial expression dynamics. For expression recognition, a probe image sequence is first projected into a low-dimensional subspace and then matched against the Bayesian temporal model on the manifold.

For constructing the Bayesian temporal model, we consider the following (Shan et al. 2006a). Given a probe image sequence, it is first mapped into an embedded subspace $Z_t, t = 0, 1, 2, \ldots$. An estimate for the likelihood of labelling this probe with certain expression class is then represented as a temporally accumulated posterior probability at time t, $p(X_t|Z_{0:t})$, where the state variable X represents the class label of a facial expression. For a model of seven expression classes including 'neutral', 'anger', 'disgust', 'fear', 'joy', 'sadness' and 'surprise', we have $X = \{x_i, i = 1, \ldots, 7\}$. From a Bayesian perspective, $p(X_t|Z_{0:t})$ is computed by

$$p(X_t|Z_{0:t}) = \frac{p(Z_t|X_t)p(X_t|Z_{0:t-1})}{p(Z_t|Z_{0:t-1})} \tag{4.20}$$

where

$$p(X_t|Z_{0:t-1}) = \int p(X_t|X_{t-1})p(X_{t-1}|Z_{0:t-1})\,dX_{t-1} \tag{4.21}$$

Hence

$$p(X_t|Z_{0:t}) = \int p(X_{t-1}|Z_{0:t-1})\frac{p(Z_t|X_t)p(X_t|X_{t-1})}{p(Z_t|Z_{0:t-1})}\,dX_{t-1} \tag{4.22}$$

Note that in (4.21), the Markov property is used to derive $p(X_t|X_{t-1}, Z_{0:t-1}) = p(X_t|X_{t-1})$. Now the problem is reduced to estimating the prior $p(X_0|Z_0)$, the transition model $p(X_t|X_{t-1})$, and the observation model $p(Z_t|X_t)$.

The prior $p(X_0|Z_0) \equiv p(X_0)$ is estimated from a gallery of expression image sequences. An expression class transition probability from time $t-1$ to t is given by $p(X_t|X_{t-1})$ and estimated as

$$p(X_t|X_{t-1}) = p(X_t = x_j|X_{t-1} = x_i) = \begin{cases} \varepsilon, & T_{i,j} = 0 \\ \alpha T_{i,j}, & \text{otherwise} \end{cases} \tag{4.23}$$

where ε is a small empirical number, α is a scale coefficient, and $T_{i,j}$ is a transition frequency measure, defined by

$$T_{i,j} = \sum I(X_{t-1} = x_i \text{ and } X_t = x_j); \quad i = 1, \ldots, 7; \ j = 1, \ldots, 7$$

where

$$I(A) = \begin{cases} 1, & A \text{ is true} \\ 0, & A \text{ is false} \end{cases} \tag{4.24}$$

$T_{i,j}$ can be easily estimated from the gallery of image sequences, and ε and α can be set such that $\sum_j p(x_j|x_i) = 1$.

The expression manifold derived by SLPP preserves optimally local neighbourhood information in the data space, as SLPP establishes essentially a k-nearest

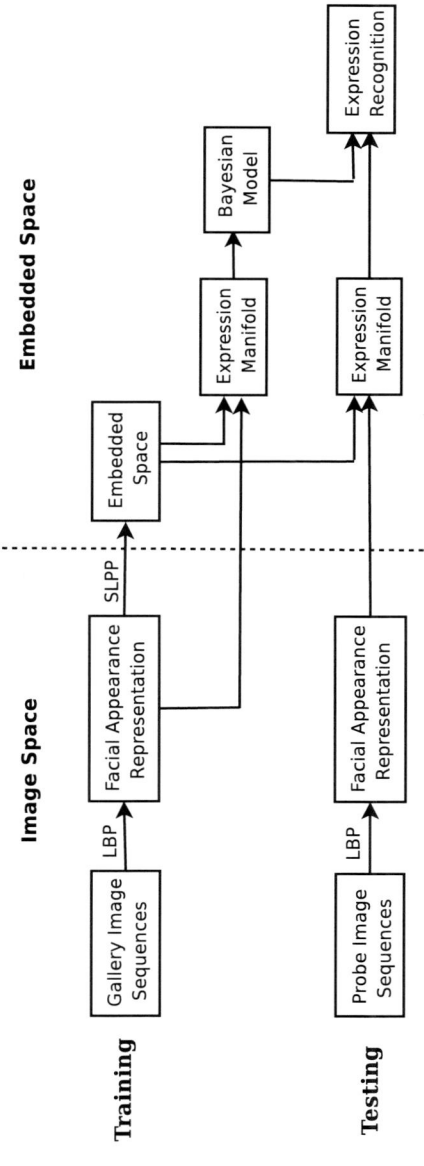

Fig. 4.15 Constructing a Bayesian temporal model of facial expression manifold in an embedded subspace

neighbour graph. To take advantage of the characteristics of such a locality pre-
serving structure, one can define a likelihood function $p(Z_t|X_t)$ according to the
nearest neighbour information. For example, given an observation Z_t at time t, if
there are other observations labelled as 'anger' in a k-nearest neighbourhood of Z_t,
there is less ambiguity for this observation Z_t to be classified as 'anger'. Therefore
let 'anger' be denoted by x_1, the observation will have a higher $p(Z_t|X_t = x_1)$.

More precisely, let $\{N_j, j = 1, \dots, k\}$ be a k-nearest neighbour of frame Z_t, a
neighbourhood distribution measure is computed as

$$M_i = \sum I(N_j = x_i), \quad j = 1, \dots, k; \ i = 1, \dots, 7$$

A neighbourhood likelihood function $p(Z_t|X_t)$ is then defined as

$$p(Z_t|X_t) = p(Z_t|X_t = x_i) = \begin{cases} \tau, & M_i = 0 \\ \beta M_i, & \text{otherwise} \end{cases} \qquad (4.25)$$

where τ is a small empirical number, β is a scale coefficient, and τ and β are
selected such that $\sum_i p(Z_t|X_t = x_i) = 1$.

Given the prior $p(X_0)$, the expression class transition probability $p(X_t|X_{t-1})$,
and the k-nearest neighbourhood likelihood function $p(Z_t|X_t)$, the posterior
$p(X_t|Z_{0:t})$ is computed by (4.22). This Bayesian temporal model of expression
provides a time cumulative probability distribution measure of all candidate expression classes in the current frame of an unseen probe expression sequence. The model
exploits explicitly temporal information represented in the expression manifold.

Some examples of facial expression recognition in image sequences using the
Bayesian temporal model are illustrated in Fig. 4.16. It can be seen that the posterior probabilities of different facial expression classes are updated over time as more
information becomes available from each observation image sequence. The model
takes into account temporal transitions of a facial expression. Shan et al. (2005a)
show through comparative experiments on benchmarking databases that, by incorporating temporal information, this Bayesian temporal model outperforms a static
image-based model for facial expression recognition when identical number of image frames are used.

4.3 Discussion

In this chapter, we considered the problem of understanding facial expression as
a form of human behaviour. In particular, we discussed two different approaches
to modelling expressions for recognition, one based on individual static face images, and the other exploiting spatio-temporal characteristics from expression image sequences. We also addressed the issue of image representation for modelling
expressions, and described in details local binary patterns (LBP) as an effective representational model. For expression image pattern classification, we considered a
number of techniques for designing a classifier, and suggest that a support vector

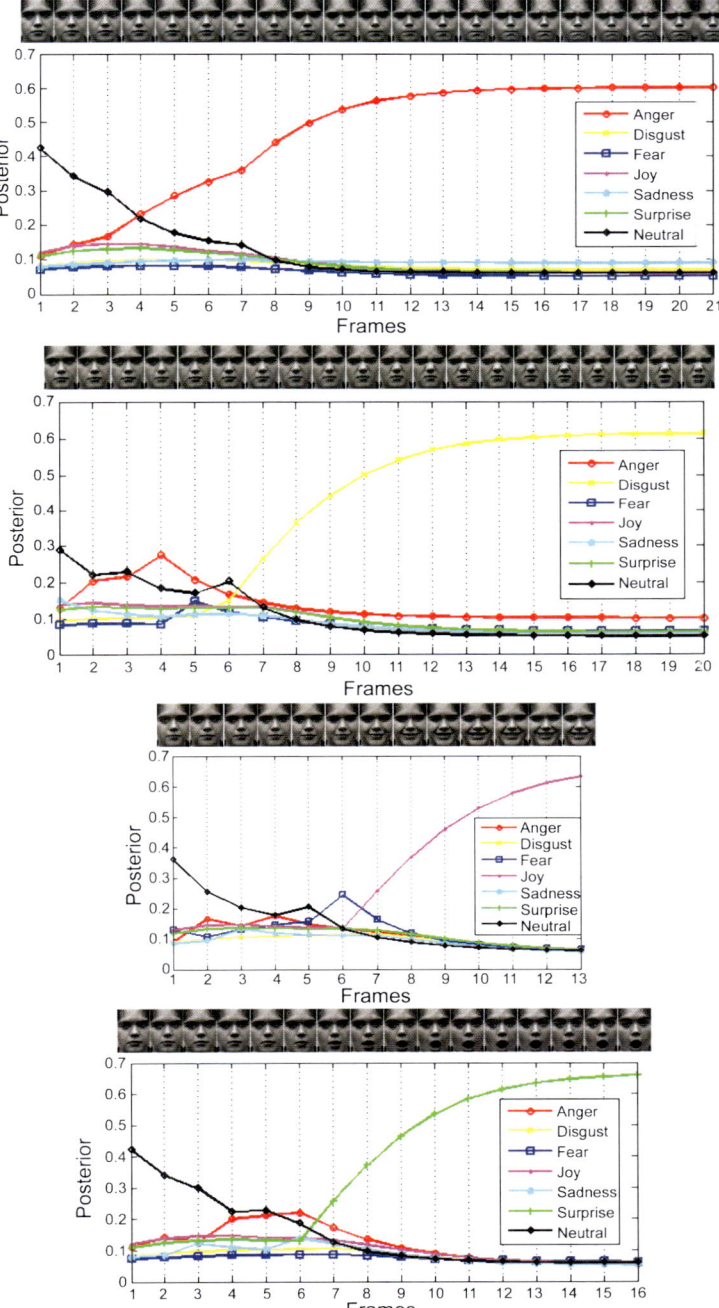

Fig. 4.16 Examples of facial expression recognition using a Bayesian temporal model. From *top* to *bottom*: Expression sequences of 'anger', 'disgust', 'joy', and 'surprise'

machine-based classifier works well with a LBP-based recognition. For modelling temporal information of facial expression, we considered the idea of constructing an expression manifold of image sequences and modelling dynamic transitions in facial expression using a Bayesian temporal model.

Intuitively, a dynamic model is able to capture richer spatio-temporal information describing facial expressions as behaviour. This is partly supported by some empirical results (Shan et al. 2005a). However, there is no clear evidence that significantly better recognition performance can be gained from a dynamic model (Pantic and Patras 2006; Shan et al. 2005b, 2009). The reasons can be two-fold. First, a facial expression is a highly constrained behaviour, with highly localised motion and a rather rigid overall structural composition of facial action components. This allows for a simple static template-based representation to achieve reasonably good recognition accuracy. Second, the benchmarking databases used by most of the studies reported in the literature, including the MMI database (Pantic et al. 2005), the JAFFE database (Lyons et al. 1999), and the Cohn–Kanade database (Kanade et al. 2000), only contain image sequences of facial expressions performed in well-controlled lab environments under good lighting conditions with high image resolution.

In a more realistic visual environment where static visual appearance information is less detailed, possibly incomplete, and subject to background clutters, a model is likely to gain greater benefit from utilising spatio-temporal information about a behaviour. In the rest of Part II, we consider visual analysis of more complex human behaviours, for which static template-based models no longer work well. We shall start with gesture recognition in the next chapter.

References

Ahonen, T., Hadid, A., Pietikäinen, M.: Face recognition with local binary patterns. In: European Conference on Computer Vision, Prague, Czech Republic, May 2004, pp. 469–481 (2004)

Bartlett, M., Littlewort, G., Fasel, I., Movellan, R.: Real time face detection and facial expression recognition: Development and application to human computer interaction. In: Workshop on Human-Computer Interaction, Madison, USA, June 2003, pp. 53–60 (2003)

Bartlett, M., Littlewort, G., Frank, M., Lainscsek, C., Fasel, I., Movellan, J.: Recognizing facial expression: machine learning and application to spontaneous behavior. In: IEEE Conference on Computer Vision and Pattern Recognition, San Diego, USA, June 2005, pp. 568–573 (2005)

Bassili, J.N.: Emotion recognition: The role of facial movement and the relative importance of upper and lower area of the face. J. Pers. Soc. Psychol. **37**(11), 2049–2058 (1979)

Belhumeur, P.N., Hespanha, J.P., Kriegman, D.J.: Eigenfaces vs. fisherfaces: recognition using class specific linear projection. IEEE Trans. Pattern Anal. Mach. Intell. **19**(7), 711–720 (1997)

Belkin, M., Niyogi, P.: Laplacian eigenmaps and spectral techniques for embedding and clustering. In: Advances in Neural Information Processing Systems, pp. 585–591 (2001)

Bishop, C.M.: Pattern Recognition and Machine Learning. Springer, Berlin (2006)

Chang, Y., Hu, C., Turk, M.: Manifold of facial expression. In: IEEE International Workshop on Analysis and Modeling of Faces and Gestures, Nice, France, October 2003, pp. 28–35 (2003)

Cohen, I., Sebe, N., Garg, A., Chen, L., Huang, T.S.: Facial expression recognition from video sequences: temporal and static modeling. Comput. Vis. Image Underst. 91, 160–187 (2003)

Donato, G., Bartlett, M., Hager, J., Ekman, P., Sejnowski, T.: Classifying facial actions. IEEE Trans. Pattern Anal. Mach. Intell. 21(10), 974–989 (1999)

Essa, I., Pentland, A.: Coding, analysis, interpretation, and recognition of facial expressions. IEEE Trans. Pattern Anal. Mach. Intell. 19(7), 757–763 (1997)

Feng, X., Hadid, A., Pietikäinen, M.: A coarse-to-fine classification scheme for facial expression recognition. In: International Conference on Image Analysis and Recognition, pp. 668–675 (2004)

Feng, X., Pietikäinen, M., Hadid, T.: Facial expression recognition with local binary patterns and linear programming. Pattern Recognit. Image Anal. 15(2), 546–548 (2005)

Fidaleo, D., Trivedi, M.: Manifold analysis of facial gestures for face recognition. In: ACM SIGMM Multimedia Biometrics Methods and Application Workshop, Berkley, USA, pp. 65–69 (2003)

Freund, Y., Schapire, R.E.: A decision-theoretic generalization of on-line learning and an application to boosting. J. Comput. Syst. Sci. 55(1), 119–139 (1997)

Gong, S., McKenna, S., Collins, J.J.: An investigation into face pose distributions. In: IEEE International Conference on Automatic Face and Gesture Recognition, Killington, USA, October 1996, pp. 265–270 (1996)

Gong, S., Psarrou, A., Romdhani, S.: Corresponding dynamic appearances. Image Vis. Comput. 20(4), 307–318 (2002)

Guo, G., Dyer, C.R.: Simultaneous feature selection and classifier training via linear programming: A case study for face expression recognition. In: IEEE Conference on Computer Vision and Pattern Recognition, Madison, USA, June 2003, pp. 346–352 (2003)

Hadid, A., Pietikäinen, M., Ahonen, T.: A discriminative feature space for detecting and recognizing faces. In: IEEE Conference on Computer Vision and Pattern Recognition, Washington, DC, USA, June 2004, pp. 797–804 (2004)

He, X., Niyogi, P.: Locality preserving projections. In: Advances in Neural Information Processing Systems, pp. 153–160 (2003)

He, X., Yan, S., Hu, Y., Niyogi, P., Zhang, H.: Face recognition using Laplacianfaces. IEEE Trans. Pattern Anal. Mach. Intell. 27(3), 328–340 (2005)

Hsu, C.W., Chang, C.C., Lin, C.J.: A practical guide to support vector classification. Technical report, National Taiwan University, Taipei (2003)

Kanade, T., Cohn, J.F., Tian, Y.: Comprehensive database for facial expression analysis. In: IEEE International Conference on Automatic Face and Gesture Recognition, pp. 46–53 (2000)

Li, Y., Gong, S., Liddell, H.: Constructing facial identity surfaces for recognition. Int. J. Comput. Vis. 53(1), 71–92 (2003)

Liao, S., Fan, W., Chung, C.S., Yeung, D.Y.: Facial expression recognition using advanced local binary patterns, Tsallis entropies and global appearance features. In: IEEE International Conference on Image Processing, pp. 665–668 (2006)

Lyons, M.J., Budynek, J., Akamatsu, S.: Automatic classification of single facial images. IEEE Trans. Pattern Anal. Mach. Intell. 21(12), 1357–1362 (1999)

McKenna, S., Gong, S., Würtz, R.P., Tanner, J., Banin, D.: Tracking facial motion using Gabor wavelets and flexible shape models. In: IAPR International Conference on Audio-Video Based Biometric Person Authentication, Crans-Montana, Switzerland, March 1997, pp. 35–43 (1997)

Ojala, T., Pietikäinen, M., Harwood, D.: A comparative study of texture measures with classification based on featured distribution. Pattern Recognit. **29**(1), 51–59 (1996)

Ojala, T., Pietikäinen, M., Mäenpää, T.: Multiresolution gray-scale and rotation invariant texture classification with local binary patterns. IEEE Trans. Pattern Anal. Mach. Intell. **24**(7), 971–987 (2002)

Padgett, C., Cottrell, G.: Representing face images for emotion classification. In: Advances in Neural Information Processing Systems, pp. 894–900 (1997)

Pantic, M., Patras, I.: Dynamics of facial expression: recognition of facial actions and their temporal segments from face profile image sequences. IEEE Trans. Syst. Man Cybern. **36**(2), 433–449 (2006)

Pantic, M., Rothkrantz, L.: Expert system for automatic analysis of facial expression. Image Vis. Comput. **18**(11), 881–905 (2000)

Pantic, M., Rothkrantz, L.: Facial action recognition for facial expression analysis from static face images. IEEE Trans. Syst. Man Cybern. **34**(3), 1449–1461 (2004)

Pantic, M., Valstar, M., Rademaker, R., Maat, L.: Web-based database for facial expression analysis. In: IEEE International Conference on Multimedia and Expo, pp. 5–10 (2005)

Romdhani, S., Gong, S., Psarrou, A.: Multi-view nonlinear active shape model using kernel pca. In: British Machine Vision Conference, Nottingham, UK, September 1999, pp. 483–492 (1999)

Saul, L., Roweis, S.: Think globally, fit locally: Unsupervised learning of low dimensional manifolds. J. Mach. Learn. Res. **4**, 119–155 (2003)

Schapire, R.E., Singer, Y.: Improved boosting algorithms using confidence-rated predictions. Mach. Learn. **37**(3), 297–336 (1999)

Shan, C., Gong, S., McOwan, P.: Appearance manifold of facial expression. In: Sebe, N., Lew, M.S., Huang, T.S. (eds.) Computer Vision in Human-Computer Interaction. Lecture Notes in Computer Science, pp. 221–230. Springer, Berlin (2005a)

Shan, C., Gong, S., McOwan, P.: Robust facial expression recognition using local binary patterns. In: IEEE International Conference on Image Processing, Genoa, Italy, September 2005, pp. 370–373 (2005b)

Shan, C., Gong, S., McOwan, P.: Dynamic facial expression recognition using a Bayesian temporal manifold model. In: British Machine Vision Conference, Edinburgh, UK, September 2006, pp. 297–306 (2006a)

Shan, C., Gong, S., McOwan, P.: A comprehensive empirical study on linear subspace methods for facial expression analysis. In: IEEE Workshop on Vision for Human-Computer Interaction, New York, USA, June 2006, pp. 153–158 (2006b)

Shan, C., Gong, S., McOwan, P.: Facial expression recognition based on local binary patterns: a comprehensive study. Image Vis. Comput. **27**(6), 803–816 (2009)

Tenenbaum, J.B., Silva, V., Langford, J.C.: A global geometric framework for nonlinear dimensionality reduction. Science **290**, 2319–2323 (2000)

Tian, Y.: Evaluation of face resolution for expression analysis. In: IEEE Workshop on Face Processing in Video, Washington, DC, USA, June 2004

Turk, M., Pentland, A.: Face recognition using eigenfaces. In: IEEE Conference on Computer Vision and Pattern Recognition, Maui, USA, June 1991, pp. 586–591 (1991)

Valstar, M., Pantic, M.: Fully automatic facial action unit detection and temporal analysis. In: Computer Vision and Pattern Recognition Workshop, pp. 149–154 (2006)

Valstar, M., Patras, I., Pantic, M.: Facial action unit detection using probabilistic actively learned support vector machines on tracked facial point data. In: Computer Vision and Pattern Recognition Workshop, pp. 76–84 (2005)

Vapnik, V.: The Nature of Statistical Learning Theory. Springer, New York (1995)

Vapnik, V.: Statistical Learning Theory. Wiley, New York (1998)

Viola, P., Jones, M.: Rapid object detection using a boosted cascade of simple features. In: IEEE Conference on Computer Vision and Pattern Recognition, Kauai, USA, December 2001, pp. 511–518 (2001)

Viola, P., Jones, M., Snow, D.: Detecting pedestrians using patterns of motion and appearance. In: IEEE International Conference on Computer Vision, Nice, France, October 2003, pp. 734–741 (2003)

Würtz, R.P.: Multilayer Dynamic Link Networks for Establishing Image Point Correspondences and Visual Object Recognition. Verlag Harri Deutsch, Frankfurt am Main (1995)

Zhang, Z., Lyons, M.J., Schuster, M., Akamatsu, S.: Comparison between geometry-based and Gabor-wavelets-based facial expression recognition using multi-layer perceptron. In: IEEE International Conference on Automatic Face and Gesture Recognition, Nara, Japan, April 1998, pp. 454–459 (1998)

Zhao, G., Pietikainen, M.: Dynamic texture recognition using local binary patterns with an application to facial expressions. IEEE Trans. Pattern Anal. Mach. Intell. 29(6), 915–928 (2007)

Chapter 5
Modelling Gesture

Gesture, particularly hand gesture, is an important part of human body language which conveys messages and reveals human intention and emotional state. Automatic interpretation of gesture provides an important means for interaction and communication between human and computer, going beyond the conventional text- and graphic-based interface. Broadly speaking, human gesture can be composed of movements from any body part of a person, although the most relevant body parts are face and hand. In this sense, facial expression is a special case of gesture. Facial expression and hand movement often act together to define a coherent gesture, and can be better understood if analysed together. This is especially true when interpreting human emotional state based on visual observation of body language.

Similar to facial expression, visual analysis of gesture starts with finding a suitable representation, followed by modelling and recognition. A gesture is a dynamic process, typically characterised by the spatio-temporal trajectory of body motion. It can be modelled as trajectories in a high-dimensional feature space, representing spatio-temporal correspondences between visual observations (Sect. 5.1). In this context, we describe plausible methods for tracking both individual body parts and overall body movement to construct trajectories for gesture representation. Unsupervised learning is considered for automatically segmenting a continuous gesture movement sequence into atomic components and discovering the number of distinctive gesture classes (Sect. 5.2). In contrast, supervised learning is studied for modelling a gesture sequence as a stochastic process, with classification of different gesture processes learned from labelled training data (Sect. 5.3). We also consider the problem of affective state recognition by analysing both facial expression and body gesture together for interpreting the emotional state of a person (Sect. 5.4).

5.1 Tracking Gesture

5.1.1 Motion Moment Trajectory

In a well-controlled environment where only a single person is present, it is possible to establish a trajectory-based gesture representation without tracking individual

S. Gong, T. Xiang, *Visual Analysis of Behaviour*,
DOI 10.1007/978-0-85729-670-2_5, © Springer-Verlag London Limited 2011

body parts. The first step is to compute holistically human motion of an entire body. Various levels of sophistication are possible for detecting human motion (McKenna and Gong 1998). A straightforward and simplistic approach is to perform two-frame temporal differencing for detecting significant changes in intensity images. More specifically, at time t, a binary image $\mathbf{B}_t$ is obtained by thresholding $|\mathbf{I}_t - \mathbf{I}_{t-1}|$ at a value T tuned to the imaging noise. A set of motion moment features can then be extracted from each image $\mathbf{B}_t$. Among these features, the motion area A is considered as the zeroth-order motion moment, the centroid coordinates $(\bar{x}, \bar{y})$ as the first-order motion moments, and the elongation E as the second-order motion moments. They are estimated as follows:

$$A_t = \sum_{x,y} \mathbf{B}_t[x, y] \tag{5.1}$$

$$\bar{x}_t = \frac{1}{A_t} \sum_{x,y} x\mathbf{B}_t[x, y] \tag{5.2}$$

$$\bar{y}_t = \frac{1}{A_t} \sum_{x,y} y\mathbf{B}_t[x, y] \tag{5.3}$$

$$E_t = \frac{\chi_{\max}}{\chi_{\min}} \tag{5.4}$$

where $\chi^2 = \frac{1}{2}(a + c) + \frac{1}{2}(a - c)\cos 2\theta + \frac{1}{2}b \sin 2\theta$

$$a = \sum_{x,y}(x - \bar{x})^2 \mathbf{B}_t[x, y]$$

$$b = 2\sum_{x,y}(x - \bar{x})(y - \bar{y})\mathbf{B}_t[x, y]$$

$$c = \sum_{x,y}(y - \bar{y})^2 \mathbf{B}_t[x, y]$$

$$\sin 2\theta = \pm \frac{b}{\sqrt{b^2 + (a - c)^2}}$$

$$\cos 2\theta = \pm \frac{a - c}{\sqrt{b^2 + (a - c)^2}}$$

The minimum and maximum values of χ are found by changing the signs of the values of $\sin 2\theta$ and $\cos 2\theta$. Elongation has a lower bound of $E = 1$ in the case of a motion image with an isotropic distribution of motion. In order to obtain a feature set with invariance to translational shifts in the image space, the displacement of the centroid is estimated as $u_t = \bar{x}_t - \bar{x}_{t-1}$, $v_t = \bar{y}_t - \bar{y}_{t-1}$. At time t, the estimated feature set is (A_t, u_t, v_t, E_t). Now a set of feature trajectories can be computed from an observed image sequence by estimating the moment features for each frame. In

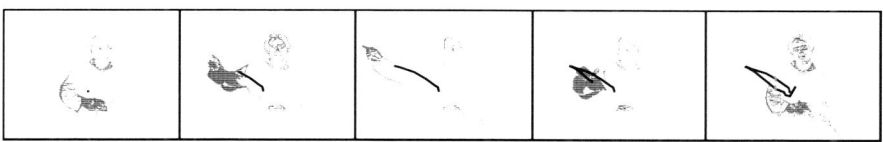

Fig. 5.1 Example temporal difference frames from a 'point right' gesture sequence. The area of detected motion is shown as *mid-grey* here, whilst the path described by the estimated centroid is overlaid in *black*

particular, at time t, a temporal trajectory $\mathbf{z}_t = (\ldots, z_{t-2}, z_{t-1}, z_t)$ is available for each feature, where z_t denotes A_t, u_t, v_t or E_t.

Such simple features are easy to compute, but will also exhibit variations due to extrinsic factors such as illumination and viewing geometry. In particular, area and centroid displacement will scale differently. The feature set will also vary between different gestures and between different instances of the same gesture. In particular, the visual texture of clothing will cause variation with more highly textured clothing increasing the motion area. Consequently identical gestures performed by the same person with different clothing will be represented by different trajectories. This undesirable variation can have an adverse effect on gesture recognition. Figure 5.1 shows an example of five temporal difference frames from a gesture sequence where a person points right. The path described by the estimated centroid travelling over time is also shown. It is evident that during periods of low motion, estimates of first and second-order moments become unreliable.

5.1.2 2D Colour-Based Tracking

When there are multiple people in a scene, association of body parts to the correct person is required for a meaningful gesture interpretation. This implies that individual body parts for each person need be tracked in order to construct a trajectory-based gesture representation. Let us first look at a simplistic two-dimensional (2D) body parts tracking method. The items of interest are the face and two hands of each person. The tracker is based on skin-colour modelling (Raja et al. 1998).

The face and hands need be detected before tracking. An easy to compute and fast detection model is based on the calculation of skin-colour probability for each pixel in the image. Using this approach, skin colour is represented by a mixture of Gaussian probability model in the hue-saturation colour space. Suppose a D-dimensional random variable $\mathbf{y}$ follows a K-component mixture distribution, then the probability density function of $\mathbf{y}$ can be written as

$$p(\mathbf{y}|\boldsymbol{\theta}) = \sum_{k=1}^{K} w_k p(\mathbf{y}|\boldsymbol{\theta}_k) \tag{5.5}$$

where w_k is the mixing probability for the kth mixture component with $0 \leq w_k \leq 1$. $\sum_{k=1}^{K} w_k = 1$, $\boldsymbol{\theta}_k$ are the internal parameters describing the kth mixture component,

(a) original image   (b) pixels classified as skin

Fig. 5.2   Example of skin-colour detection

and $\boldsymbol{\theta} = \{\boldsymbol{\theta}_1, \ldots, \boldsymbol{\theta}_K; w_1, \ldots, w_K\}$ is a C_K-dimensional vector describing the complete set of parameters for the mixture model. Let us denote N independent and identically distributed examples of $\mathbf{y}$ as $\mathcal{Y} = \{\mathbf{y}^{(1)}, \ldots, \mathbf{y}^{(N)}\}$. The log-likelihood of observing $\mathcal{Y}$ given a K-component mixture model is

$$\log p(\mathcal{Y}|\boldsymbol{\theta}) = \sum_{n=1}^{N}\left(\log \sum_{k=1}^{K} w_k\, p\left(\mathbf{y}^{(n)}|\boldsymbol{\theta}_k\right)\right) \tag{5.6}$$

where $p(\mathbf{y}^{(n)}|\boldsymbol{\theta}_k)$ defines the model kernels, that is, the form of the probability distribution function for the kth component. The model kernel functions for different mixture components are assumed to have the same form of Gaussian. If the number of mixture components K is known, the maximum likelihood estimation (MLE) of model parameters, as given by

$$\hat{\boldsymbol{\theta}} = \arg\max_{\boldsymbol{\theta}}\left\{\log p(\mathcal{Y}|\boldsymbol{\theta})\right\}$$

This can be computed using the expectation-maximisation (EM) algorithm (Dempster et al. 1977).

A mixture-of-Gaussian-based skin-colour model is trained from example pixels as a foreground model. A background model can also be built (Russell and Gong 2006), and combined with the foreground skin model using the Bayes' rule. The skin-colour probabilities are thresholded to give a binary image of skin and non-skin pixels. This model can also be updated on-the-fly to adapt to changes in lighting and view angle resulting in an adaptive skin-colour model (Raja et al. 1998). An example of detecting skin-colour pixels is shown in Fig. 5.2.

To form skin-coloured body parts, detected skin pixels are grouped by clustering to define blobs in the image. This can be computed using a connected component algorithm (Ballard and Brown 1982; Sherrah and Gong 2000b). The output of this clustering process is a set of rectangular bounding boxes of skin-colour pixel groups (Fig. 5.2(a)), which could be faces, hands, or skin-coloured furniture or clothing. To find faces, one solution is to use a multi-pose and multi-scale support vector machine to scan inside these boxes at multiple positions (Li et al. 2000; Ng and Gong 2002). When a face is found, it initialises a person tracker. An example of face detection is shown in Fig. 5.3.

Fig. 5.3 Examples of
detecting faces using a
support vector machine. All
outer boxes are skin clusters,
while *inner boxes* are
detected faces

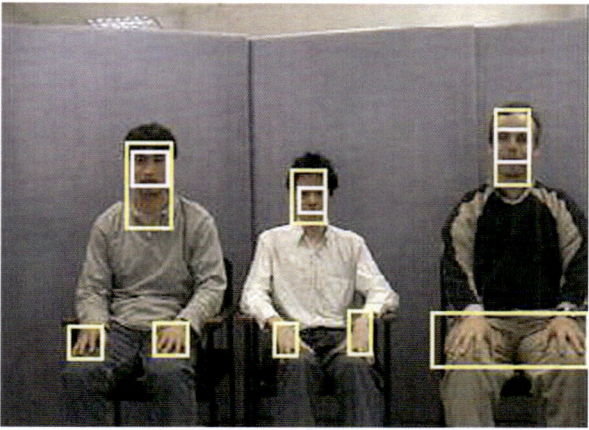

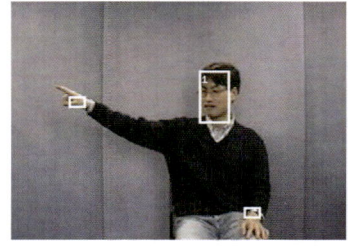   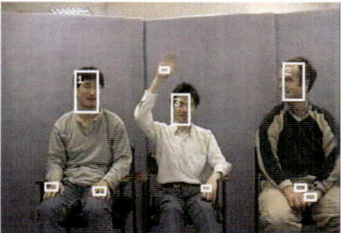

(a) single person scenario   (b) multiple people scenario

Fig. 5.4 Examples of 2D body part tracking

Given a detected face, the task of a person tracker is to track the subject's head
and hands. A generic box tracker can be readily used to track each body part based
on an iterative mean shift algorithm (Bradski 1998). A person tracker can consist
of three box trackers for the head and both hands, respectively, all initialised with
different parameters. The head box tracker is initialised using the support vector
machine-based face detection. The hand box trackers are initialised heuristically
with respect to the head position. An example of 2D body part tracking is shown in
Fig. 5.4. With tracked hands, the trajectory of the centroid of a hand can be used to
model gesture. Although trajectories of hands are more directly related to body gesture, head tracking can assist in yielding more accurate hand tracking by providing
a contextual constraint on the likely hand location (Sherrah and Gong 2000a).

5.1.3 Bayesian Association

Colour-based tracking of body parts is a relatively robust and inexpensive approach.
Nevertheless, there are a number of problems that can cause tracking failure. They

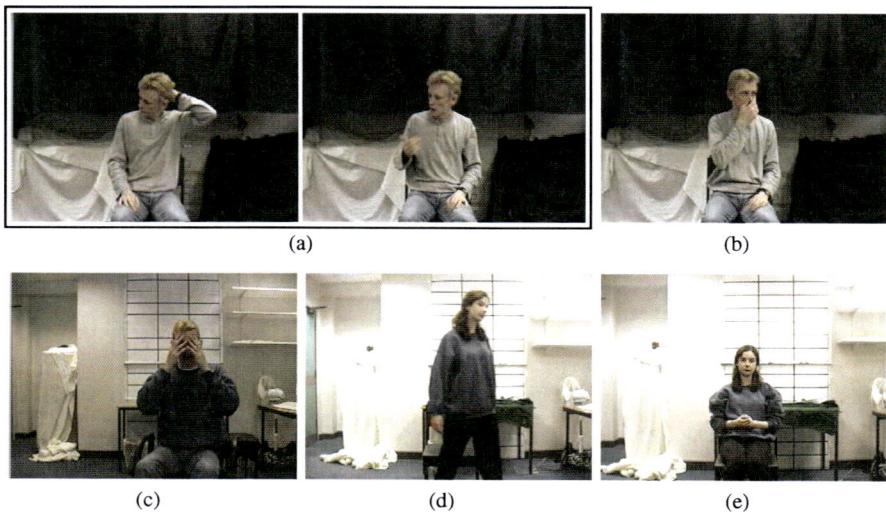

(a)   (b)

(c)   (d)   (e)

Fig. 5.5 Examples of the difficulties associated with tracking the body. (**a**) Motion is discontinuous between frames; (**b**) one hand occludes the face; (**c**) both hands occlude the face; (**d**) a hand is invisible in the image; and (**e**) the hands occlude each other

include noise, uncertainty, and ambiguity due to occlusion and distracting skin-coloured background clutters. The two most difficult problems to deal with when tracking a person's head and hands are occlusion and hand association. Occlusion occurs when a hand passes in front of the face or intersects with the other hand. This can happen rather regularly in natural conversational and interactive gestures. Hand association requires that the hands found in the current image frame are matched correctly to those in the last frame. Another challenge for 2D body part tracking is that the movements of body parts captured in an image sequence are often discontinuous. Some examples of these difficulties are shown in Fig 5.5.

Hand movements can also be very rapid during a gesture resulting in large displacement of hands in two consecutive image frames. To illustrate the nature of discontinuous body motion under these conditions, Fig. 5.6 shows the head and hands positions and accelerations, visualised as vectors, for two image sequences, along with example frames. The video frames were sampled at 18 frames per second (FPS). There are significant and sudden variations in both the magnitude and orientation of the acceleration of the hands.

A plausible method for object tracking is to explicitly model the dynamics of object movement. Good examples of such models include Kalman filtering and conditional density propagation (CONDENSATION) (Isard and Blake 1998). However, given large displacements in gesture movement and sudden change of speed and direction typically observed in a hand gesture, it can be problematic to model the dynamics of body parts. Under these conditions, explicit modelling of body dynamics inevitably makes too strong an assumption about image data. Rather, the tracking can be performed better and more robustly through a process of deduction.

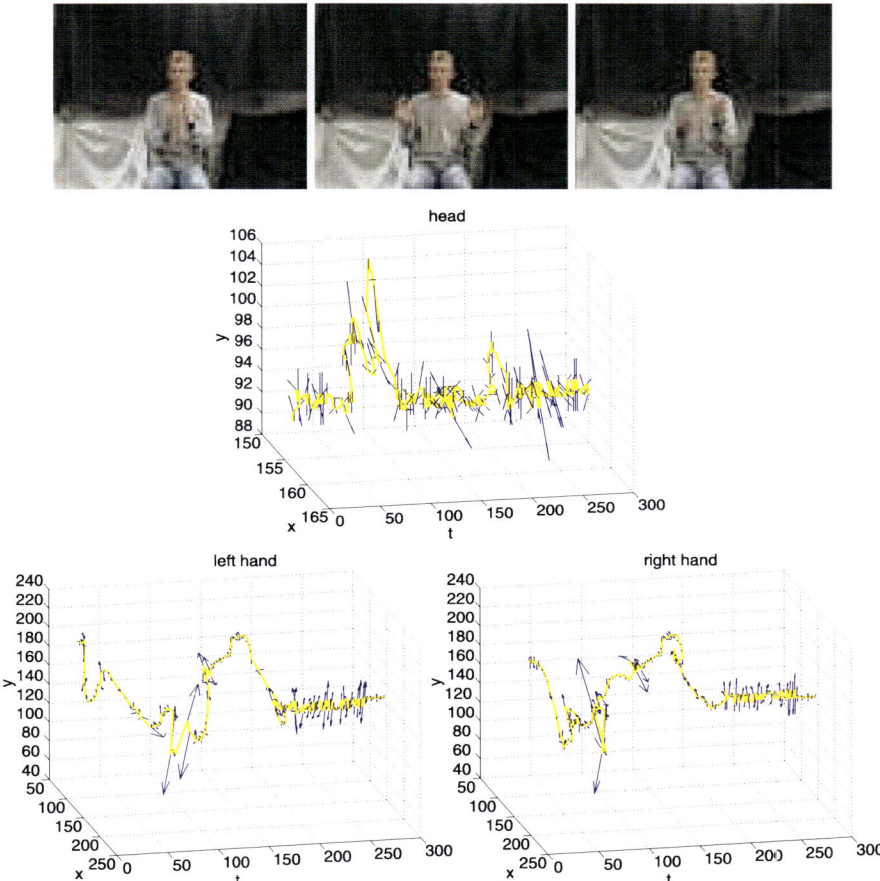

Fig. 5.6 An example gesture sequence with tracked head and hand positions and accelerations. At each time frame, the 2D acceleration is shown as an *arrow* with arrowhead size proportional to the acceleration magnitude

This requires full exploitation of both visual cues and high-level contextual knowledge. For instance, one knows that at any given time either (1) a hand is associated with a skin-colour cluster, or (2) it occludes the face, rendering it therefore invisible using only skin colour (Fig. 5.5(b), (c)), or (3) it has disappeared from the image (Fig. 5.5(d)). When considering both hands, a possibility arises that both hands are associated with the same skin-colour cluster, as when one claps the hands together for example (Fig. 5.5(e)).

To overcome these problems in tracking multiple visually similar objects, one plausible solution is Bayesian data association (Sherrah and Gong 200Cb). In particular, for associating body parts from gesture motion, this method exploits the following contextual knowledge as constraints:

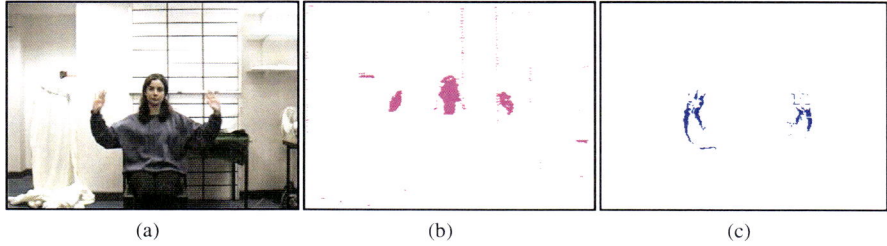

(a)   (b)   (c)

Fig. 5.7 Different simple visual cues for tracking body parts. (**a**) Original image, (**b**) binary skin–colour map, and (**c**) binary motion map

1. A person is oriented roughly towards the camera for most of the time.
2. A person is wearing long sleeves, and no artificial gloves are worn if skin colour is used for identifying hands.
3. Sufficiently good head detection is possible.
4. The head and hands are the dominant moving skin-colour clusters in the view.

Computing Visual Cues

To overcome body part tracking difficulties, a model should also utilise multiple visual cues, preferably from independent sources. Visual cues that are simple and fast to compute for body part tracking include skin colour, image motion and hand orientation. Skin-colour probabilities are computed for pixels in an image to generate a binary skin image (Fig. 5.7(b)). Image motion is computed by frame-differencing (Fig. 5.7(c)).

Skin colour and motion are natural cues for focusing attention on regions of interest in images, especially in the case of tracking body parts from gesture movement. One may consider suppressing distracting skin colour alike background clutters by applying a logic AND operator to the binary skin and motion maps directly. However, such a crude approach to fusing these cues is premature due to potential loss of useful information. For example, the motion information generally occurs only at the edges of a moving object, making the fused information too sparse.

The problem of associating the correct hands over time can usually be solved using spatial constraints. However, certain situations arise under occlusion in which choosing the nearest skin-coloured part to the previous hand position can result in incorrect hand assignment. This problem cannot be solved purely using colour and motion information. In the absence of depth information or 3D skeletal constraints on body parts, image intensity information can be exploited to assist in resolving incorrect assignment. Specifically, the intensity image of each hand can be used to obtain a very coarse estimate of hand orientation which is robust even in low resolution imagery. Restricted kinematics of the human body are loosely modelled to exploit the fact that only certain hand orientations are likely at a position in the image relative to the head of the same person.

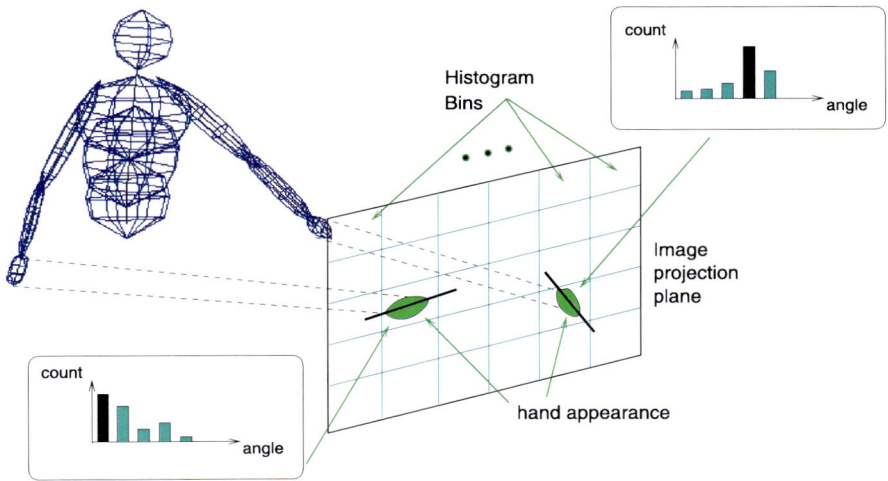

Fig. 5.8   Schematic diagram of a hand orientation histogram process

A statistical hand orientation model can be estimated (Sherrah and Gong 2000b), as illustrated in Fig. 5.8. Assuming that the subject is facing the camera, the image is divided coarsely into a grid of histogram bins. A histogram of likely hand orientations can be synthesised exhaustively for each 2D position of the hand in the image projection relative to the head position. To do this, a three-dimensional (3D) avatar model of the human body is used to exhaustively sample the range of possible arm joint angles in upright posture. Assuming that the hand extends parallel to the forearm, the 2D projection is made to obtain the appearance of hand orientation and position in the image plane, and the corresponding histogram bin is updated. During tracking, the quantised hand orientation is obtained according to the maximum response from a bank of oriented Gabor filters. The tracked hand position relative to the tracked head position is used to index the histogram and to obtain the likelihood of the hand orientation given the position.

Reasoning about Body-Parts Association

Given multiple visual cues described above, the problem is now to determine a model for the association of the left and right hands. One can consider this situation to be equivalent to watching a mime artist wearing a white face mask and white gloves in black clothing and a black background (Fig. 5.7(b)). Further, only discontinuous information is available as though a strobe light were operating, creating a 'jerky' effect (Fig. 5.5(a)). A robust mechanism is required for reasoning about the situation. One such mechanism is Bayesian networks.

A Bayesian network provides a rigorous framework for combining semantic and sensor-level reasoning under conditions of uncertainty (Charniak 1991; Cowell et al.

1999; Pearl 1988). Given a set of variables **W** representing a scenario, the assumption is that all our knowledge of the current state of affairs is encoded in the joint distribution of the variables conditioned on the existing evidence, $P(\mathbf{w}|\mathbf{e})$. Explicit modelling of this distribution is unintuitive and often infeasible. Instead, conditional independencies between variables can be exploited to sparsely specify the joint distribution in terms of more tangible conditional distributions between variables.

A Bayesian network model is a directed acyclic graph that explicitly defines the statistical or causal dependencies between all variables.[1] These dependencies are known a priori and used to create the network structure. Nodes in the network represent random variables, while directed links point from conditioning to dependent variables. For a link between two variables, $X \rightarrow Y$, the conditional distribution of their instantiations $P(y|x)$ in the absence of evidence must be specified beforehand from contextual knowledge. As evidence is presented to the network over time through variable instantiation, a set of beliefs are established which reflect both prior and observed information:

$$\mathrm{BEL}(x) = P(x|\mathbf{e}) \tag{5.7}$$

where $\mathrm{BEL}(x)$ is the belief in the value of variable X given the evidence **e**. Updating of beliefs occurs through a distributed message-passing process that is made possible by the exploitation of local dependencies and global independencies. Hence, dissemination of evidence to update currently held beliefs can be performed in a tractable manner to arrive at a globally consistent evaluation of the situation.

A Bayesian network can subsequently be used for prediction and queries regarding values of single variables given current evidence. However, if the most probable joint configuration of several variables given the evidence is required, then a process of belief revision, as opposed to belief updating,[2] must be applied to obtain the most probable explanation of the evidence at hand, $\mathbf{w}^*$, defined by the following criterion:

$$P(\mathbf{w}^*|\mathbf{e}) = \overset{\max}{\mathbf{w}} \; P(\mathbf{w}|\mathbf{e}) \tag{5.8}$$

where **w** is any instantiation of the variables **W** consistent with the evidence **e**, known as an explanation or extension of **e**, and $\mathbf{w}^*$ is the most probable explanation or extension. This corresponds to the locally computed function expressing the local belief in the extension:

$$\mathrm{BEL}^*(x) = \overset{\max}{\mathbf{w}'_X} \; P(x, \mathbf{w}'_X|\mathbf{e}) \tag{5.9}$$

where $\mathbf{W}'_X = \mathbf{W} - X$.

Sherrah and Gong (2000b) designed a Bayesian network data association model for tracking hands, as shown in Fig. 5.9. The abbreviations are: LH = left hand,

---

[1]Therefore the statistical independencies are implicitly defined as well.

[2]There is a difference between the notions of belief updating and belief revision. This is because in general, the values for variables X and Y that maximise their joint distribution are not the values that maximise their individual marginal distributions.

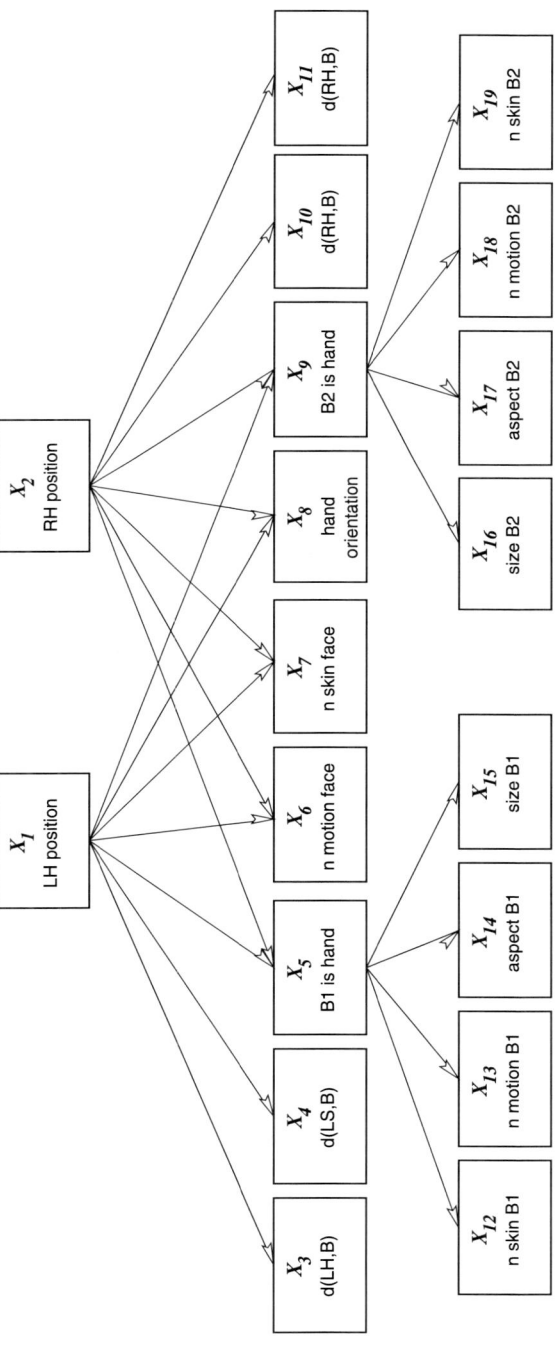

Fig. 5.9 A Bayesian network-based data association model for representing dependencies amongst variables in tracking human body parts

RH = right hand, LS = left shoulder, RS = right shoulder, B1 = skin cluster 1, B2 = skin cluster 2. There are 19 variables, $\mathbf{W} = \{X_1, X_2, \ldots, X_{19}\}$. The first point to note is that some of the variables are conceptual, namely X_1, X_2, X_5 and X_9, while the remaining variables correspond to image-measurable quantities, $\mathbf{e} = \{X_3, X_4, X_6, X_7, X_8, X_{10}, \ldots, X_{19}\}$. All quantities in the network are or have been transformed to discrete variables. The conditional probability distributions attributed to each variable in the network are specified beforehand using either domain knowledge or statistical sampling. At each time step, all of the measurement variables are instantiated from observations. B1 and B2 refer to the two largest skin clusters in the image apart from the head. Absence of clusters is handled by setting the variables X_5 and X_9 to have zero probability of being a hand.

The localised belief revision method is employed until the network stabilises. The most probable joint explanation of the observations is obtained as

$$P\left(\mathbf{w}^* | \{x_3, x_4, x_6, x_7, x_8, x_{10}, \ldots, x_{19}\}\right) = \overset{\max}{\mathbf{w}} \ P\left(\mathbf{w} | \{x_3, x_4, x_6, x_7, x_8, x_{10}, \ldots, x_{19}\}\right)$$
(5.10)

This yields the most likely joint values of X_1 and X_2, which can be used to set the left- and right-hand box position. This network is not singly connected, due to the loops formed through X_1 and X_2. Consequently, the simple belief revision algorithm of Pearl (1988) cannot be used due to non-convergence. Instead, the more general inference algorithm of Lauritzen and Spiegelhalter (1990) needs be applied. This inference method transforms the network to a join tree, each node of which contains a sub-set of variables called a clique. The transformation to the join tree needs be performed only once off-line. Inference then proceeds on the join tree by a message-passing mechanism similar to the method proposed by Pearl (1988).

The complexity of the propagation algorithm is proportional to the span of the join tree and the largest state space size amongst the cliques. The variables and their dependencies are as follows.

X_1 and X_2: the primary hypotheses regarding the left- and right-hand positions, respectively. These variables are discrete with values {CLUSTER1, CLUSTER2, HEAD} which represent skin cluster 1, skin cluster 2 and occlusion of the head, respectively. Note that disappearance of the hands is not modelled here for simplicity.

X_3; X_{10}: the distance in pixels of the previous left/right-hand box position from the currently hypothesised cluster. The dependency imposes a weak spatio-temporal constraint that hands are more likely to have moved a small distance than a large distance from one frame to the next.

X_4; X_{11}: the distance in pixels of the hypothesised cluster from the left/right shoulder. The shoulder position is estimated from the tracked head box. This dependency specifies that the hypothesised cluster should lie within a certain distance of the shoulder as defined by the length of the arm.

X_5, X_{12}, X_{13}, X_{14}, X_{15}; X_9, X_{16}, X_{17}, X_{18}, X_{19}: these variables determine whether each cluster is a hand. X_5 and X_9 are Boolean variables specifying whether or not their respective clusters are hands or noise. The variables have an obvious dependency on X_1 and X_2: if either hand is a cluster, then that cluster must be a hand. The descendants of X_5 and X_9 provide evidence that the clusters are hands. X_{12} and X_{19} are the number of skin pixels in each cluster, which have some distribution depending on whether or not the cluster is a hand. X_{13} and X_{18} are the number of motion pixels in each cluster, expected to be high if the cluster is a hand. Note that these values can still be non-zero for non-hands due to shadows, highlights and noise on skin-coloured background objects. X_{14} and X_{17} are the aspect ratios of the clusters which will have a certain distribution if the cluster is a hand, but no constraints if the cluster is not a hand. X_{15} and X_{16} are the spatial areas of the enclosing rectangles of the clusters. For hands, these values have a distribution in terms relative to the size of the head box, but for non-hands there are no expectations.

X_6 and X_7: the number of moving pixels and number of skin-coloured pixels in the head exclusion box, respectively. If either of the hands is hypothesised to occlude the head, one expects more skin pixels and some motion.

X_8: orientation of the respective hand, which depends to some extent on its spatial position in the screen relative to the head box. This orientation is calculated for each hypothesised hand position, and the likely hand orientation histogram (Fig. 5.8) is used to assign a conditional probability.

Under this Bayesian data association framework, all of the visual cues can be considered simultaneously and consistently to arrive at a most probable explanation for the positions of both hands. A Bayesian network lends the benefit of being able to *explain away* evidence, which can be of use in the network. For example, if the belief that the right hand occludes the face increases, this decreases the belief that the left hand also occludes the face because it explains any motion of growth in the number of skin pixels in the head region. This comes about through the indirect coupling of the hypotheses X_1 and X_2 and the fixed amount of probability attributable to any single piece of evidence. Hence probabilities are consistent and evidence is not double counted (Pearl 1988).

Sherrah and Gong (2000b) show comparative experiments to demonstrate that a Bayesian data association-based tracker can improve the robustness and consistency of tracking discontinuous hand movements during a gesture. This is a good example of a model that is able to combine all available information at all levels, including non-visual contextual information. Some examples of body part tracking based on Bayesian data association are shown in Fig. 5.10. Each sub-figure shows a gesture sequence from left to right. The frames shown are not consecutive. In each image a bounding box locates the head and each of the two hands. The hand boxes are labelled left and right, showing the correct assignments. In the first example in Fig. 5.10(a), the hands are accurately tracked before, during and after mutual occlusion. In Fig. 5.10(b), typical coughing and nose-scratching movements bring about occlusion of the head by a single hand. In this sequence, the two frames marked with "A" are adjacent frames, exhibiting the significant motion discontinuity that can be encountered. Nevertheless, the tracker is able to correctly follow the hands. In Fig. 5.10(c), the subject undergoes significant whole body motion and the head is constantly moving. With the hands alternately occluding each other and the face, the tracker is still able to follow the body parts. In the second last frame, both hands

Fig. 5.10 Examples of tracking body parts by Bayesian data association

simultaneously occlude the face. The tracker recovers well into the next frame. The example of Fig. 5.10(d) has the subject partially leaving the screen twice to fetch objects. Note that in the frames marked "M", one hand is not visible in the image. Since this case is not explicitly modelled by the Bayesian data association model, occlusion with the head or the other hand is deduced. After these periods of disappearance, the hand is once again correctly associated and tracked.

5.1.4 3D Model-Based Tracking

So far we considered tracking body parts based on visual cues in 2D views. Alternatively, a 3D model of the skeleton of a human body has the potential to overcome tracking ambiguities arisen from losing the depth information in a 2D image plane. The basic parameters of a 3D body model include the positions of its 3D vertices. Additional parameters such as relative angles between bones can also be introduced into the model.

A 3D model can be used to track human body parts from a single camera view. Methods for tracking 3D models from a single view include the use of inverse kinematics for estimating body joint rotation parameters (Regh 1995), and modelling bones using primitive solids such as superquadrics (Gavrila and Davis 1995; Pentland 1990) or cylinders (Hogg 1983). A search procedure together with a similarity measure is used to compute the parameters of a 3D skeleton (Gavrila and Davis 1995; Hogg 1983). Another approach is to learn explicitly any correlations between 2D visual observations and an underlying 3D model (Bowden et al. 1998).

However, body part self-occlusions make tracking a 3D model from a single view difficult, due to missing visual features resulting in model matching ambiguities. To tackle this problem, a multi-view approach can be adopted, which utilises information from different views to constrain the 3D pose of a model. In the following, we consider a method for estimating the degree of ambiguity in 2D visual cues and combining them with a 3D skeleton model for tracking human body parts (Ong and Gong 2002).

Learning a 3D Skeleton Model

A 3D skeleton of a human body can be defined as a collection of 3D vertices together with a hierarchical arrangement of bones. The bones are used to link two 3D vertices and constrain their possible movements. One can start with an example-based point distribution model, commonly used for modelling and tracking 2D shapes. This technique can be extended to track 3D skeletons whereby a hybrid 2D–3D representation of a human model is constructed (Bowden et al. 1998). In this model, a state vector is defined to represent an instance of a similar hybrid model consisting of observable 2D data and its corresponding 3D skeleton. The observable data are 2D image features, including the 2D shape of a person's body and image positions of body parts, all directly measured from images. The shape is represented

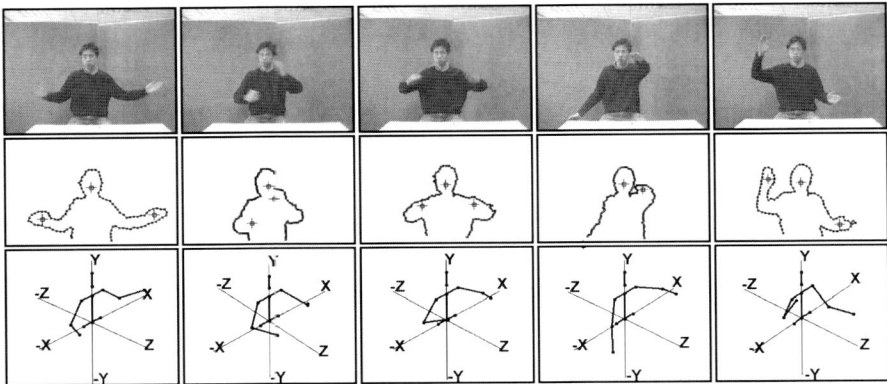

Fig. 5.11 Examples of a joint 2D–3D state vector representing a human body using hybrid 2D and 3D information. The *top row* shows 2D images of a gesture sequence. The *middle row* shows to the 2D body contours ($\mathbf{v}_S$) and 2D body parts positions ($\mathbf{v}_C$). The *bottom row* shows the corresponding 3D skeleton ($\mathbf{v}_T$)

by the contour of a person's silhouette, consisting of N_C number of 2D vertices, $\mathbf{v}_S = (x_1, y_1, \ldots, x_{N_C}, y_{N_C})$. The body parts consist of the image positions of the left hand (x_L, y_L), the right hand (x_R, y_R) and the head (x_H, y_H). The head is used as an alignment reference point for both the body parts and the contour. The positions of left and right hands are concatenated into a vector, $\mathbf{v}_C$. Finally, a corresponding 3D representation of a human body, which consists of N_T number of 3D vertices of a body skeleton, is similarly concatenated into a vector ($\mathbf{v}_T$). A joint state vector is defined as $\mathbf{v} = (\mathbf{v}_S, \mathbf{v}_C, \mathbf{v}_T)$, which consists of a hybrid representation of both 2D and 3D point distribution information about the human body: 2D shape contour positions, 2D body parts positions and their 3D skeleton vertices (Fig. 5.11).

The state space of this hybrid point distribution model is inherently nonlinear. This is due to both the articulated nature of the skeleton model, and the restriction that body segments are rigid. Learning a 3D skeleton model requires some means of capturing this nonlinear space from some training data. Hierarchical principal component analysis (PCA) is an effective model for representing high-dimensional nonlinear spaces (Heap and Hogg 1997). This method consists of two main components:

1. Dimensionality reduction by PCA is carried out to remove redundant dimensions in a point distribution space. This lower dimensional space is defined by a global eigenspace into which all training data are projected. Since the projection to this global eigenspace is linear, a distribution subspace occupied by the projected training examples remains nonlinear.
2. The nonlinear subspace describing the projected training examples can be modelled by a group of N_{clust} clusters denoted as $\{\mathbf{c}_1, \ldots, \mathbf{c}_{N_{\text{clust}}}\}$. Each cluster ($\mathbf{c}_i$) is defined as $\{\boldsymbol{\mu}_i, \Sigma_i, \mathbf{P}_i, \Lambda_i\}$, where the mean vector ($\boldsymbol{\mu}_i$) denotes the cluster's location, and the covariance matrix (Σ_i) represents the cluster's orientation and size. Each cluster can be further represented in a local eigenspace by the prin-

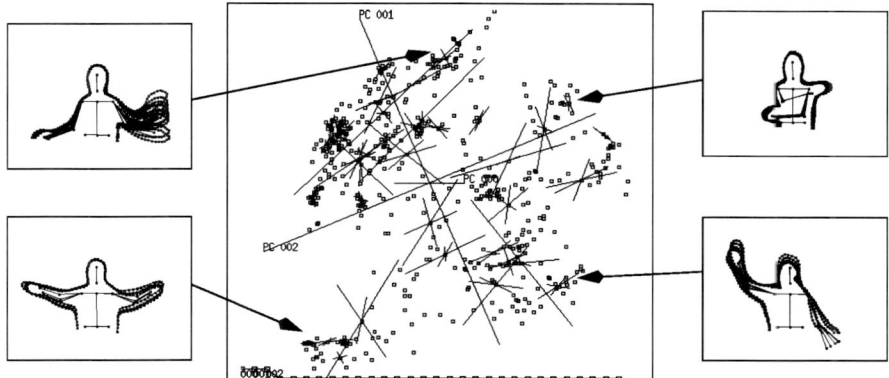

Fig. 5.12 Some examples of projecting human body joint 2D–3D state vectors into a global eigenspace, where the three largest eigen-dimensions are shown. Local principal components of different clusters formed from the training data are also shown. Each side figure shows multiple instances of reconstructed body shape, part positions and skeleton along the largest local principal component from the mean of a cluster (Ong and Gong 2002)

cipal components of its covariance matrix, where the eigenvectors of Σ_i is $\mathbf{P}_i$, and the eigenvalues are the diagonal elements of matrix (Λ_i). One can now consider these clusters as a set of localised principal components (see examples in Fig. 5.12).

Having learned a distribution of hybrid state vectors of 2D body shape, 2D part position and 3D skeleton extracted from some training data, a model can be built to track the state vectors of a human body in motion. To make use of the learned distribution, tracking is performed by projecting instances of 2D state vector components extracted from gesture images into the global eigenspace, and reconstructing the hybrid state vectors in the eigenspace from the learned distribution of the training data.

For this tracking process to be successful, there are potentially two problems need be solved. First, the possibility of discontinuous 2D shape contours occurring between views needs be addressed. In certain views, some feature points of a 2D shape contour have no correspondences to previous views. This can be caused by small 3D movements resulting in large changes in the state vector (see examples in Fig. 5.13). Such phenomenon can cause a huge jump in the projections of state vectors into the global eigenspace. To address this problem, a technique introduced by Heap (1997) can be adopted, which models a matrix of transition probabilities between different subspaces in a global eigenspace. The subspaces are represented by clusters in the global eigenspace.

Second, to reconstruct the 3D skeleton component of the state vector, a prediction model on state transition must cope with non-deterministic and discontinuous temporal change. The CONDENSATION algorithm (Isard and Blake 1998) is a plausible candidate for this task. The stochastic nature of CONDENSATION allows certain degree of discontinuous changes to be handled by modelling the dynamics of

Fig. 5.13 Small 3D body
movement can cause
discontinuities in the feature
points of its corresponding
2D shape contour

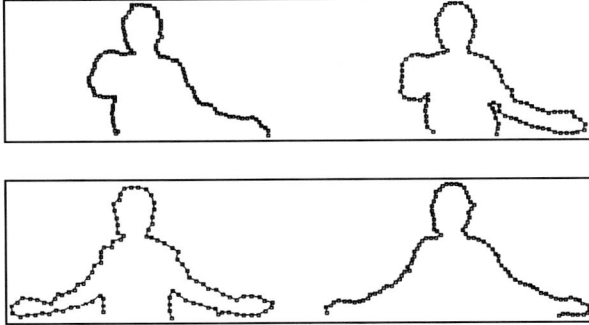

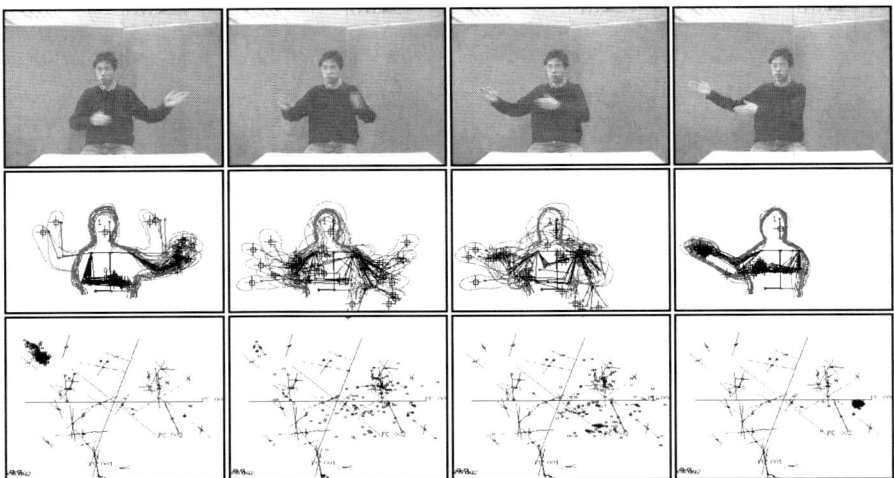

Fig. 5.14 An example of tracking a population of state samples by CONDENSATION. The *top row* shows images of a gesture sequence. The *bottom row* shows the global eigenspace with clusters of local principal components. The examples are *highlighted* by *small dots*. The *middle row* shows reconstructed state vectors using 10 out of 100 samples propagated over time (Ong and Gong 2002)

propagating multiple sample states therefore enabling a tracker to recover from failure. The ability to have multiple hypotheses also provides the potential for coping with ambiguities in the estimated skeleton in certain views. Although this does not resolve ambiguities directly, tracking is made more robust when the problem arises. The CONDENSATION algorithm is used to track a population of projected state vectors as a probability density function in the global eigenspace (see Fig. 5.14). The sample prediction step is modified to make use of the transitional probability matrix. This allows the tracker to propagate samples across different subspaces, therefore coping with the discontinuous nature of the data.

More precisely, an estimated state vector ($\mathbf{v}$) is reconstructed by taking a linear combination of N_{gev} number of *global* eigenvectors ($\mathbf{g}_1, \ldots, \mathbf{g}_{N_{\text{gev}}}$):

$$\mathbf{v} = \sum_{i=1}^{N_{\text{gev}}} s_i \mathbf{g}_i \tag{5.11}$$

In the reconstructed vector $\mathbf{v} = (\mathbf{v}_S, \mathbf{v}_C, \mathbf{v}_T)$, the accuracy of both the shape contour ($\mathbf{v}_S$) and the body parts ($\mathbf{v}_C$) are measured from the image before being combined to yield a final fitness value. More specifically, a prediction accuracy value (f_S) for the contour is computed as follows:

1. Assign the prediction accuracy value, $f_S = 0$.
2. For N_C number of vertices of a contour:
   a. Find the distance (s) from each 2D shape vertex position to the pixel of greatest intensity gradient by searching along its normal.
   b. Compute $f_S = f_S + s$.

We now consider how to measure the accuracy of the predictions of state vector on the positions of body parts, (x_{p1}, y_{p1}) and (x_{p2}, y_{p2}). In each image frame, three skin-coloured regions of interests are tracked corresponding to positions of the hands and the face. Two of these three positions are taken and ordered into a four-dimensional vector $\mathbf{m}_b = (x_{m1}, y_{m1}, x_{m2}, y_{m2})$ where (x_{m1}, y_{m1}) and (x_{m2}, y_{m2}) are the coordinates of the first and second position, respectively. The accuracy of the predictions on the positions of two body parts is defined as

$$f_C = \sqrt{(x_{p1} - x_{m1})^2 + (y_{p1} - y_{m1})^2} + \sqrt{(x_{p2} - x_{m2})^2 + (y_{p2} - y_{m2})^2} \tag{5.12}$$

A final fitness value (f_n) for $\mathbf{v}$ of the nth sample, $s_n^{(t+1)}$, is given by the individual fitness measurements as follows:

$$f_n = O \exp\left(\frac{-f_C}{2P}\right) + R \exp\left(\frac{-f_S}{2Q}\right) \tag{5.13}$$

where O and R are scale constants used to even out differences in scale between the two weighted fitness measurements, f_C and f_S. Constants P and Q represent the amount of variance or tolerance allowed between the predicted and observed values for the body parts and the shape contour, respectively.

View Selection

By selecting a different viewpoint to minimise self-occlusion, visual observations can be made less ambiguous. How does a model measure the goodness of a viewpoint when tracking articulated motion of a human body? Here we describe a method for estimating the degree of ambiguities in 2D image measurements at a given viewpoint. When measured at different views, its value allows us to select the least ambiguous information from one of the available views. Visual observations become ambiguous when more than one projections of a 3D skeleton can be matched. Figure 5.15 illustrates an example of this phenomenon. One considers that

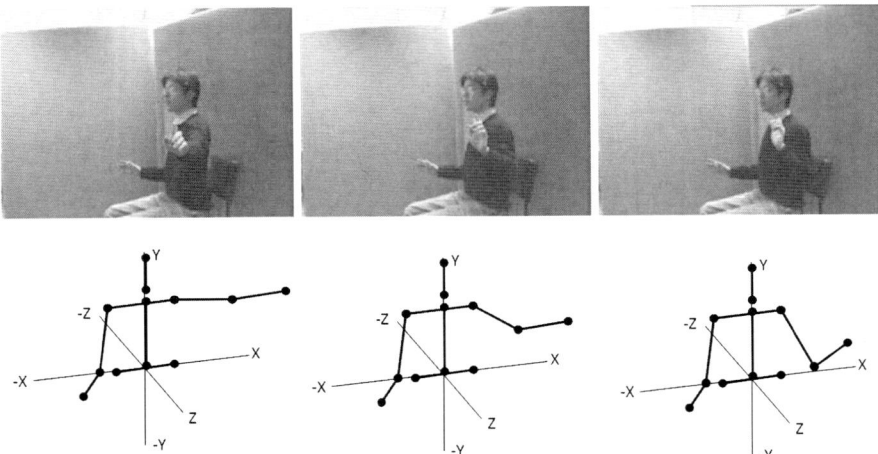

Fig. 5.15 2D visual observation of a 3D skeleton model can be ambiguous from certain viewpoints. This example shows that despite a 3D skeleton is changing, there is little difference in the 2D shape contour and the positions of the hands in the images

a state vector's visual observation is ambiguous if there are more than one state vectors which have similar 2D shapes and part positions but dissimilar underlying 3D skeletons.

Let us define an observation subspace of a state vector $(\mathbf{v}_C, \mathbf{v}_S, \mathbf{v}_T)$ to be $\mathbf{v}_S$ and $\mathbf{v}_C$. There are N_C number of vertices for a shape contour $(\mathbf{v}_S)$ while the number of tracked body part positions is two. Thus, the observation subspace spans from dimension 0 through dimension $2N_C + 4$ in the state vector. We also define a skeleton subspace to be the dimensions of a state vector for all the 3D skeleton vertices. There are N_T number of 3D vertices for a 3D skeleton. To measure how close the 2D observations in two state vectors are, we define an observation-distance (d_{ob}) between two state vectors $\mathbf{x}$ and $\mathbf{y}$ as

$$d_{ob}(\mathbf{x}, \mathbf{y}) = \sum_{i=1}^{2N_C+4} \sqrt{dx_{ob}^2 + dy_{ob}^2} \qquad (5.14)$$

where $dx_{ob} = x_{2i} - y_{2i}$ and $dy_{ob} = x_{2i+1} - y_{2i+1}$.

Similarly, to measure how similar the 3D skeleton component in two state vectors are, we define a skeleton-distance (d_s) between two state vectors $\mathbf{x}$ and $\mathbf{y}$ as

$$d_s(\mathbf{x}, \mathbf{y}) = \sum_{i=1}^{3N_T} \sqrt{dx_s^2 + dy_s^2 + dz_s^2} \qquad (5.15)$$

where $dx_s = x_{3i} - y_{3i}, dy_s = x_{3i+1} - y_{3i+1}$ and $dz_s = x_{3i+2} - y_{3i+2}$. Given K number of example state vectors of the 2D–3D hybrid representation, $\{\mathbf{y}_1, \mathbf{y}_2, \ldots, \mathbf{y}_K\}$, we define the ambiguity of an sample state vector as

$$a = \sum_{j=1}^{K} G(s_{2D}, \sigma) f(s_{3D}, \beta, t) \tag{5.16}$$

where $s_{2D} = d_o(\mathbf{w}, \mathbf{y}_j)$ and $s_{3D} = d_s(\mathbf{w}, \mathbf{y}_j)$. Equation (5.16) measures the ambiguity between a 3D skeleton and its 2D observations in a single view.

If two state vectors have a very similar visual observation, the model estimates the difference in their 3D skeletons. The role of the one-dimensional Gaussian kernel $G(x, \sigma)$ is to return the similarity between observations of two sample state vectors. The smaller the observation-distance, the greater the similarity. The rate at which this value reduces as the observation-distance increases is determined by the standard deviation (σ) of the Gaussian kernel.

5.2 Segmentation and Atomic Action

Most natural gestures are expressive body motions without clear spatio-temporal trajectories. In order to interpret gestures by their trajectories, continuous body movement patterns need be segmented into fragments corresponding to meaningful gestures, where starting and ending points of gestures can be determined. Visually segmenting a gesture from body motion is difficult due to image measurement noise, complex temporal scaling caused by variations in speed, and variations in performing a gesture by different people, or the same person at a different time.

For segmenting continuous visual observation of natural gestures, McNeill (1992) identifies five basic categories of hand gestures: iconic, metaphoric, cohesive, deictic and beat gestures. All gestures have their temporal signature in common. Gestures are typically embedded by the hands being in a rest position and can be divided into either bi-phasic or tri-phasic gestures. Beat and deictic gestures are examples for bi-phasic gestures. They have two movement phases, away from the rest position into gesture space and back again, while iconic metaphoric and cohesive gestures have three including preparation, stroke and retraction. They are executed by transitioning from a rest position into gesture space (preparation), this is followed by a small movement with hold (stroke) and a movement back to the rest position (retraction). Some examples are shown in Figs. 5.16, 5.17 and 5.18. Given these observations, gestures can be viewed as a recurrent sequence of atomic components, similar to phonemes in speech, starting and ending in rest positions and governed by a high level structure controlling the temporal sequence. A process for automatic segmentation of gestures can be designed to start with the extraction of atomic components, followed by the identification of rest positions, and the finding of an ordered sequence of enclosed components.

5.2.1 Temporal Segmentation

Temporal segmentation partitions a continuous observation sequence into plausible atomic gesture components. A gesture can be represented by a continuous trajectory

Fig. 5.16 Deictic gestures: 'pointing left' (*top row*), 'pointing right' (*bottom row*)

Fig. 5.17 Metaphoric gestures: 'he bent a tree' (*top row*), 'there was a big explosion' (*bottom row*)

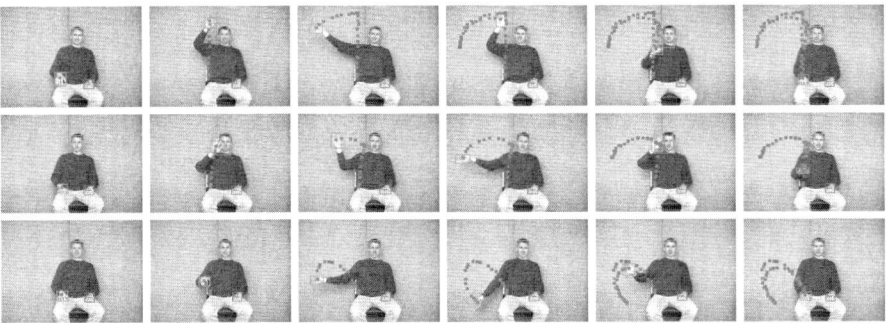

Fig. 5.18 Communicative gestures: 'waving high' (*top row*), 'waving low' (*middle row*) and 'please sit down' (*bottom row*)

of 2D (x, y) positions of a person's moving hand tracked in the image space. Segmentation can be performed in two steps. In the first step, the complete observation sequence is analysed and segments are obtained by detecting vertices where velocity drops below a given threshold. These vertices correspond to rest positions and pause positions that typically occur in bi-phasic gestures between transition into and out of a gesture, and in tri-phasic gestures between stroke and retraction. The second step analyses the segments for discontinuities in orientation to recover strokes. A method adopted by Walter et al. (2001) exploits the curvature primal sketch of Asada and Brady (1986) for segmenting continuous hand gesture trajectories into atomic actions (see Fig. 5.19).

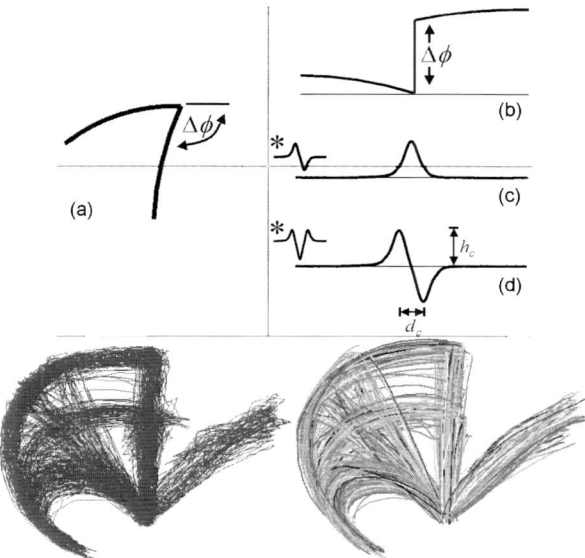

Fig. 5.19 Segmentation of continuous gesture trajectories into atomic components (Walter et al. 2001). *Bottom left*: An example of a continuously recorded hand gesture trajectory. *Bottom right*: The hand gesture trajectory is segmented into 701 atomic actions (*bottom right*). (**a**) A trajectory segment with discontinuity $\Delta\Phi$. (**b**) The trajectory in orientation space, relating the orientation of the curve to the arc length along the curve. (**c**) Filter response $N'_\sigma * f$ of the orientation of the trajectory $f(t)$ convoluted with the first derivative of a Gaussian $N_\sigma(t)$. (**d**) Filter response $N''_\sigma * f$ of the orientation of the trajectory $f(t)$ convoluted with the second derivative of a Gaussian $N_\sigma(t)$

5.2.2 Atomic Actions

An atomic action in the image space, extracted from a person's continuous hand motion, is a short trajectory segment consisting of c 2D discrete vertices $v_c = [x_1, y_1, x_2, y_2, \ldots, x_c, y_c]$ as its temporal duration. Each instance of an atomic action can also have a different number of vertices c. For clustering, it requires that all atomic actions are represented in a common feature space, here referred to as the gesture component space. This requires their temporal duration being normalised by interpolation into a fixed number of vertices, and each atomic action being represented by a feature vector. Dimensionality reduction may also be applied to the component space before clustering.

Not every segmented trajectory fragment is a genuine atomic action for a gesture feature. Some can be simply noise or clutter (see examples in Fig 5.20). Clutters in a distribution can be filtered to some extent (Byers and Raftery 1998). It is assumed that features and clutters can be described by two super-imposed random processes. They approximate the distribution of the distance d_k from a randomly chosen sample to its k_{th} nearest neighbours by a mixture of two Gaussian components with model

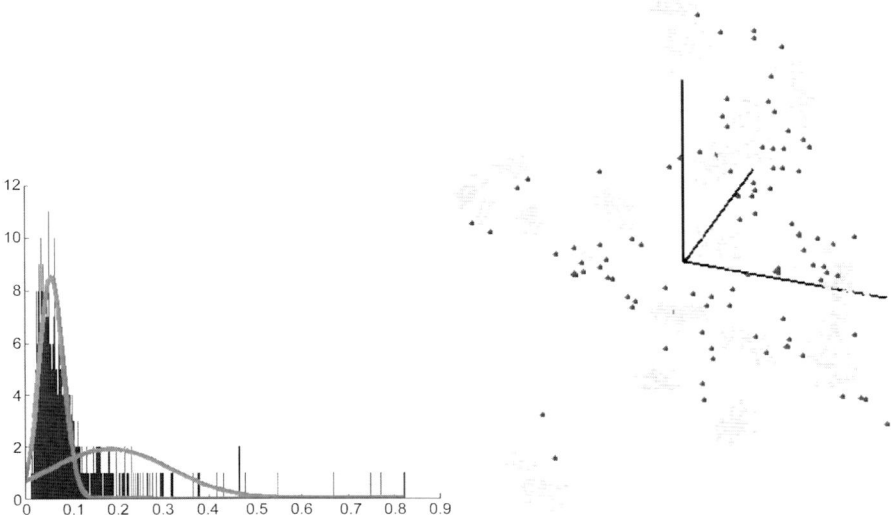

Fig. 5.20 *Left*: Average Euclidean distances calculated from the atomic actions in Fig. 5.19 to their nearest 15 neighbours, approximated by a mixture of two Gaussian components. *Right*: The distribution of the atomic actions visualised by the three largest principal components of the gesture component space. Each atomic action can also be classified as a gesture feature (*light colour*) or clutter (*dark colour*)

parameters w and $\Theta_d = [\lambda_1, \lambda_2, \sigma_1, \sigma_2]$, defined as

$$f(d_k | w, \Theta_d) = wN(d_k, \lambda_1, \sigma_1) + (1 - w)N(d_k, \lambda_2, \sigma_2) \tag{5.17}$$

where λ_1 and λ_2 are the average distances of a component to its k_{th} nearest neighbours, σ_1 and σ_2 the corresponding variances, w and $(1 - w)$ the mixture probabilities for feature and noise, respectively. The mixture parameters are estimated using the EM algorithm. More precisely, segmented atomic action components are filtered by applying the following procedure:

1. For K randomly chosen components
   a. Calculate the average Euclidean distance to its kth nearest neighbours.
   b. Estimate the model parameters $[w, \lambda_1, \lambda_2, \sigma_1, \sigma_2]$ using the EM algorithm.
2. Classify each component according to whether the average distance to its kth nearest neighbours has a higher probability for feature or clutter.

An example is shown in Fig. 5.20. Using the atomic actions extracted from some gesture sequences (Fig. 5.19), one can calculate the average Euclidean distance for all components to the nearest 15 neighbours and approximate the resulting distribution by a mixture of two Gaussian components. A clutter is represented by the distribution with the largest variance and all atomic actions whose average distance to their kth nearest neighbours falls within this distribution is classified as clutter, or noise.

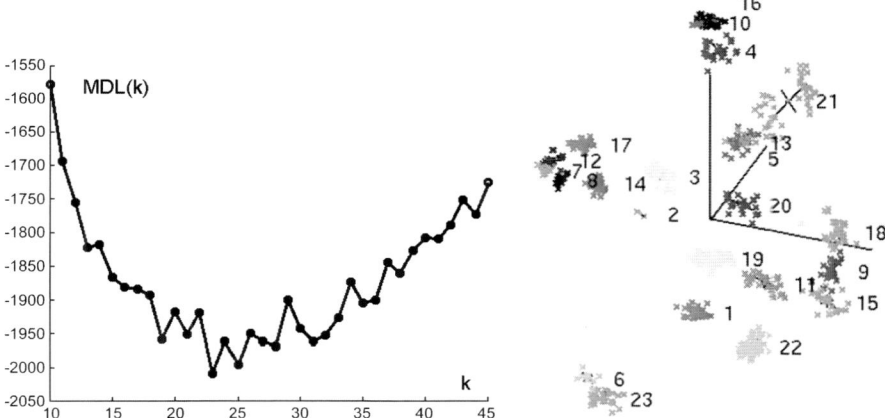

Fig. 5.21 Automatic discovery of typical atomic action groups from the gesture sequences of Fig. 5.19 using minimum description length. *Left*: Computed minimum description length (MDL(k)) values suggest that the optimal number of typical atomic action groups for this particular gesture data set is 23. *Right*: The gesture component space for these training data is represented by 23 clusters, shown by the three largest principle components of that space

Atomic actions are considered as shared building blocks for all the gesture trajectories. Moreover, they conform to a small set of typical groups, as not all atomic actions are unique. Such groups can be represented by a mixture model, such as a mixture of Gaussian, where each Gaussian component represents the distribution of one group of typical atomic actions. To determine automatically how many typical groups exist and where they are in a gesture component space, clustering techniques can be used. In clustering, two problems need be solved in order to determine a mixture model. For estimating a mixture of Gaussian, a method is needed to both discover the optimal number of mixture components with different weights, known as the problem of model order selection, and to estimate the parameters of all the Gaussian components. Walter et al. (2001) show how minimum description length (Rissanen 1978) can be exploited to automatically segment gestures into a small set of clusters of typical atomic actions, without any a priori knowledge on the number of clusters and where they may lie in a gesture component space.

Figure 5.21 shows an example of automatic discovery of typical atomic action groups from the gesture sequences of Fig. 5.19 using minimum description length. A global minimum in minimum description length values indicates that the optimal number of clusters is 23, resulting in the gesture component space for these gesture sequences being represented by 23 clusters. The corresponding localised gesture trajectory segments are shown in Fig. 5.22.

5.3 Markov Processes

If natural gesture motion patterns are temporally segmented into gesture components, such as atomic action groups, known gesture classes can be represented by

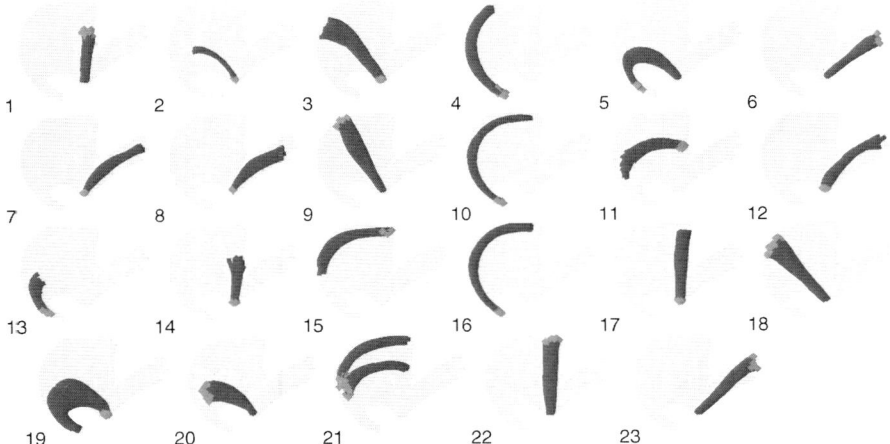

Fig. 5.22 Localised gesture trajectory segments corresponding to the automatically discovered 23 typical groups of atomic actions shown in Fig. 5.21

different distributions of unique characteristics in a gesture component space. For instance, different gestures types are represented by different gesture clusters. The problem of classifying an unknown gesture sequence can then be solved by measuring a similarity distance metric between a set of gesture components extracted from the unknown gesture sequence and different known gesture distributions. Based on this general concept, classification of human gestures can be treated as a problem of template matching (McKenna and Gong 1998). However, modelling gestures as templates does not take into account temporal information that may define unique characteristics of different gestures. Human gestures are intrinsically spatio-temporal trajectories with uncertainty, and can be more suitably modelled as stochastic processes either using Markov processes (Gong et al. 1999), or by propagating conditional state density samples using the CONDENSATION algorithm (Black and Jepson 1998). In the following, we consider in more details a model for interpreting gestures by a Markov process guided continuous propagation of conditional state density functions.

In general, a Markov process describes a time varying random process of Markov property as a series of either continuous or discrete states in a state space.[3] Suppose that human gesture sequences are temporally ordered, and can be represented as a finite set of ordered visual observations $\mathcal{O}_T = \{\mathbf{o}_1, \ldots, \mathbf{o}_t, \ldots, \mathbf{o}_T\}$, where $\mathbf{o}_t$ denotes the observation vector $\mathbf{o}$ at time t. Let us assume that the spatio-temporal trajectory of a gesture can be modelled as a time varying random process described by a first-order Markov process.[4] In this case, the conditional probability of state

---

[3] By Markov property, one assumes that the future states of a time varying random process is conditionally dependent on the current state, and independent from the past.

[4] A first-order Markov process has the property such that the next state of a time varying random process is dependent only on the current state. Similarly, a second-order Markov process has the

$\mathbf{q}_t$ given $\mathbf{q}_{t-1}$ is independent of its former history $\mathcal{Q}_{t-2} = \{\mathbf{q}_1, \mathbf{q}_2, \dots, \mathbf{q}_{t-2}\}$, and defined as

$$p(\mathbf{q}_t|\mathbf{q}_{t-1}) = p(\mathbf{q}_t|\mathcal{Q}_{t-1}) \tag{5.18}$$

Furthermore, it is assumed that a conditional, multi-modal observation probability $p(\mathbf{o}_t|\mathbf{q}_t)$ is independent of its observation history $\mathcal{O}_{t-1} = \{\mathbf{o}_1, \mathbf{o}_2, \dots, \mathbf{o}_{t-1}\}$, and therefore equal to the conditional observation probability given the history. It is defined as

$$p(\mathbf{o}_t|\mathbf{q}_t) = p(\mathbf{o}_t|\mathbf{q}_t, \mathcal{O}_{t-1}) \tag{5.19}$$

Therefore, the state probability $p(\mathbf{q}_t|\mathcal{O}_t)$ can be propagated based on the Bayes' rule as follows:

$$p(\mathbf{q}_t|\mathcal{O}_t) = k_t\, p(\mathbf{o}_t|\mathbf{q}_t) p(\mathbf{q}_t|\mathcal{O}_{t-1}) \tag{5.20}$$

where $p(\mathbf{q}_t|\mathcal{O}_{t-1})$ is the prior from the accumulated observation history up to time $t-1$, $p(\mathbf{o}_t|\mathbf{q}_t)$ is the conditional observation density and k_t is a normalisation factor. The prior density $p(\mathbf{q}_t|\mathcal{O}_{t-1})$ for accumulated observation history can be regarded as a prediction taken from the posterior at the previous time $p(\mathbf{q}_{t-1}|\mathcal{O}_{t-1})$ with the state transition probability $p(\mathbf{q}_t|\mathbf{q}_{t-1})$:

$$p(\mathbf{q}_t|\mathcal{O}_{t-1}) = \int_{q_{t-1}} p(\mathbf{q}_t|\mathbf{q}_{t-1}) p(\mathbf{q}_{t-1}|\mathcal{O}_{t-1}) \tag{5.21}$$

Equation (5.21) can be estimated by factored sampling using the CONDENSA-TION algorithm (Isard and Blake 1996). By doing so, the posterior $p(\mathbf{q}_{t-1}|\mathcal{O}_{t-1})$ is approximated by a fixed number of state density samples. The prediction is more accurate if the number of samples increases, but there is a corresponding increase in computational cost.

The CONDENSATION algorithm does not make any strong parametric assumptions about the form of the state density, $p(\mathbf{q}_t)$, and can therefore model potentially different object trajectories and build cumulatively the best interpretation (expectation) as more observations become available from tracking an object over time. However, the model does not utilise prior information from past experience, for instance, learning prior from past trajectories of different typical gestures in order to explore richer contextual information for interpreting current observations. In other words, it only accumulates information from current observations up to time t, $\mathcal{O}_t$. The state propagation density $p(\mathbf{q}_t|\mathbf{q}_{t-1})$ at the current state is given by the density estimation from the previous state plus Gaussian noise. Consequently, to estimate the prior $p(\mathbf{q}_t|\mathcal{O}_{t-1})$, the model must propagate a very large pool of conditional densities over time. As a result, the prediction can be both expensive and sensitive to observation noise. To significantly reduce the number of density samples to be

---

property such that the next state is dependent on both the current state and the previous state but independent from any other past states.

propagated over time and to better cope with noise and large variations in observations, a model needs to explore prior knowledge that can be extracted from typical gesture motion patterns observed in the past.

One idea is to learn and impose a priori knowledge of both observation covariance and the underlying state transition expectation from some training data of typical gestures. By doing so, the model effectively utilises additional contextual information beyond current observations in order to minimise ambiguities in sampling and propagating observation conditional state densities. This can be considered as a Markov process of propagating conditional densities with strong priors. A hidden Markov model (HMM) serves this purpose well. More specifically, an HMM can be used to learn the prior knowledge of both observation covariance and state transition probabilities for a finite set of discrete landmark states in the state space. Once an HMM is learned from past gesture examples, the continuous propagation of conditional state densities for explaining the current observations over time is then 'channelled' through those landmark states as contextual constraints.

An HMM model $\lambda = (\mathbf{A}, \mathbf{b}, \pi)$, see Fig. 3.4(a), can be fully described by a set of probabilistic parameters as follows:

1. $\mathbf{A}$ is a matrix of state transition probabilities where element a_{ij} describes the probability $p(\mathbf{q}_{t+1} = j | \mathbf{q}_t = i)$ and $\sum_{j=1}^{N} a_{ij} = 1$.
2. $\mathbf{b}$ is a vector of observation-density functions $b_j(\mathbf{o}_t)$ for each state j where $b_j(\mathbf{o}_t) = p(\mathbf{o}_t | \mathbf{q}_t = j)$. The observation density $b_j(\mathbf{o}_t)$ can be discrete or continuous, for example, a Gaussian mixture $b(\mathbf{o}_t) = \sum_{k=1}^{K} c_k \mathcal{G}(\mathbf{o}_t, \mu_k, \Sigma_k)$ with mixture coefficient c_k, mean μ_k and covariance Σ_k for the kth mixture in a given state.
3. π is a vector of initial probabilities of being in state j at time $t = 1$, where $\sum_{i=1}^{N} \pi_i = 1$.

Let us define that a CONDENSATION state vector at time t is $\mathbf{q}_t = (\mathbf{q}_t, \lambda)$, given by the current hidden Markov state $\mathbf{q}_t$ for a model λ. By training HMMs on some example gesture trajectories, a priori knowledge on both state propagation and conditional observation density are learned by assigning the hidden Markov state transition probabilities $p(\mathbf{q}_t = j | \mathbf{q}_{t-1} = i)$ of a learned model λ to the CONDENSATION state propagation densities as follows:

$$p(\mathbf{q}_t | \mathbf{q}_{t-1}) = p(\mathbf{q}_t = j | \mathbf{q}_{t-1} = i, \lambda) = a_{ij} \qquad (5.22)$$

Similarly, the prior on the observation conditional density $p(\mathbf{o}_t | \mathbf{q}_t)$ is given by the Markov observation densities at each hidden state as

$$p(\mathbf{o}_t | \mathbf{q}_t) = p(\mathbf{o}_t | \mathbf{q}_t = j, \lambda) = b_j(\mathbf{o}_t) \qquad (5.23)$$

The Markov observation density at each Markov state $b_j(\mathbf{o}_t)$ gives the prior on the observation covariance. As a result, the process of both sampling and propagating CONDENSATION states is not only more guided, but also more robust against current observational noise and for coping with large variations.

Learning an HMM for modelling gesture examples involves automatic gesture component segmentation through clustering, as discussed in Sect. 5.2. It also requires the estimation of HMM state transition distributions and a conditional observation-density distribution at each HMM state. They are computed using the Baum–Welch algorithm (Baum and Petrie 1966).

A model for gesture classification can be made more robust if the current observation is taken into account before prediction. Let us consider the state propagation density $p(\mathbf{q}_t|\mathbf{q}_{t-1})$ in (5.21) to be augmented by the current observation, $p(\mathbf{q}_t|\mathbf{q}_{t-1}, \mathbf{o}_t) = p(\mathbf{q}_t|\mathbf{q}_{t-1}, \mathcal{O}_t)$. Assuming observations are independent over time and future observations have no effect on past states $p(\mathbf{q}_{t-1}|\mathcal{O}_t) = p(\mathbf{q}_{t-1}|\mathcal{O}_{t-1})$, the prediction process of (5.21) is then replaced by

$$p(\mathbf{q}_t|\mathcal{O}_t) = \sum_{\mathbf{q}_{t-1}} p(\mathbf{q}_t|\mathbf{q}_{t-1}, \mathbf{o}_t) p(\mathbf{q}_{t-1}|\mathcal{O}_{t-1})$$

$$= \sum_{\mathbf{q}_{t-1}} k_t p(\mathbf{o}_t|\mathbf{q}_t) p(\mathbf{q}_t|\mathbf{q}_{t-1}) p(\mathbf{q}_{t-1}|\mathbf{o}_{t-1}) \qquad (5.24)$$

where $k_t = \frac{1}{p(\mathbf{o}_t|\mathbf{q}_{t-1})}$ and

$$p(\mathbf{q}_t|\mathbf{q}_{t-1}, \mathbf{o}_t) = \frac{p(\mathbf{o}_t, \mathbf{q}_t|\mathbf{q}_{t-1})}{p(\mathbf{o}_t|\mathbf{q}_{t-1})} = \frac{p(\mathbf{o}_t|\mathbf{q}_t, \mathbf{q}_{t-1}) p(\mathbf{q}_t|\mathbf{q}_{t-1})}{p(\mathbf{o}_t|\mathbf{q}_{t-1})}$$

$$= \frac{p(\mathbf{o}_t|\mathbf{q}_t) p(\mathbf{q}_t|\mathbf{q}_{t-1})}{p(\mathbf{o}_t|\mathbf{q}_{t-1})} \qquad (5.25)$$

Given that the observation and state transitions are constrained by the underlying HMM, the state transition density is then given by

$$p(\mathbf{q}_t|\mathbf{q}_{t-1}, \mathbf{o}_t) = p(\mathbf{q}_t = j|\mathbf{q}_{t-1} = i, \mathbf{o}_t) = \frac{a_{ij}^\lambda b_j^\lambda(\mathbf{o}_t)}{\sum_{n=1}^N a_{in}^\lambda b_n^\lambda(\mathbf{o}_t)} \qquad (5.26)$$

This on-line observation augmented prediction unifies the processes of innovation and prediction in a CONDENSATION process given by (5.20) and (5.21). Without augmentation, a CONDENSATION process performs a blind search based on current observation history alone. An on-line observation augmented prediction takes the current observation into account and adapts the prior to perform a guided search in prediction. Using this Markov process guided conditional density propagation model to interpret human gestures both improves classification accuracy and reduces computational complexity and cost (Gong et al. 1999).

5.4 Affective State Analysis

To be able to recognise the affective state, namely the emotion, of a person is important for successful interpersonal social interaction. Affective arousal modulates all nonverbal communication cues such as facial expression, body moment, posture, gesture, and the tone of voice. Design and development of an automated system able to detect and interpret human affective behaviour is a challenging problem (Pantic et al. 2007; Picard 1997).

Gesture alone can be ambiguous for interpreting the affective state of a person. For example, the gestures captured in three different upper body behaviours shown in Fig. 2.3 are very similar. However, it is clear that the affective states expressed by these three body behaviours are different, as can be seen from their facial expressions. Both human face and body can reveal emotional state of an individual. Psychological studies suggest that the combined visual channels of facial expression and body gesture are the most informative, and their integration is a mandatory process occurring early in the human processing stream (Ambady and Rosenthal 1992; Meeren et al. 2005). For automatic interpretation of the affective state of a person, combining visual information extracted from both facial expression and body gesture behaviours has been shown to be effective (Gunes and Piccardi 2007; Kapoor and Picard 2005; Pantic et al. 2005).

In the following, we consider a model for representing and interpreting affective body behaviour in video data. First, let us consider a representational scheme that exploits space-time interest points computed from image sequences (Dollár et al. 2005). Different from representing human gestures by their motion trajectories, this representation does not rely on robust segmentation and tracking, therefore relaxing the viewing conditions under which behaviours are observed, such as background clutters and occlusion.

5.4.1 Space-Time Interest Points

There are different types of spatio-temporal imagery features for behaviour analysis in video data (Dollár et al. 2005; Efros et al. 2003; Ke et al. 2005; Laptev and Lindeberg 2003; Niebles et al. 2006). Among them, Dollár et al. (2005) suggest a model for detecting space-time interest points by a response function computed using separable linear filters. This assumes that the camera is stationary, or a mechanism to remove camera motion is available. The response function has the form of

$$R = (I * g * h_{ev})^2 + (I * g * h_{od})^2 \tag{5.27}$$

where $I(x, y, t)$ denotes image frames from an image sequence; $*$ denotes a convolution operator; $g(x, y; \sigma)$ is a 2D Gaussian smoothing kernel, applied spatially in the image space (x, y); h_{ev} and h_{od} are a quadrature pair of one-dimensional Gabor filters applied temporally over time, where

$$h_{ev}(t; \tau, \omega) = -\cos(2\pi t\omega)e^{-t^2/\tau^2} \tag{5.28}$$

$$h_{od}(t; \tau, \omega) = -\sin(2\pi t\omega)e^{-t^2/\tau^2} \tag{5.29}$$

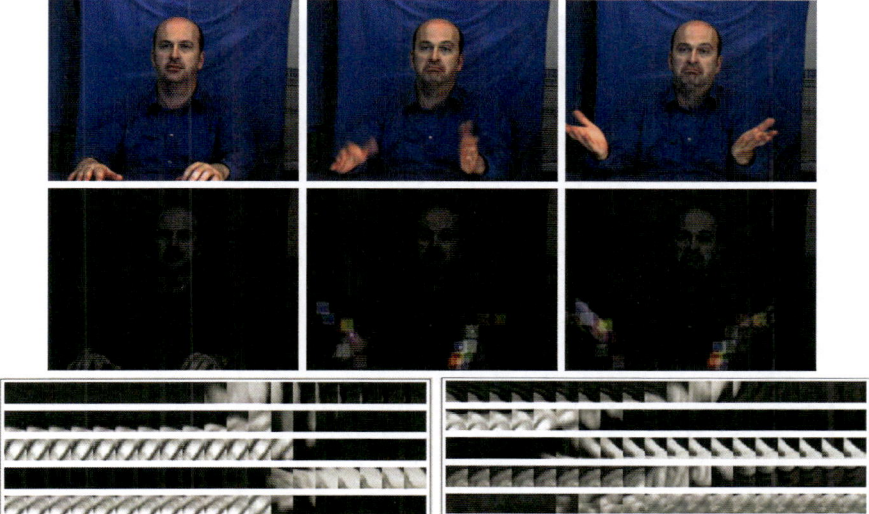

Fig. 5.23 Extracting local space-time interest point cuboids. *Top row*: Example image frames from three different upper body behaviours. *Middle row*: Local space-time cuboids extracted from the image frames where each cuboid is labelled by a *different colour*. *Bottom row*: Space-time local cuboids are shown when flattened over time

The two parameters σ and τ correspond approximately to the spatial and temporal scales of a space-time interest point detector. Each interest point is extracted as a local maximum of the above response function. It can be considered that any image region with spatially distinctive characteristics undergoing a complex motion will induce a strong response, whilst regions undergoing pure translational motion, or without spatially distinctive features, will generate a weak response.

At each detected interest point, a local space-time cuboid is extracted centred at the interest point. This space-time cuboid defines a spatio-temporal volume of pixels. Some examples of space-time interest point cuboids are shown in Fig. 5.23. The temporal dimension in t of each cuboid is typically set empirically to six times of the size of the spatial dimensions in x and y. After extracting these local space-time cuboids, a gesture sequence is represented as a collection of cuboids. A descriptor is designed to represent all the pixels in each cuboid. For example, by computing some local pixel-wise features including the average pixel intensity value, an intensity gradient map or optical flows for a cuboid, a descriptor can be a histogram with a chosen number of bins that represents a distribution of the values of those local features spreading across the bins.[5]

Given some training image sequences of upper human body motion capturing both facial expressions and body gestures, a large number of space-time interest

---

[5]The number of histogram bins for a cuboid descriptor is considered as the dimensionality of the descriptor. If this dimensionality is too large, it can be further reduced by projecting the descriptors of all the cuboids into a lower dimensional space using PCA.

point cuboids can be extracted, and represented by their corresponding descriptors. These local video descriptors of human body behaviour can be clustered into a smaller set of groups. This is because not all descriptors of local space-time interest point cuboids are unique. After clustering, each group is considered as a prototype. The set of prototypes forms a library. Each interest point cuboid is assigned a unique prototype label as the result of the clustering process. A human body behaviour image sequence can then be represented by a histogram of the number of different prototype labels assigned to all the space-time cuboids computed from this image sequence. The bin size for this histogram is the size of the prototype library, namely the number of clusters of the video descriptors. This is simply a bag-of-words representation.

5.4.2 *Expression and Gesture Correlation*

It is considered that integration of facial expression and body gesture visual cues is a mandatory process occurring early in the human processing stream (Ambady and Rosenthal 1992; Meeren et al. 2005). This suggests that their patterns of variation are inherently correlated. Moreover, if this latent correlation occurs at an early stage of a recognition process, a model is needed to map feature spaces of behaviours from different bodily parts into a common space so that visual interpretation of affective state can be performed with all the information available in a common context. To model the correlation between two feature spaces of facial expression and body gesture, we exploit a particular statistical method known as canonical correlation analysis (Hotelling 1936).

Canonical correlation analysis (CCA) is a statistical technique developed for measuring linear relationships between two multi-dimensional variables. It finds pairs of base vectors, known as canonical factors, such that the correlations between the projections of the variables onto these canonical factors are mutually maximised. CCA is applied to solve a number of computer vision problems. For instance, Borga (1998) considers CCA for finding corresponding points in stereo images. Melzer et al. (2003) exploit CCA for modelling correlations between 2D image intensity properties and object 3D pose. Hardoon et al. (2004) study a CCA-based method for learning semantic correlation between web images and text.

Given two zero-mean random variables $\mathbf{x} \in R^m$ and $\mathbf{y} \in R^n$, CCA finds pairs of directions $\mathbf{w}_x$ and $\mathbf{w}_y$ that maximise the correlation between the projections $x = \mathbf{w}_x^T \mathbf{x}$ and $y = \mathbf{w}_y^T \mathbf{y}$. The projections x and y are called canonical variates. More precisely, CCA maximises the following function:

$$\rho = \frac{\bar{E}[xy]}{\sqrt{E[x^2]E[y^2]}} = \frac{E[\mathbf{w}_x^T \mathbf{x}\mathbf{y}^T \mathbf{w}_y]}{\sqrt{E[\mathbf{w}_x^T \mathbf{x}\mathbf{x}^T \mathbf{w}_x]E[\mathbf{w}_y^T \mathbf{y}\mathbf{y}^T \mathbf{w}_y]}}$$

$$= \frac{\mathbf{w}_x^T \mathbf{C}_{xy} \mathbf{w}_y}{\sqrt{\mathbf{w}_x^T \mathbf{C}_{xx} \mathbf{w}_x \mathbf{w}_y^T \mathbf{C}_{yy} \mathbf{w}_y}} \tag{5.30}$$

where $\mathbf{C}_{xx} \in R^{m \times m}$ and $\mathbf{C}_{yy} \in R^{n \times n}$ are the within-set covariance matrices of $\mathbf{x}$ and $\mathbf{y}$, respectively, whilst $\mathbf{C}_{xy} \in R^{m \times n}$ denotes their between-sets covariance matrix. A number of up to $k = \min(m, n)$ canonical factor pairs $\langle \mathbf{w}_x^i, \mathbf{w}_y^i \rangle, i = 1, \ldots, k$ can be obtained by successively solving

$$
\begin{aligned}
&\arg \max_{\mathbf{w}_x^i, \mathbf{w}_y^i} \{\rho\}, \\
&\text{subject to} \quad \rho\left(\mathbf{w}_x^j, \mathbf{w}_x^i\right) = \rho\left(\mathbf{w}_y^j, \mathbf{w}_y^i\right) = 0, \\
&\text{for } j = 1, \ldots, i - 1
\end{aligned}
\tag{5.31}
$$

This optimisation problem can be solved by setting the derivatives of (5.30) equal to zero, with respect to $\mathbf{w}_x$ and $\mathbf{w}_y$, and resulting in the following eigenequations:

$$
\begin{cases}
\mathbf{C}_{xx}^{-1} \mathbf{C}_{xy} \mathbf{C}_{yy}^{-1} \mathbf{C}_{yx} \mathbf{w}_x = \rho^2 \mathbf{w}_x \\
\mathbf{C}_{yy}^{-1} \mathbf{C}_{yx} \mathbf{C}_{xx}^{-1} \mathbf{C}_{xy} \mathbf{w}_y = \rho^2 \mathbf{w}_y
\end{cases}
\tag{5.32}
$$

Matrix inversions need be performed in (5.32), leading to numerical instability if $\mathbf{C}_{xx}$ and $\mathbf{C}_{yy}$ are rank deficient. Alternatively, $\mathbf{w}_x$ and $\mathbf{w}_y$ can be computed by principal angles, as CCA is a statistical interpretation of principal angles between two linear subspaces (Golub and Zha 1992).

Given $B = \{\mathbf{x} | \mathbf{x} \in R^m\}$ and $F = \{\mathbf{y} | \mathbf{y} \in R^n\}$, where $\mathbf{x}$ and $\mathbf{y}$ are the feature vectors extracted from body images and face images, respectively, CCA is applied to discover the relationship between $\mathbf{x}$ and $\mathbf{y}$. Suppose $\langle \mathbf{w}_x^i, \mathbf{w}_y^i \rangle, i = 1, \ldots, k$ are the canonical factor pairs found, d $(1 \leq d \leq k)$ factor pairs can be used to represent their correlation information. With $\mathbf{W}_x = [\mathbf{w}_x^1, \ldots, \mathbf{w}_x^d]$ and $\mathbf{W}_y = [\mathbf{w}_y^1, \ldots, \mathbf{w}_y^d]$, the original feature vectors are projected as $\mathbf{x}' = \mathbf{W}_x^T \mathbf{x} = [x_1, \ldots, x_d]^T$ and $\mathbf{y}' = \mathbf{W}_y^T \mathbf{y} = [y_1, \ldots, y_d]^T$ in a lower dimensional correlation space, where x_i and y_i are uncorrelated with the previous pairs x_j and y_j, $j = 1, \ldots, i - 1$. The projected feature vectors $\mathbf{x}'$ and $\mathbf{y}'$ are then combined to form a new feature vector as

$$
\mathbf{z} = \begin{pmatrix} \mathbf{x}' \\ \mathbf{y}' \end{pmatrix} = \begin{pmatrix} \mathbf{W}_x^T \mathbf{x} \\ \mathbf{W}_y^T \mathbf{y} \end{pmatrix} = \begin{pmatrix} \mathbf{W}_x & 0 \\ 0 & \mathbf{W}_y \end{pmatrix}^T \begin{pmatrix} \mathbf{x} \\ \mathbf{y} \end{pmatrix}
\tag{5.33}
$$

This joint feature vector represents visual information from both facial and body parts of a human body in a single common feature space. For interpreting the affective state of a person given a joint facial and body parts CCA feature space representation, a support vector machine classifier can be employed.

Shan et al. (2007) show that interpreting affective state using facial expression is more accurate than by body gesture. This is because there are more variations in body gesture, and different gestures are often associated with the same affective state. It is also shown that a joint CCA facial and body gesture feature space representation improves the accuracy of affective state interpretation when compared to using either facial expression or gesture alone.

5.5 Discussion

We considered a number of core problems for representing and interpreting human gestures. They include tracking body parts in motion caused by gestures, segmenting natural gesture sequences into typical gesture components as atomic actions, modelling gestures as Markov processes, and interpreting human affective state by learning correlations between facial expression and body gesture in a common space-time feature space.

Structured motion patterns caused by facial expressions and hand gestures show many common characteristics. Both are spatio-temporal visual patterns with correlated and constrained movements of sub-components, for example, different articulated body parts during a gesture and different facial parts contributing towards an overall facial expression. However, there are also a number of important differences, suggesting the adoption of different visual feature representations. First, the movements of different body parts in gesturing are less constrained than in the case of facial expression. For instance, both hands can move much more freely compared to the inter-relationships between eyes and mouth. Secondly, the temporal ordering in gesture sequences is more important for interpretation than in the case of facial expression. For the former, different ordering of the same atomic action components can easily give rise to a different meaning. Whilst for the latter, changes in the temporal order of local facial parts mostly result in localised changes in imagery intensity properties with little overall structural change. Consequently, whilst a stochastic process based model is deemed to be necessary for overcoming visual ambiguities in the interpretation of gestures, static image template-based matching can perform rather well for the interpretation of facial expression patterns.

Computer-based automatic visual interpretation of human gestures is most naturally applicable in a human-computer interaction situation when it is reasonable to assume that a person is standing or sitting with mostly localised body movement. In this situation, one can also assume that a camera is not too far away from the interacting subject so that sufficient imagery details are available for body parts tracking. In the next chapter, we shall consider the problem of visual interpretation of a more general kind of human actions, from which both global and local human body movements are present and visual observations are often undertaken from a distance.

References

Ambady, N., Rosenthal, R.: Thin slices of expressive behaviour as predictors of interpersonal consequences: a meta-analysis. Psychol. Bull. **111**(2), 256–274 (1992)

Asada, H., Brady, M.: The curvature primal sketch. IEEE Trans. Pattern Anal. Mach. Intell. **8**(1), 2–14 (1986)

Ballard, D., Brown, C.: Computer Vision. Prentice Hall, New York (1982)

Baum, L.E., Petrie, T.: Statistical inference for probabilistic functions of finite state Markov chains. Annu. Math. Stat. **37**, 1554–1563 (1966)

Black, M., Jepson, A.: Recognizing temporal trajectories using the condensation algorithm. In: IEEE International Conference on Automatic Face and Gesture Recognition, Nara, Japan, April 1998, pp. 16–21 (1998)

Borga, M.: Learning multidimensional signal processing. PhD thesis, Linkoping University, SE-581 83 Linkoping, Sweden. Dissertation No. 531 (1998)

Bowden, R., Mitchell, T., Sarhadi, M.: Reconstructing 3D pose and motion from a single camera view. In: British Machine Vision Conference, Southhampton, UK, pp. 904–913 (1998)

Bradski, G.R.: Computer vision face tracking for use in a perceptual user interface. Intel Technol. J., 2nd Quarter (1998)

Byers, S., Raftery, A.: Nearest-neighbour clutter removal for estimating features in spatial point processes. J. Am. Stat. Assoc. **93**(442), 577–584 (1998)

Charniak, E.: Bayesian networks without tears. AI Mag. **12**(4), 50–63 (1991)

Cowell, R., Philip, A., Lauritzen, S., Spiegelhalter, D.: Probabilistic Networks and Expert Systems. Springer, Berlin (1999)

Dempster, A.P., Laird, N.M., Rubin, D.B.: Maximum likelihood from incomplete data via the EM algorithm. J. R. Stat. Soc. **39**(1), 1–38 (1977)

Dollár, P., Rabaud, V., Cottrell, G., Belongie, S.: Behavior recognition via sparse spatio-temporal features. In: IEEE International Workshop on Visual Surveillance and Performance Evaluation of Tracking and Surveillance, pp. 65–72 (2005)

Efros, A., Berg, A., Mori, G., Malik, J.: Recognizing action at a distance. In: IEEE International Conference on Computer Vision, pp. 726–733 (2003)

Gavrila, D., Davis, L.S.: Towards 3-D model based tracking and recognition of human movement: a multi-view approach. In: IEEE International Conference on Automatic Face and Gesture Recognition, pp. 272–277 (1995)

Golub, G.H., Zha, H.: The canonical correlations of matrix pairs and their numerical computation. Technical report, Stanford University, Stanford, USA (1992)

Gong, S., Walter, M., Psarrou, A.: Recognition of temporal structures: Learning prior and propagating observation augmented densities via hidden Markov states. In: IEEE International Conference on Computer Vision, Corfu, Greece, pp. 157–162 (1999)

Gunes, H., Piccardi, M.: Bi-modal emotion recognition from expressive face and body gestures. J. Netw. Comput. Appl. **30**(4), 1334–1345 (2007)

Hardoon, D., Szedmak, S., Shawe-Taylor, J.: Canonical correlation analysis: an overview with application to learning methods. Neural Comput. **16**(12), 2639–2664 (2004)

Heap, T.: Learning deformable shape models for object tracking. PhD thesis, School of Computer Studies, University of Leeds, UK (1997)

Heap, T., Hogg, D.C.: Improving specificity in pdms using a hierarchical approach. In: British Machine Vision Conference, pp. 80–89 (1997)

Hogg, D.C.: Model based vision: A program to see a walking person. Image Vis. Comput. **1**(1), 5–20 (1983)

Hotelling, H.: Relations between two sets of variates. Biometrika **8**, 321–377 (1936)

Isard, M., Blake, A.: Contour tracking by stochastic propagation of conditional density. In: European Conference on Computer Vision, Cambridge, UK, pp. 343–357 (1996)

Isard, M., Blake, A.: CONDENSATION—conditional density propagation for visual tracking. Int. J. Comput. Vis. **29**(1), 5–28 (1998)

Kapoor, A., Picard, R.W.: Multimodal affect recognition in learning environments. In: ACM International Conference on Multimedia, pp. 677–682 (2005)

Ke, Y., Sukthankar, R., Hebert, M.: Efficient visual event detection using volumetric features. In: IEEE International Conference on Computer Vision, pp. 166–173 (2005)

Laptev, I., Lindeberg, T.: Space-time interest points. In: IEEE International Conference on Computer Vision, pp. 432–439 (2003)

Lauritzen, S., Spiegelhalter, D.: Local Computations with probabilities on graphical structures and their application to expert systems. In: Shafer, G., Pearl, J. (eds.) Readings in Uncertain Reasoning, pp 415–448. Morgan Kaufmann, San Mateo (1990)

Li, Y., Gong, S., Liddell, H.: Support vector regression and classification based multi-view face detection and recognition. In: IEEE International Conference on Automatic Face and Gesture Recognition, Grenoble, France, March 2000, pp. 300–305 (2000)

McKenna, S., Gong, S.: Gesture recognition for visually mediated interaction using probabilistic event trajectories In: British Machine Vision Conference, Southampton, UK, pp. 498–508 (1998)

McNeill, D.: Hand and Mind: What Gestures Reveal About Thought. University of Chicago Press, Chicago (1992)

Meeren, H., Heijnsbergen, C., Gelder, B.: Rapid perceptual integration of facial expression and emotional body language. Proc. Natl. Acad. Sci. USA **102**(45), 16518–16523 (2005)

Melzer, T., Reiter, M., Bischof, H.: Appearance models based on kernel canonical correlation analysis. Pattern Recognit. **39**(9), 1961–1973 (2003)

Ng, J., Gong, S.: Composite support vector machines for the detection of faces across views and pose estimation. Image Vis. Comput. **20**(5–6), 359–368 (2002)

Niebles, J., Wang, H., Fei-Fei, L.: Unsupervised learning of human action categories using spatial-temporal words. In: British Machine Vision Conference, Edinburgh, UK, September 2006

Ong, E.-J., Gong, S.: The dynamics of linear combinations. Image Vis. Comput. **20**(5–6), 397–414 (2002)

Pantic, M., Sebe, N., Cohn, J., Huang, T.S.: Affective multimodal human-computer interaction. In: ACM International Conference on Multimedia, pp. 669–676 (2005)

Pantic, M., Pentland, A., Nijholt, A., Huang, T.S.: Human computing and machine understanding of human behavior: a survey. In: Huang, T.S., Nijholt, A., Pantic, M., Pentland, A. (eds.) Artificial Intelligence for Human Computing, vol. 4451, pp. 47–71. Springer, Berlin (2007)

Pearl, J.: Probabilistic Reasoning in Intelligent Systems: Networks of Plausible Inference. Morgan Kaufmann, San Mateo (1988)

Pentland, A.: Automatic extraction of deformable models. Int. J. Comput. Vis. **4**, 107–126 (1990)

Picard, R.W.: Affective Computing. MIT Press, Cambridge (1997)

Raja, Y., McKenna, S., Gong, S.: Tracking and segmenting people in varying lighting conditions using colour. In: IEEE International Conference on Automatic Face and Gesture Recognition, Nara, Japan, pp. 228–233 (1998)

Regh, J.M.: Visual analysis of high DOF articulated objects with application to hand tracking. PhD thesis, Carnegie Mellon University, Pittsburgh, USA (1995)

Rissanen, J.: Modelling by shortest data description. Automatica **14**, 465–471 (1978)

Russell, D., Gong, S.: Minimum cuts of a time-varying background. In: British Machine Vision Conference, Edinburgh, UK, September 2006, pp. 809–818 (2006)

Shan, C., Gong, S., McOwan, P.: Beyond facial expressions: Learning human emotion from body gestures. In: British Machine Vision Conference, Warwick, UK, September 2007

Sherrah, J., Gong, S.: Resolving visual uncertainty and occlusion through probabilistic reasoning. In: British Machine Vision Conference, Bristol, UK, September 2000, pp. 252–261 (2000a)

Sherrah, J., Gong, S.: Tracking discontinuous motion using Bayesian inference. In: European Conference on Computer Vision, Dublin, Ireland, June 2000, pp. 150–166 (2000b)

Walter, M., Psarrou, A., Gong, S.: Data driven model acquisition using minimum description length. In: British Machine Vision Conference, Manchester, UK, pp. 673–683 (2001)

Chapter 6
Action Recognition

Understanding the meaning of actions is essential for human social communication (Decety et al. 1997). Automatic visual analysis of actions is important for visual surveillance, video indexing and browsing, and analysis of sporting events. Facial expression and gesture are mainly associated with movement of one or a number of individual body parts. Actions are typically associated with whole body movement, for instance, walking, sitting down, riding a horse. Actions also involve possibly multiple people interacting, such as hugging or shaking hands. To observe holistic body movement associated with actions, the viewpoint is necessarily further away, typically from a medium distance of 5 to 20 metres, and it may also be from a non-stationary viewpoint in order to keep the subject of interest within the view. Furthermore, as the field of view becomes wider, it is more likely to have moving objects in the background, making the segmentation and tracking of the actions of interest harder still. Similar in nature to understanding gestures, action modelling is challenging because the same action performed by different people can be visually different due to the variations in the visual appearance of the people, and changes in viewing angle, position, occlusion, illumination, and shadow.

Given a different visual environment under which action recognition is likely to be performed, representation and modelling of actions differ from those for facial expression and gesture. In the following, we consider three different approaches to modelling and interpreting action image sequences. First, suppose that an action is observed in a relatively static scene against non-cluttered background from a not too distant view, a model can be built to extract imagery features from segmented human body silhouettes for action representation (Sect. 6.1). A graphical model based on hidden conditional random field is exploited for treating action recognition as a graph classification problem (Sect. 6.2). Second, when actions are observed from further away with a non-stationary viewpoint, reliable extraction of detailed silhouettes is no longer feasible. A model based on space-time interest points becomes more suitable for action representation (Sect. 6.3). Such a model is designed to utilise both local appearance information and global distribution information of spatio-temporal imagery features. Third, realistic human actions are often observed continuously against a background of many distractors in an uncontrolled

S. Gong, T. Xiang, *Visual Analysis of Behaviour*,
DOI 10.1007/978-0-85729-670-2_6, © Springer-Verlag London Limited 2011

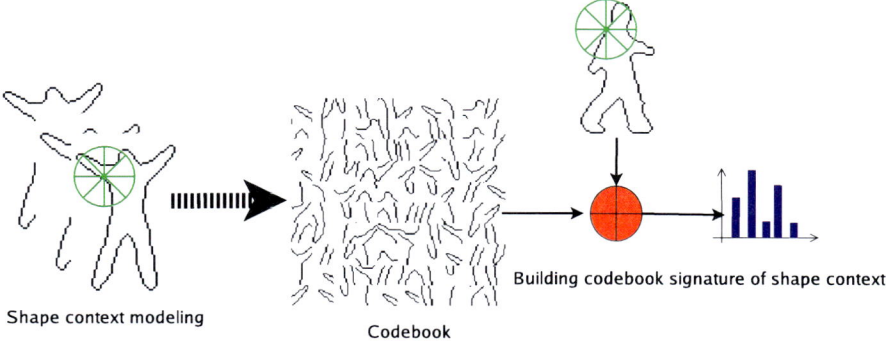

Fig. 6.1 Constructing a codebook of human silhouette shapes

and crowded environment. Action recognition under such conditions requires more than classification of action sequences. The latter assumes that segmented image sequences of isolated individual actions are readily available. This assumption is no longer valid in a crowded environment when a model must also localise and detect actions of interest in the presence of distracting objects of irrelevance. To address this problem, a model is considered to represent actions by combining space-time interest point features with local trajectory features (Sect. 6.4). Given this representation, a space-time sliding volume based search is exploited for detecting actions of known classes. The acute problem of learning an action detector with a minimal amount of annotated training data is also addressed.

6.1 Human Silhouette

A human action can be represented by computing body silhouette shapes from an image sequence. A silhouette is a binary image representing human body shape in each image frame (Blank et al. 2005). Given a silhouette, shape features are extracted by computing a chain code. Describing the silhouette of objects by their chain codes is adopted for shape retrieval and matching (Freeman 1961). However, the chain code itself is not invariant to changes in shape orientation caused by changes in 3D pose of a human body undergoing action. Specifically, if a human body shape is rotated by θ degrees, the corresponding chain code $C(p)$ will be shifted by an offset, $\Delta p(\theta)$. Let $C(p + \Delta p(\theta))$ be the resultant chain code after rotation. To obtain rotation invariance, a Fourier transform can be performed on $C(p)$, resulting in abs$(F(C(p)) =$ abs$(F(C(p + \Delta p(\theta))))$. The first n components of the Fourier spectrum can be used as action features, known as spectrum features, to form a vector descriptor of each silhouette shape. These shape descriptors can be used to construct a codebook based representation of human silhouette shapes (Fig 6.1), similar to the concept of building a prototype library for a gesture representation discussed in Sect. 5.4.

Although spectrum feature descriptors are good for capturing actions that cause body shape change, such as bending or walking, human body actions are not always necessarily associated with significant body shape changes. For instance, in a jumping sequence, the human silhouette does not change a great deal over time. As a result, its spectrum features cannot capture discriminative information. To overcome this problem, motion moment features can be exploited to form descriptors of action. Specifically, for each silhouette, inter-frame differencing can be used to detect motion changes resulting in a binary image. A set of moment features can be extracted from each binary image and used to form a vector descriptor of each inter-frame difference binary image. These features can include the motion area, representing the zeroth-order moment; the centroid coordinates, giving the first-order moments; and the elongation, derived from the second-order moments. These features are similar to those used for hand gesture representation as described in Sect. 5.1.

Figure 6.2(a) shows an example in which a motion moment based descriptor of human silhouettes is more distinctive than a spectrum feature based descriptor in a jumping action sequence. Figure 6.2(b) gives an example in which the spectrum feature descriptor is more distinctive than a motion moment descriptor in a waving action sequence. This suggests that different types of features are complimentary, and that they can be utilised jointly for action representation, for instance. by forming a concatenated joint feature vector descriptor.

6.2 Hidden Conditional Random Fields

Given a sequence of frame-wise descriptors representing both chain code shape and inter-frame motion moment features extracted from an action image sequence, a model is needed to capture the temporal dependencies between consecutive frame-descriptors over time. A hidden Markov model (HMM) can be considered as a candidate for action modelling, as in the case of gesture modelling, due to its generative power to predict and explain visual observations, often partial and incomplete. As a generative model, an HMM estimates and assigns a joint probability distribution over an observation and class label sequence. On the other hand, for learning a classification function, statistical learning theory suggests that a model optimised for maximising a discrimination function is more effective for discriminating class labels without also having to explain observations, especially when only a small quantity of training data are available (Fukunaga 1990).

A conditional random field (CRF) is a discriminative undirected graphical model, designed for classifying sequential data, not necessarily temporally ordered. A CRF is a Markov random field trained discriminatively given observational sequences. Unlike an HMM, it does not model the joint distribution over both class label and observations (Lafferty et al. 2001). CRFs have been exploited for modelling action sequences (Sminchisescu et al. 2006; Sutton and McCallum 2007). There remains the need for modelling latent relationships between observations and class labels. To address this problem, hidden state variables are introduced to construct a hidden conditional random field (HCRF) (Gunawardana et al. 2005; Quattoni et al.

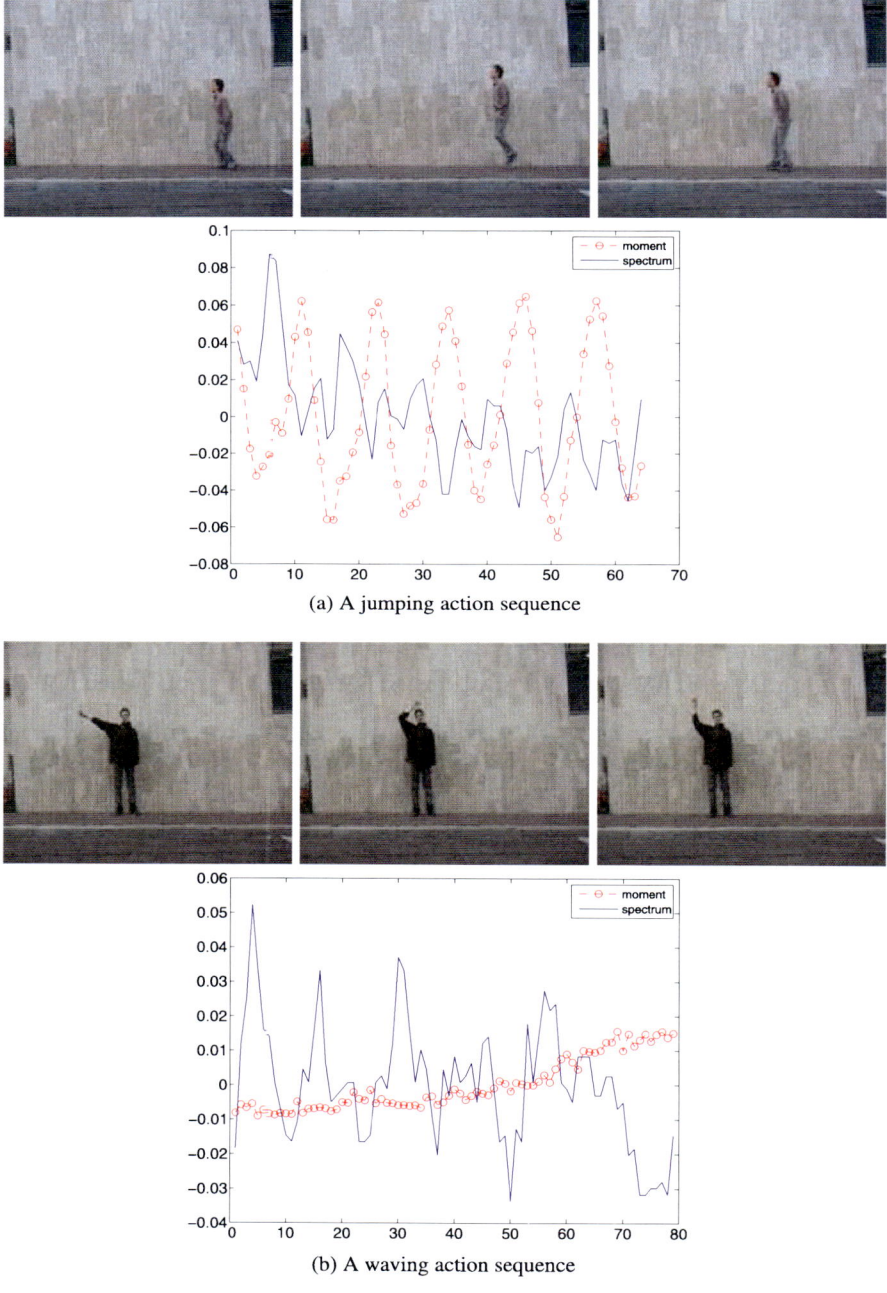

(a) A jumping action sequence

(b) A waving action sequence

Fig. 6.2 Comparing time series of a spectrum feature descriptor with a motion moment feature descriptor extracted from different actions

2006; Wang et al. 2006). Compared to a conventional CRF, an HCRF incorporates a sequence class label into the optimisation of observation conditional probabilities. However, due to the non-convexity nature of the objective function of an HCRF, it is sensitive to parameter initialisation and unstable with noisy data. For action classification, we consider an observable HCRF using an HMM pathing (Zhang and Gong 2010). This model has a convex objective function which leads to a global optimum.

Given a sequence composed of a set of n local observations $\{x_1, x_2, x_3, \ldots, x_n\}$, denoted by $\mathbf{X}$, and its class labels $y \in Y$, one aims to find a model that defines a mapping $p(y|\mathbf{X})$ between them, where y is conditioned on $\mathbf{X}$. An HCRF is defined as

$$p(y|\mathbf{X}; \Theta) = \frac{p(y, \mathbf{X}; \Theta)}{p(\mathbf{X}; \Theta)} = \frac{\sum_{\mathbf{H}} p(y, \mathbf{H}, \mathbf{X}; \Theta)}{\sum_{y, \mathbf{H}} p(y, \mathbf{H}, \mathbf{X}; \Theta)} = \frac{\sum_{\mathbf{H}} e^{\phi(y, \mathbf{H}, \mathbf{X}; \Theta)}}{\sum_{y, \mathbf{H}} e^{\phi(y, \mathbf{H}, \mathbf{X}; \Theta)}} \quad (6.1)$$

where Θ is a set of parameters of the model, $\mathbf{H} = \{h_1, h_2, \ldots, h_n\}$. Each $h_i \in \hat{H}$ captures a certain underlying structure of a class and $\hat{H}$ is a set of hidden states in the model. $\phi(y, \mathbf{H}, \mathbf{X}; \Theta)$ is a potential function which measures the compatibility between a label, a set of observations and a configuration of hidden variables.

Based on maximum likelihood estimation by minimising negative log-likelihood, a regularised objective function of HCRF is

$$L(\Theta) = -\sum_{i=1}^{s} \log p(y_i|\mathbf{X}_i, \Theta) + \frac{\|\Theta\|^2}{2\sigma^2}$$

$$= \underbrace{-\sum_{i=1}^{s} \log \sum_{\mathbf{H}} e^{\phi(y_i, \mathbf{H}, \mathbf{X}_i; \Theta)}}_{(1)} + \underbrace{\sum_{i=1}^{s} \log \sum_{y, \mathbf{H}} e^{\phi(y_i, \mathbf{H}, \mathbf{X}_i; \Theta)}}_{(2)} + \underbrace{\frac{\|\Theta\|^2}{2\sigma^2}}_{(3)} \quad (6.2)$$

where s is the total number of training sequences with known class labels. The first term and the second term are the log-likelihood of the data. The third term is the log of a Gaussian prior with variance σ^2, $p(\Theta) \sim \exp(\frac{\|\Theta\|^2}{2\sigma^2})$, similar to regularisation of a conditional random field (Sutton and McCallum 2007). The optimal parameters, $\Theta^* = \arg\min_{\Theta} L(\Theta)$, can be found by a gradient descent algorithm using Quasi-Newton optimisation. Note that the objective function (6.2) and its gradient can be written in terms of marginal distributions over the hidden variables. These distributions can be computed exactly using an inference method such as belief propagation when the model graph structure is a chain (Sutton and McCallum 2007). The objective function of an HCRF in (6.2) is not convex. This is because a non-negative sum of a concave function (term 1) and two convex (terms 2 and 3) functions does not guarantee convexity. Thus its global convergence depends heavily on the initialisation. Moreover, it is also important to normalise data first, because the sum of potential can give infinite values in the inference process for the gradient calculation. This will cause numerical instability.

Fig. 6.3 The graph model of
a hidden conditional random
field

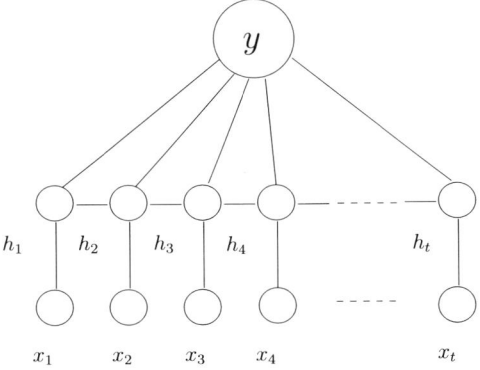

6.2.1 HCRF Potential Function

Figure 6.3 shows an HCRF as an undirected graph. A potential function can be
defined according to both the observation sequence interacting with the hidden states
of a model, and the class label sequence interacting with individual hidden states
represented by hidden graph nodes and the edges between hidden nodes:

$$\phi(y, \mathbf{H}, \mathbf{X}; \theta) = \sum_{j} f(x_j)\theta(h_j) + \sum_{j} f(h_j)\theta(y, h_j) + \sum_{e_k \in E} f(e_k)\theta(y, e_k) \quad (6.3)$$

where e_k is the edge between a pair of nodes j and j'. For action sequence classification, an HCRF graph model is defined as a chain where each node corresponds
to a hidden state variable at time t, $f(x_j)$ is an observation feature vector of node
j, $f(h_j)$ is the feature vector of a hidden node j, $f(e_k)$ is the feature vector of an
edge between node j and j'.

6.2.2 Observable HCRF

Due to the non-convexity nature of the objective function defined by (6.2), the parameter initialisation of an HCRF needs be carefully selected. This limits its usefulness. To overcome this problem, an alternative approach can be considered as
suggested by Zhang and Gong (2010). The idea is to make those hidden variables
observable by learning a class-specific HMM. Once the hidden variables become
observable to the HCRF, the objective function can be shown to be convex.

More precisely, an HMM is first learned for each action sequence class. The number of hidden states is automatically selected by a mixture of Gaussian model using
minimum description length (Rissanen 1983). Second, a Viterbi path is computed
for each training sequence. This step is referred to as HMM pathing.[1] The node of

---

[1]This shares the same principle with the concept of HMM channelling conditional density propagation in modelling gesture sequences (Gong et al. 1999), as discussed in Sect. 5.3.

each training sequence is labelled by the learned class-specific HMMs. This procedure makes the hidden states observable. The observed feature vector is continuous in $\mathbb{R}^L$, and a mixture of Gaussian based HMM is chosen for pathing. A class-specific HMM is learned by maximising the following criterion:

$$\Theta^*(c) = \arg\max_{\Theta} \sum_{i=1}^{s(c)} p(\mathbf{X}_i | y = c; \Theta)$$

$$= \arg\max_{\Theta} \log \sum_{i=1}^{s(c)} \sum_{\mathbf{H}} p(\mathbf{X}_i, \mathbf{H} | y = c; \Theta) \tag{6.4}$$

where c is the class label. The observation model is

$$p(X_t = x | h_t = i) = N(x; u_i, \Sigma_i)$$

$$= \frac{1}{(2\pi)^{\frac{d}{2}} \|\Sigma_i\|^{\frac{1}{2}}} \exp\left(-\frac{1}{2}(x - u_i)' \Sigma_i (x - u_i)\right) \tag{6.5}$$

where u_i and Σ_i are the mean and variance of the ith Gaussian mixture component. A Viterbi path is inferred as

$$h_{1:t}^* = \arg\max_{h_{1:t}} p(h_{1:t} | x_{1:t}) \tag{6.6}$$

This enables the hidden states of each training sequence observable which are then used for learning an HCRF. Given HMM pathing, the model function of an HCRF defined in (6.1) becomes

$$p(y | \mathbf{X}, \hat{\Theta}) = \frac{p(y, \mathbf{X}; \hat{\Theta})}{p(\mathbf{X}; \hat{\Theta})} \stackrel{\text{def}}{=} \frac{p(y, \mathbf{H}(\mathbf{X}); \hat{\Theta})}{\sum_y p(y, \mathbf{H}(\mathbf{X}); \hat{\Theta})} = \frac{e^{\phi(y, \mathbf{H}(\mathbf{X}); \hat{\Theta})}}{\sum_y e^{\phi(y, \mathbf{H}(\mathbf{X}); \hat{\Theta})}} \tag{6.7}$$

and the HCRF objective function given by (6.2) becomes

$$L(\hat{\Theta}) = -\sum_{i=1}^{s} \log p(y_i | \mathbf{X}_i, \hat{\Theta}) + \frac{\|\hat{\Theta}\|^2}{2\sigma^2}$$

$$= -\sum_{i=1}^{s} \phi(y, \mathbf{H}(\mathbf{X}_i); \hat{\Theta}) + \sum_{i=1}^{s} \log \sum_y e^{\phi(y, \mathbf{H}(\mathbf{X}_i); \hat{\Theta})} + \frac{\|\hat{\Theta}\|^2}{2\sigma^2} \tag{6.8}$$

The effect of HMM pathing is to change the hidden relationship between H, X in (6.1) to observable in (6.7), so that the hidden variable is a direct function of observation by $H(X)$. This transforms the objective function $L(\hat{\Theta})$ to be convex. Accordingly, the local potential of (6.3) is redefined as

$$\phi(y, \mathbf{H}, \mathbf{X}; \hat{\Theta}) \stackrel{\text{def}}{=} \sum_j f(h_j) \theta_h(y, h_j) + \sum_{e_k} f(e_k) \theta_e(y, e_k) \tag{6.9}$$

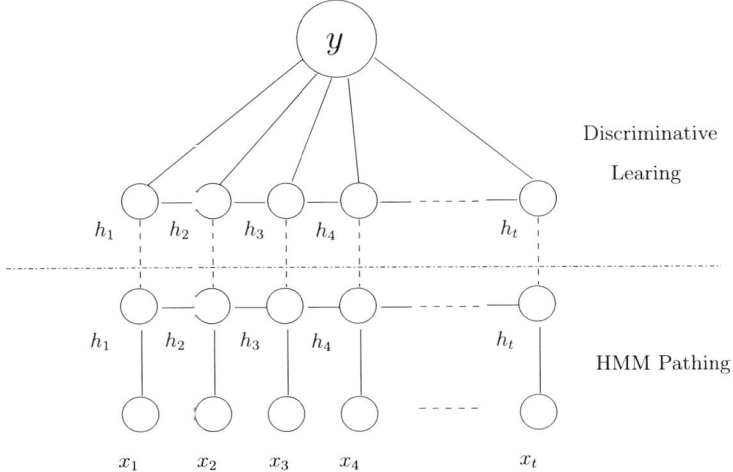

Fig. 6.4 The graph model of an observable hidden conditional random field

Fig. 6.5 The graph model of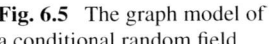
a conditional random field

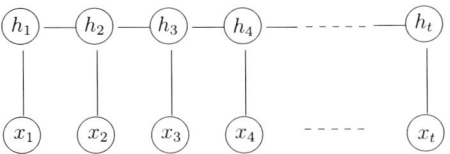

In this observable HCRF, the parameter set $\hat{\Theta}$ contains two components: $\hat{\Theta} = [\theta_h, \theta_e]$. Whilst $\theta_h[y, h_j]$ is used to refer to a parameter that measures the compatibility between a state h_j and an action label y, $\theta_e[y, e_k]$ corresponds to a parameter for compatibility between action label y and the edge between nodes j and j'.

This model still retains the advantage of a general HCRF for discriminative learning of the parameters of the edge potentials, representing temporal dependencies, interacted with the action sequence label. Moreover, (6.8) guarantees a global optimum. It is possible that the learning of class-specific HMMs for HMM pathing is not globally optimal. However, it provides a much improved initialisation for overcoming the non-convex objective function problem suffered by a conventional HCRF. The model complexity is also reduced by enabling the hidden variables observable for model learning. For estimating a conventional HCRF using gradient descent, the gradient needs be computed by inference (Wang et al. 2006), whose complexity is usually exponential to the number of hidden variables. In contrast, such inference is not necessary for this observable HCRF with the summation over the hidden variables being avoided. It should be noted that the objective function of a CRF is $\log \sum_{i=1}^{s} p(\mathbf{H}|\mathbf{X}_i)$, where $\mathbf{H}$ is the observed node label instead of the whole sequence class label. This is inherently different from that of an observable HCRF as defined in (6.8). Figure 6.4 shows the graph of an observable HCRF, whilst the graph for a CRF is shown in Fig. 6.5.

Given a learned model, action sequence classification is straightforward by maximising the posterior probability of the learned model with parameters $\hat{\Theta}^*$, which defines a classification decision rule as

$$y^* = \arg\max_y p(y|\mathbf{X}, \hat{\Theta}^*) \tag{6.10}$$

Zhang and Gong (2010) show that a hidden conditional random field based action classification model outperforms alternative action models using a mixture of Gaussian (Figueiredo and Jain 2002), an HMM (Ahmad and Lee 2006; Lee and Kim 1999), logistic regression (Brzezinski 1999), a support vector machine (Chapelle et al. 1999), and a conditional random field (Sminchisescu et al. 2006).

6.3 Space-Time Clouds

Computing object-centred shape and motion moment features for action recognition is only viable in a well-controlled environment with no moving background clutter nor viewpoint changes. For action recognition in a more realistic visual environment, object segmentation becomes challenging and is mostly unreliable. To overcome this problem, space-time interest point based models are a suitable alternative (Dollár et al. 2005; Niebles et al. 2008; Savarese et al. 2008; Schüldt et al. 2004). These models are based on a bag-of-words feature representation. Similar techniques have also been exploited for 2D object categorisation and detection (Dalal and Triggs 2005). Compared to object tracking and spatio-temporal shape-based models, they are more robust to noise, viewpoint changes, and the lack of visual details. However, a bag-of-words representation relies solely on the discriminative power of individual local space-time descriptors in that only local appearance information centred at each individual interest point is utilised. More holistic spatial, and in particular temporal, information about the distribution of all the interest points from an action sequence is not considered.

To address this limitation of a conventional bag-of-words action representation, we consider the idea of a cloud-of-points representation such that an action image sequence is represented by clouds of space-time interest points at different temporal scales. Specifically, extracted interest points are accumulated over time at different temporal scales to form point clouds (see Fig. 6.6). Holistic features are computed from these point clouds to represent explicitly global spatial and temporal distributions of local space-time interest point descriptors. A feature selection process is considered for filtering the most informative features for representing different action classes.

6.3.1 Clouds of Space-Time Interest Points

Space-time interest points are local spatio-temporal features that can be used to describe image sequences of human body behaviour such as gestures (Sect. 5.4). For

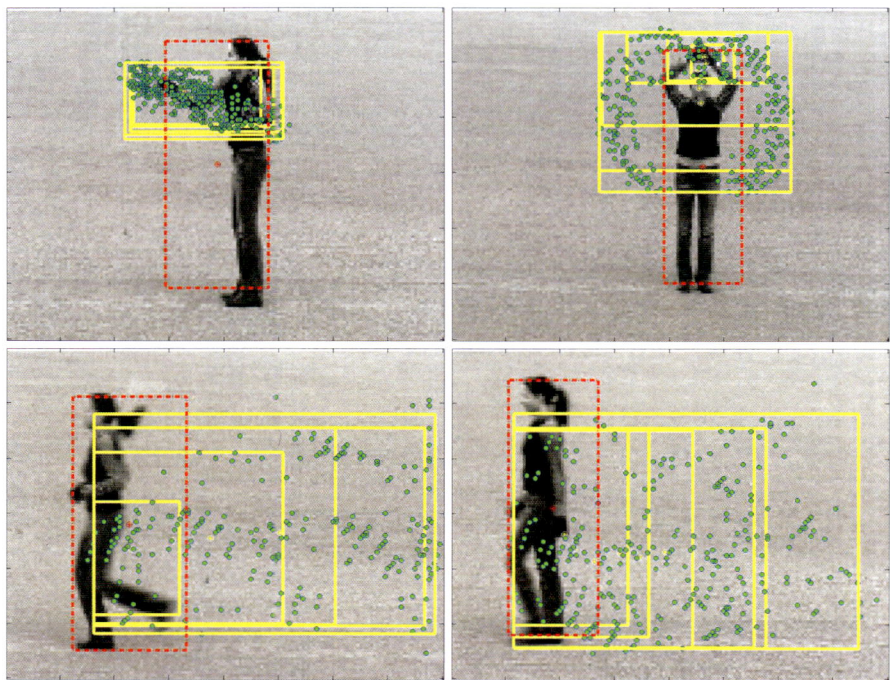

Fig. 6.6 Clouds of space-time interest points from action sequences. The *red rectangles* indicate the detected foreground areas The *green points* are the extracted interest points, whilst the *yellow rectangles* indicate clouds at different scales

less constrained action recognition, this type of representation has its limitations. For instance, the one-dimensional Gabor filter applied in the temporal domain by Dollár et al. (2005) is sensitive to both background clutter and textured foreground details. To overcome this problem, Bregonzio et al. (2009) explores different filters for detecting salient local space-time image volumes undergoing complex motions.

More specifically, space-time interest points are detected by two steps: (1) frame differencing for focus of attention and region of interest detection, and (2) Gabor filtering on the detected regions of interest using 2D Gabor filters of different orientations. This two-steps approach facilitates saliency detection in both the temporal and spatial domains. The 2D Gabor filters are composed of two parts. The first part, $s(x, y, i)$, represents the real part of a complex sinusoid, known as the carrier:

$$s(x, y, i) = \cos\big(2\pi(\mu_0 x + \upsilon_0 y) + \theta_i\big) \tag{6.11}$$

where θ_i defines the orientation of the filter with eight orientations being considered: $\theta_{i=1,\dots,8} = \{0°, 22.5°, 45°, 67.5°, 90°, 112.5°, 135°, 157.5°\}$, and μ_0 and υ_0 are the spatial frequencies of the sinusoid controlling the spatial scale of the filter. The second part of the filter, $G(x, y)$, represents a 2D Gaussian-shaped function, known

(a) Boxing   (b) Hand waving   (c) Running

Fig. 6.7 Comparison between space-time interest points of Bregonzio et al. (2009) (*green circle points*) and those of Dollár et al. (2005) (*red square points*)

as the envelope:

$$G(x, y) = \exp\left(-\frac{\frac{x^2}{\rho^2} + \frac{y^2}{\rho^2}}{2}\right) \tag{6.12}$$

where ρ is the parameter that controls the width of $G(x, y)$. Let $\mu_0 = \upsilon_0 = \frac{1}{2\rho}$, so the only parameter controlling the spatial scale is ρ.

Figure 6.7 shows some examples of space-time interest points detected using this method. It is evident that they are more descriptive compared to those detected by the method of Dollár et al. (2005). In particular, these interest points are more likely to correspond to body parts undergoing temporal change without drifting to either irrelevant static textured foreground regions or static background textures with strong local intensity gradients.

Multi-scale space-time clouds of interest points are constructed as follows. Suppose an action sequence $\mathbf{V}$ consists of T image frames, represented by $\mathbf{V} = [\mathbf{I}_1, \ldots, \mathbf{I}_t, \ldots, \mathbf{I}_T]$, where $\mathbf{I}_t$ is the tth image frame. For an image frame $\mathbf{I}_t$, a total of S interest point clouds at different temporal scales can be formed, denoted by $[\mathbf{C}_t^1, \ldots, \mathbf{C}_t^s, \ldots, \mathbf{C}_t^S]$. A cloud at the sth scale is constructed by accumulating the interest points detected over the past $s \times N_s$ frames, where N_s is the difference between two consecutive time scales, measured by the number of frames. Examples of multi-scale space-time clouds of interest points are shown in Fig. 6.8. It is evident that different types of actions exhibit space-time clouds of distinctively different shape, relative location, and distribution.

Feature Extraction

To represent space-time clouds more effectively, two sets of cloud features are extracted from S clouds $[\mathbf{C}_t^1, \ldots, \mathbf{C}_t^s, \ldots, \mathbf{C}_t^S]$ constructed for the tth time frame. The first set of features aim to describe the shape and speed characteristics of the foreground object. They are computed as follows:

1. Regions of interest are detected by image frame differencing.
2. A series of 2D Gabor filters are applied to the detected regions of interest.
3. A Prewitt edge detector (Parker 1997) is employed to coarsely segment an object from the detected foreground area.
4. Two features are computed from the segmented object bounding box: O_t^r measuring the height and width ratio of the object, and O_t^{Sp} measuring the absolute speed of the object.

The second set of features are extracted from clouds of different scales, so they are time scale specific. More precisely, eight features are computed for the sth time scale cloud,[2] and denoted as

$$\left\{ C_s^r, C_s^{Sp}, C_s^D, C_s^{Vd}, C_s^{Hd}, C_s^{Hr}, C_s^{Wr}, C_s^{Or} \right\} \tag{6.13}$$

where

1. C_s^r is the height and width ratio of the cloud.
2. C_s^{Sp} is the absolute speed of the cloud.
3. C_s^D is the density of the interest point within the cloud, which is computed as the total number of points normalised by the area of the cloud.
4. C_s^{Vd} and C_s^{Hd} measure the spatial relationship between the cloud and the detected object area. Whilst C_s^{Vd} is the vertical distance between the geometrical centroid of the object area and the cloud, C_s^{Hd} is the horizontal distance.
5. C_s^{Hr} and C_s^{Wr} are the height and width ratios between the object area and the cloud, respectively.
6. C_s^{Or} measures how much the two areas overlap.

Overall, these eight features can be put into two categories: C_s^r, C_s^{Sp} and C_s^D measure the shape, speed, and density of each space-time cloud. The five remaining features capture the relative shape and location information between the object foreground bounding box region and the individual space-time cloud areas.

---

[2]Subscript t is omitted for clarity.

(a) Boxing   (b) Hand clapping   (c) Hand waving

(e) Running   (f) Walking   (g) Galloping sideways

Fig. 6.8 Examples of clouds of space-time interest points at multiple scales

To minimise the sensitivity of these cloud features to outliers in the detected space-time interest points, removing some of the outliers can be considered before computing these cloud features. This problem is similar to that of removing clutters from detected atomic actions when constructing a gesture representation space (Fig. 5.20 in Sect. 5.2), where clutters in a distribution can be filtered out by modelling them as a random process (Byers and Raftery 1998). A simpler approach can also be adopted to evaluate the interest point distribution over a number of consecutive time frames and removing those points that are too far away from the distribution centroid. An even simpler model entails computing both the centroid of all the points in each frame and the average distance from each point to the centroid. If the distance between an interest point and the centroid is above a threshold, it is removed.

Feature Selection by Variance

Now each frame is represented by $8S + 2$ features, where S is the total number of temporal scales. Using a total of $(8S + 2) \times T$ cloud features to represent an action sequence, where T is the length of the sequence, can lead to a feature space of high dimension. This is sensitive to model over-fitting and poor recognition accuracy (Bishop 2006). To reduce the dimensionality of the feature space, and also making the representation less sensitive to both feature noise and variation in the length of an action sequence, a histogram of N_b bins is constructed for each of the $8S + 2$ cloud features collected over time by linear quantisation. As a result, each action sequence regardless of length is represented by $8S + 2$ histograms, which can be considered as $(8S + 2) \times N_b$ global scalar features with $N_b \ll T$. A space of $(8S + 2) \times N_b$ features remains to have a high dimension, although it is now action duration independent. Taking into account that there are also uninformative and redundant features, further dimensionality reduction by feature selection is exploited as follows.

Action cloud feature selection is based on measuring the relevance of each feature according to how much its value varies within each action class and across different classes. Specifically, a feature is deemed as being informative about a specific class of actions if its value varies little for actions of the same class but varies significantly for actions of different classes. To computer this class relevance measure from some training data, we consider the following steps. First, given a training data set of A action classes, for the ith feature f_i in the ath class, its mean and standard deviation within the class are computed as $\mu_{f_i}^a$ and $\sigma_{f_i}^a$, respectively. A class relevance measure for feature f_i is then denoted as R_{f_i} and computed as

$$R_{f_i} = \frac{\sqrt{\frac{1}{A} \sum_{a=1}^{A} (\mu_{f_i}^a - \hat{\mu}_{f_i})^2}}{\frac{1}{A} \sum_{a=1}^{A} \sigma_{f_i}^a} \tag{6.14}$$

where $\hat{\mu}_{f_i} = \frac{1}{A} \sum_{a=1}^{A} \mu_{f_i}^a$ is the inter-class mean of the A intra-class feature means. The numerator and denominator of (6.14) correspond, respectively, to the standard deviation of the intra-class means, and the inter-class mean of the intra-class standard deviations. The former measures how the feature value varies across different action classes, where a higher value signifies a more informative feature. The latter informs on how the feature value varies within each action class, where a lower value indicates a more informative feature. Overall, cloud features with higher R_{f_i} values are good and need be selected. For selection, all features are ranked according to their R_{f_i} values and a decision can be made simply based on the percentage of the cloud features to be kept.

Feature Selection by Kernel Learning

Feature selection can also be considered from another perspective. The cloud features are from multiple (S) time scales. Features of different temporal scales are not equally informative in representing different action classes. This is because actions of the same class share certain common temporal characteristics including time scale. For instance, different types of actions are likely to be performed at different speeds. Most actions are also periodic with repetitive cycles, such as running, walking, and hand clapping. These periodic cycles characterise their temporal scales. It can be considered that features of longer time scales are more useful in describing long or slow actions, and vice versa. Measuring a time scale relevance for each cloud feature can be useful for their selection in representing different action classes. Such measurements can be learned from training data.

Multiple kernel learning is introduced by Bach et al. (2004) to address the problem of selecting the optimal combination of kernel functions for a specific feature in classification. This technique can be used for visual feature selection when designing classification functions (Gehler and Nowozin 2009; Sun et al. 2009). For selecting cloud features from different time scales so that collectively they best describe an action class, multiple kernel learning can be exploited to learn the optimal combination of features for a multi-class classification problem.

Let us formally define the multi-class action classification problem as follows. Taking the one-versus-rest strategy, C binary classifiers are learned to classify an action sequence into one of the C classes. Suppose we have a training data set $(x_i, y_i)_{i=1,\dots,N}$ of N instances, whilst each training example x_i is an action sequence with a class label y_i. To represent this action, S histogram features are extracted with each histogram representing cloud features at one temporal scale. Let the sth time scale histogram feature be $f_s(x)$, where $f_s(\cdot)$ denotes a histogram feature extraction function. Using multiple kernel learning, a set of kernel functions is to be computed, each of which is a distance similarity measure. More precisely, a kernel function

$$k_s(x, x') = k\big(f_s(x), f_s(x')\big) \tag{6.15}$$

measures the similarity between a pair of action sequences represented by their corresponding sth time scale histogram features. Given an action sequence x, we denote a set of kernel responses from comparing its sth histogram feature against those of N action sequences from a training data set as

$$K_s(x) = \left[k_s(x, x_1), k_s(x, x_2), \ldots, k_s(x, x_N) \right]^T \qquad (6.16)$$

Let us now consider how to combine kernel responses from different histogram features. The aim of the multiple kernel learning here is to learn optimal weightings on different histogram features so to give a combined kernel function as

$$k^*(x, x') = \sum_{s=1}^{S} \beta_s k_s(x, x') \qquad (6.17)$$

where β_s is the weight associated to the sth temporal scale to be learned. For learning these weights, a support vector machine is constructed to discriminate one action class against the rest. That requires to solve an optimisation problem with the following objective function:

$$\min_{\alpha, \beta, b} \frac{1}{2} \sum_{s=1}^{S} \beta_s \alpha^T K_s \alpha + C \sum_{i=1}^{N} L \left(y_i, b + \sum_{s=1}^{S} \beta_s K_s(x)^T \alpha \right) \qquad (6.18)$$

$$\text{subject to} \quad \sum_{s=1}^{S} \beta_s = 1, \quad \beta_s \geq 0, \ s = 1, \ldots, S$$

where

1. S is the number of histogram features used to represent an action sequence,
2. α is a N-dimensional weight vector applied to all the action training sequences,
3. b is a scalar value,
4. K_s are the kernel responses to a histogram feature at a particular temporal scale, as defined by (6.16),
5. $L(y, z)$ denotes a hinge loss function (Bishop 2006).

The two constraints put on β_s are to ensure that the estimated value of β_s is sparse and interpretable. That is, as weights, they should be either zero or a positive number, and the sum of all weights should be 1. There are a number of techniques that can be used to solve this optimisation problem. Conventionally, the multiple kernel learning problem is formulated as a convex quadratically constrained program, and solved by a local descent algorithm such as sequential minimisation optimisation. However, such a scheme is only computationally tractable for small scale problems. Sonnenburg et al. (2006) reformulate the multiple kernel learning problem as a semi-infinite linear program that can be solved by a standard linear program and a standard support vector machine implementation.

Given the learned parameters β_s, α, and b, a binary decision function is given as

$$F_{\text{MKL}}(x) = \text{sgn} \left(\sum_{s=1}^{S} \beta_s \left(K_s(x)^T \alpha + b \right) \right) \qquad (6.19)$$

where the function sgn($\cdot$) returns a value 1 if its parameter is positive, and -1 otherwise. $K_s(x)$ measures the similarity between a new action sequence x and all N training action sequences, as defined by (6.16). If $F_{\mathrm{MKL}}(x)$ assumes the value 1, this new action sequence x is deemed as being likely to be a member of an action class from the training data. Positive response values of $\sum_{s=1}^{S} \beta_s (K_s(x)^T \alpha + b)$ can result from multiple different action classes. In this case, the action sequence is classified by the action class with the highest value.

6.3.2 Joint Local and Global Feature Representation

A cloud-of-points representation exploits different and complementary information about an action as compared to that of a bag-of-words representation. Whilst the former computes global point distributions, the latter extracts local point appearance descriptions. To utilise both global and local information, feature selection based on multiple kernel learning can be exploited as a mechanism for feature fusion. Specifically, to learn a multi-class classifier for action classification, a support vector machine is employed with multiple kernels, each of them being computed using either the bag-of-words features or the cloud-of-points features at a certain temporal scale. Multiple kernel learning is designed to learn the optimal linear combination of these kernels which maximises action classification accuracy.

More specifically, a kernel function, denoted as k_B, is computed given some extracted interest points represented as a histogram of bag-of-words features. Similarly, a kernel function for S cloud-of-points features can also be computed (6.15). Now a linear combination of multiple kernel functions (6.17) can be re-written as follows:

$$k^*(x, x') = \sum_{s=1}^{S} \beta_s k_s(x, x') + \beta_B k_B(x, x') \qquad (6.20)$$

The objective function to be optimised becomes

$$\min_{\alpha, \beta, b} \frac{1}{2} \left(\sum_{s=1}^{S} \beta_s \alpha^T K_s \alpha + \beta_B \alpha^T K_B \alpha \right)$$

$$+ C \sum_{i=1}^{N} L \left(y_i, b + \sum_{s=1}^{S} \beta_s K_s(x)^T \alpha + \beta_B K_B(x)^T \alpha \right) \qquad (6.21)$$

$$\text{subject to} \quad \sum_{s=1}^{S} \beta_s + \beta_B = 1, \quad \beta_s \geq 0, \ s = 1, \dots, S, \ \beta_B \geq 0$$

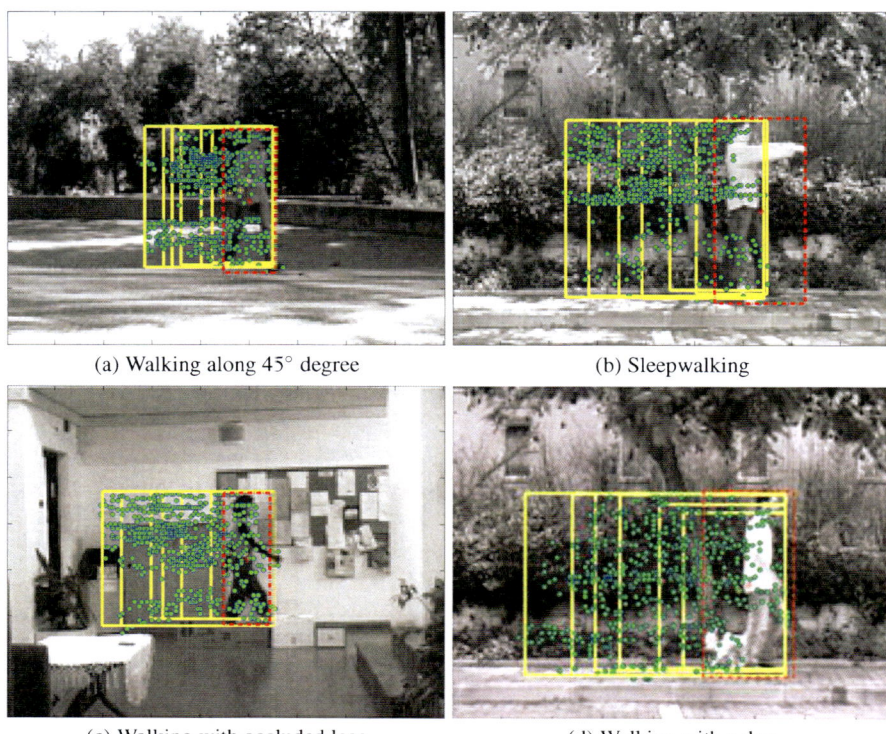

(a) Walking along $45°$ degree   (b) Sleepwalking

(c) Walking with occluded legs   (d) Walking with a dog

Fig. 6.9   Using a cloud-of-points model to classify walking actions observed under different viewing conditions

where β_B is the weight for the bag-of-words features. After parameter estimation, the final binary decision function is

$$F_{\mathrm{MKL}}(x) = \mathrm{sgn}\left(\sum_{s=1}^{S} \beta_s\left(K_s(x)^T\alpha + b\right) + \beta_B\left(K_B(x)^T\alpha + b\right)\right) \qquad (6.22)$$

Bregonzio et al. (2009) show that a cloud-of-points based action classification is robust against image noise and changes in viewpoint. It is shown that a fusion of both local and global space-time interest points can further improve classification accuracy, in particular under occlusion, changes in viewpoint and clothing (Fig. 6.9).

6.4 Localisation and Detection

To be able to automatically detect and recognise human actions is important to camera systems designed for monitoring large areas of unconstrained public spaces. For instance, this ability can assist in detecting either a public disorder and violent incident involving fighting that requires police attention, or a member of the public falls that requires medical attention (Fig. 6.10(b)). Models developed for automatic action classification often assume that a human action has already been detected and isolated in the viewpoint (Fig. 6.10(a)). For action recognition in a crowded space, a model is required to first locate and detect an action of interest, whilst such an action may well be dwarfed by other activities in the scene (Fig. 6.10(b)–(d)).

To address this problem, we consider a model that learns a discriminative classifier based on support vector machines from annotated space-time action cuboids examples. The model is used for sliding a space-time search window for action localisation and detection. The concept is similar to 2D sliding-window based object detection that both localises and recognises object of a certain class against background clutter and other distracting objects in the scene (Dalal and Triggs 2005; Felzenszwalb et al. 2010; Viola and Jones 2004; Viola et al. 2003). Similar to learning a 2D object detector, a large number of space-time action cuboids needs be annotated to learn a discriminative action detector. This is more tedious and unreliable compared to 2D object annotation. To overcome this problem, Siva and Xiang (2010) suggest a method that adopts a greedy k-nearest neighbour algorithm for automated annotation of positive training data, by which an action detector can be learned with only a single manually annotated training video. This method can greatly reduce the amount of manual annotation required. It can also be seen as a solution to the multi-instance learning problem encountered in weakly supervised learning (Andrews et al. 2003). Compared to a conventional multi-instance learning method, this approach is more capable of coping with large amounts of background activities that are visually similar to the action of interest.

6.4.1 Tracking Salient Points

Spatially salient points are 2D interest points extracted in each image frame and tracked over consecutive frames to form tracks $\{\mathbf{T}_1, \ldots, \mathbf{T}_j, \ldots, \mathbf{T}_N\}$ (Sun et al. 2009). Figure 6.11(a) shows an example. These tracks are formed by a 1-to-1 pairwise matching of scale-invariant feature transform (SIFT) descriptors (Lowe 2004) over consecutive frames. Tracks are terminated if between consecutive frames, a point travels more than 20 pixels in the x or y axes. The length of acceptable tracks is restricted to between $L_{\min} = 5$ and $L_{\max} = 30$ frames.

The 2D local appearance information along a track is described using the average SIFT descriptor. For a track $\mathbf{T}_j$ of length k frames, we have a SIFT descriptor at each frame $\{S_1, S_2, \ldots, S_k\}$. The static appearance information associated with the salient points tracked by $\mathbf{T}_j$ is then represented as an average SIFT descriptor computed as

$$S_j = \frac{1}{k} \sum_{i=1}^{k} S_i \qquad (6.23)$$

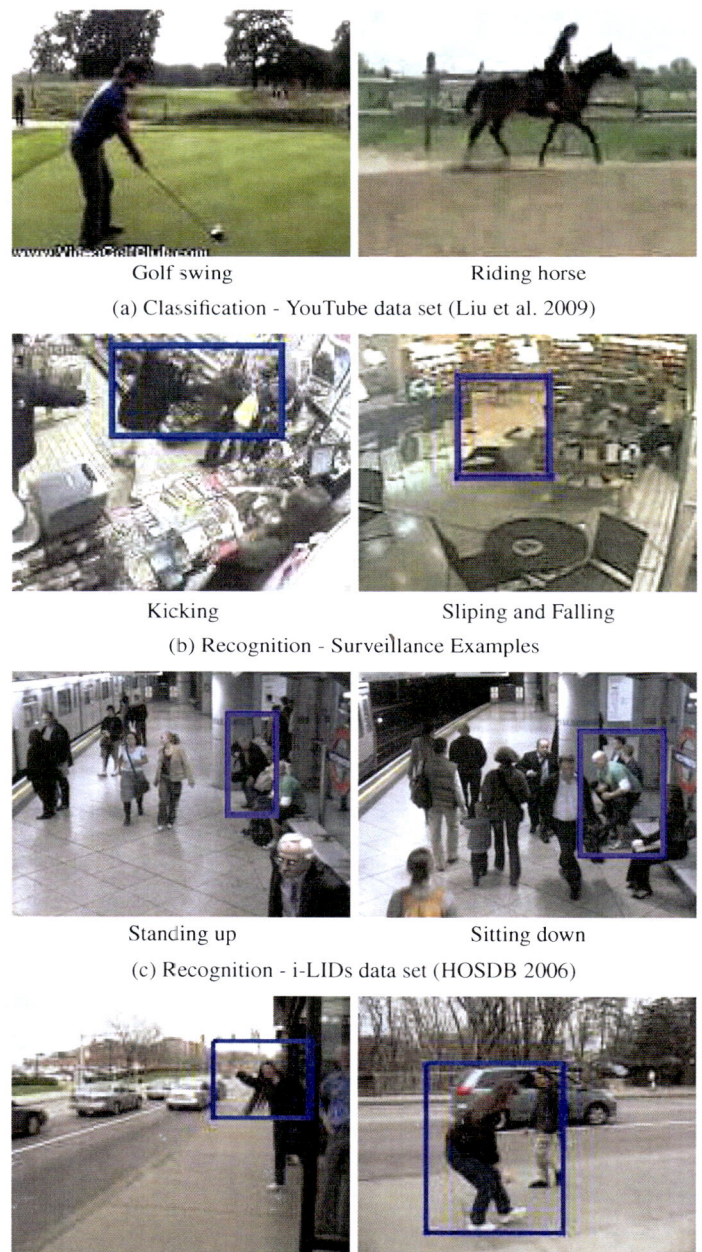

Golf swing   Riding horse

(a) Classification - YouTube data set (Liu et al. 2009)

Kicking   Sliping and Falling

(b) Recognition - Surveillance Examples

Standing up   Sitting down

(c) Recognition - i-LIDs data set (HOSDB 2006)

One hand waving   Picking up

(d) Recognition - CMU data set (Ke et al. 2007)

Fig. 6.10 Action classification versus recognition. (**a**) Classification: a viewpoint is centred on the action of interest with minimal background distractions. (**b**)–(**d**) Recognition: an action of interest is amid other background activities

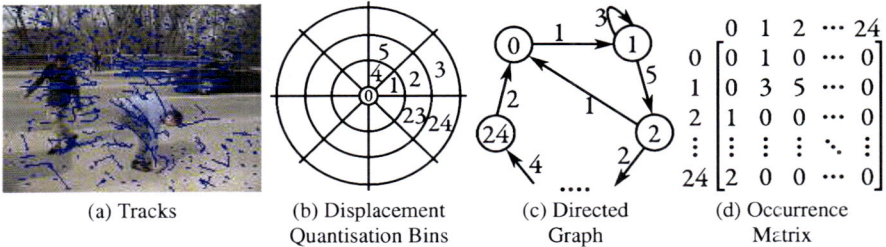

(a) Tracks	(b) Displacement Quantisation Bins	(c) Directed Graph	(d) Occurrence Matrix

Fig. 6.11 Computing a trajectory transition descriptor. (**a**) Extracting tracks. (**a**)→(**b**) Quantising track displacements into bins. (**b**)→(**c**) Sequential bin transitions along a track is transformed into a directed graph. (**c**)→(**d**) Constructing an occurrence matrix of the directed graph

The motion information captured by each track is represented using a trajectory transition descriptor. Specifically, a trajectory vector $\mathbf{T} = [(x_1, y_1, f_1), (x_2, y_2, f_2), \ldots,$ $(x_k, y_k, f_k)]$ is transformed to a displacement vector

$$\mathbf{D} = \left[(x_2 - x_1, y_2 - y_1), (x_3 - x_2, y_3 - y_2), \ldots, (x_k - x_{k-1}, y_k - y_{k-1}) \right] \quad (6.24)$$

where (x_i, y_i) is the spatial location of this track $\mathbf{T}$ at frame f_i. To achieve scale invariance, the magnitude of the displacement vector is normalised by the largest displacement magnitude $\|\mathbf{D}\|_{\max}$ along the same trajectory.

Each of the $k - 1$ displacements in vector $\mathbf{D}$ is then quantised in magnitude and angle and assigned to one of the 25 bins depicted in Fig. 6.11(b). After quantisation, the displacement vector becomes $\hat{D} = [b_1, b_2, \ldots, b_{k-1}]$, where b_i is the quantisation bin number. The sequential bin numbers in vector $\hat{D}$ are used to construct a directed graph (Fig. 6.11(c)), which is then modelled as an ergodic Markov chain. Each of the nodes of the graph corresponds to one of the 25 quantisation states and each of the edge weights corresponds to the number of transitions between the states. To follow the ergodic Markov chain formulation all edge weights in the directed graph are initialised with a negligible weight of 0.15.

The occurrence matrix representation of the connected graph (Fig. 6.11(d)) is row-normalised to construct the transition matrix $\mathbf{P}$ of the ergodic Markov chain. The motion descriptor of track $\mathbf{T}_j$ is then computed as the stationary distribution vector π_j of the Markov chain, which can be obtained as the column average of matrix $\mathbf{A}_n$, defined as

$$\mathbf{A}_n = \frac{1}{n+1} \left(\mathbf{I} + \mathbf{P} + \cdots + \mathbf{P}^n \right) \quad (6.25)$$

when $n \to \infty$ and $\mathbf{I}$ is the identity matrix.

The average SIFT S_j given by (6.23) and motion descriptor π_j are then used to construct a bag-of-words model for action representation with 1000 words for each type of descriptor. To take advantage of the spatial distribution of the tracks, one can use all six spatial grids (1×1, 2×2, 3×3, $h3 \times 1$, $v3 \times 1$, and $o2 \times 2$) and the four temporal grids ($t1$, $t2$, $t3$, and $ot2$), as suggested by Laptev et al. (2008).

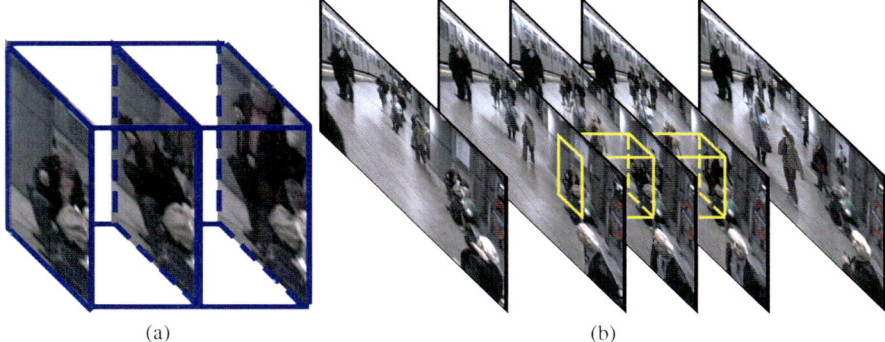

Fig. 6.12 (a) Action cuboid. (b) Action localised spatially and temporally using sliding-windows

The combination of the spatial and temporal grids gives us 24 channels each for S_j and π_j, yielding a total of 48 channels. A greedy channel selection routine is then employed to select at most 5 of the 48 channels, because these 48 channels contain redundant information. Specifically, starting with the null set ($C_0 = [\,]$), a channel c is added such that the cross-validation average precision (AP) of $C_{i+1} = C_i \bigcup c$ is the maximum over all available c. A channel c is then removed from C_{i+1} if AP of C_{i+1} is less than $C_{i+1} \setminus c$. This process is repeated till the size of C_i reaches 5. The channel combination C_i with the best overall AP is then selected as the best channel combination.

6.4.2 Automated Annotation

An action is contained within a space-time cuboid or volume (Fig. 6.12(a)). This can be called an action cuboid. An action cuboid is represented by the multi-channel bag-of-words histogram of all features contained within the cuboid. The process of sliding a window to detect an action is to move around a cuboid spatially and temporally over a video and locate all occurrences of the targeted action in the video (Fig. 6.12(b)). To determine whether a candidate action cuboid contains the action of interest, a support vector machine based classifier is learned and deployed inside the cuboid as it is shifted in space and over time.

   More specifically, one can construct a positive training set of manually annotated action cuboids and a negative training set of videos without the actions from which random cuboids can be selected. A support vector machine is then trained using a multi-channel χ^2 kernel (Laptev et al. 2008):

$$K(H_i, H_j) = \exp\left[-\sum_{c \in \mathsf{C}} \frac{1}{A_c} D_c\left(H_i^c, H_j^c\right)\right] \qquad (6.26)$$

where $H_i^c = \{h_{in}^c\}$ and $H_j^c = \{h_{jn}^c\}$ are two histograms extracted in channel c. A_c is a normalisation parameter and

$$D_c\left(H_i^c, H_j^c\right) = \frac{1}{2} \sum_{n=1}^{N^c} \frac{(h_{in}^c - h_{jn}^c)^2}{h_{in}^c + h_{jn}^c} \tag{6.27}$$

A support vector machine can be trained using a hard negative mining procedure (Felzenszwalb et al. 2010). More specifically, at each iteration, a support vector machine is trained to select the optimal channel combination. The positive cuboids are scored using the trained support vector machine and the scores are normalised, by logistic regression, to between -1 and 1. Subsequently, the trained support vector machine is run on all the negative training videos that contain no occurrence of the action of interest. Any cuboids with a normalised support vector machine score greater than a threshold, typically set to -0.75, are added to the negative training set. In addition to the hard negative mining, a positive mining can also be performed to relax the reliance on the ground truth annotation. Specifically, at each iteration, the trained support vector machine is run on all the positive locations within a search region created by increasing a ground truth action cuboid by 25% in all three dimensions. The cuboid with the maximum score at each search region is then added to the training set. With the increased positive and negative sets, the support vector machine is retrained. This iterative mining and retraining process is terminated when the detection performance on a validation set stops improving.

To apply the learned model for action detection, a cuboid is slid along the spatial and temporal axis, with overlapping, to obtain a score for each cuboid using the trained support vector machine. Once the scores at all locations are obtained, the maximum score in the scanned video can be found iteratively as an action cuboid. To avoid repeated detection of the same cuboid, all scores within a found action cuboid are set to $-\infty$. In addition, if two detected cuboids overlaps more than 25%, the cuboid with the lower score is removed as a duplicate detection. The overlap percentage is obtained as the intersection volume divided by the union volume.

A training data set consists of a set of negative video clips without the action $V_{i=1,\ldots,N_1}^-$, positive video clips known to contain at least one instance of the action $V_{i=1,\ldots,N_2}^+$, and one manually annotated clip with a labelled cuboid $Q_{k=1}$ around the action. From V_i^- and V_i^+, a set of spatially overlapping instances or cuboids $X_{i,j}^-$ and $X_{i,j}^+$ are extracted, respectively, where $j = 1, \ldots, M_i$ are the number of cuboids in video i. The aim is to select automatically positive cuboids $X_{i,j=j^*}^+$ from each video i in the positive clip set V_i^+ to add to the positive cuboid set Q_k which initially contains only the annotated positive cuboid $Q_{k=1}$. This is effectively a multi-instance learning problem (Andrews et al. 2003). However, the nature of the action recognition problem poses difficulties to a conventional multi-instance learning technique. That is, within each positive clip, there are much more negative instances, typically in the order of thousands, than positive ones, typically less than half a dozen. Furthermore, these negative instances contain many background activities that are visually similar to the action of interest. They can be easily misdetected as positive instances by a conventional multi-instance learning model. To solve this problem, one can consider a simple but effective solution to the multi-instance learning problem with an assumption that a single positive instance has been annotated.

Greedy k-Nearest Neighbour Classification

The solution is based on a greedy k-nearest neighbour algorithm. A k-nearest neighbour classifier (Bishop 2006) represents classes by a set of known examples, Q_k for action and $X_{i,j}^-$ for non-action, and classifies a new example, $X_{i,j}^+$, by checking the class of the closest known example. The accuracy of a k-nearest neighbour classifier increases as the number of known examples increases. Therefore, one wants to iteratively grow the positive action cuboid set Q_k, such that at each iteration, more of the intra-class variation can be captured and the k-nearest neighbour classifier becomes stronger. To that end, at each iteration, the current k-nearest neighbour classifier is used to select one cuboid X_{i^*,j^*}^+ from the set of all available cuboids $X_{i,j}^+$ which is the closest to the action class Q_k. To do this, (i^*, j^*) is selected as

$$\{i^*, j^*\} = \arg\min_{i,j} \left[\min_k d\left(X_{i,j}^+, Q_k\right) - \min_{l,m} d\left(X_{i,j}^+, X_{l,m}^-\right) \right] \qquad (6.28)$$

where $d(X, Y)$ is a distance function:

$$d(X, Y) = \sum_{c=\{c_{1\times1\times t1}^{\mathrm{SIFT}}, c_{1\times1\times t1}^{\mathrm{TTD}}\}} D_c\left(H_X^c, H_Y^c\right) \qquad (6.29)$$

where $D_c(H_X^c, H_Y^c)$ is given by (6.27). $\{c_{1\times1\times t1}^{\mathrm{SIFT}}, c_{1\times1\times t1}^{\mathrm{TTD}}\}$ are the bag-of-words histogram for both the average SIFT and trajectory transition descriptor, obtained using a $1 \times 1 \times t1$ grid representation. After adding X_{i^*,j^*} to Q_k, the k-nearest neighbour classifier is automatically updated. In order to reduce the risk of including false positive cuboids into the positive training set Q_k, one can take a very conservative approach. That is, from each positive clip, only one positive cuboid is selected. Specifically, all other cuboids $X_{i^*,j\neq j^*}^+$ from video i^* are removed from set $X_{i,j}^+$ before the next iteration.

Figure 6.13 shows some examples of action detection and recognition using the automated annotation training model. It is evident that action recognition in crowded scenes is challenging due to constant movement of people in the background. An action of interest is easily confused visually with some of the background actions leading to both mis-detection and false recognition. Siva and Xiang (2010) show that a model trained with minimal manual annotation can still achieve comparable recognition accuracy to that from a model learned with full annotation, and that it outperforms existing multi-instance learning methods.

Fig. 6.13 Recognising people sitting down and standing up on a crowded underground platform: *Green*—ground truth, *blue*—detected and recognised actions. (**a**)–(**d**) True recognition. (**e**) and (**g**) Mis-detection. (**f**) and (**h**) False recognition

6.5 Discussion

For automatic recognition of human actions in a visual environment, a model needs to not only classify action image sequences into a set of known categories, but also locate and detect actions of interest. For action classification, a model typically makes the assumption that the action of interest is located and centred in the viewpoint, with minimal distraction from other objects in the background. Two types of action classification techniques are considered depending on the choice of representation given the viewing condition. When a person can be reliably segmented from a visual scene, the actions of this person can be represented by silhouette shapes. When extracting a silhouette becomes difficult with increasing background clutter, greater viewing distance and a non-stationary viewpoint, a more robust part-based representation is required. To that end, space-time interest points can be exploited to extract both local spatio-temporal features and global distribution features for representing action sequences. When human actions in a crowded scene are observed from a distance, an action of interest can well be immersed into the background activities. Under such viewing conditions, action recognition is critically dependent on locating and detecting actions of interest with sufficient accuracy. A space-time action cuboid can be designed to serve as a sliding search window for locating and recognising certain types of actions from continuous visual observations of an unconstrained scene.

Overall, the focus in Part II is on the problems of visual interpretation of single object behaviours, either in isolation or in the presence of others. In particular, we consider building plausible and tractable computational models for the representation and recognition of human behaviours, ranging from facial expression, hand gesture, body affective state, to whole body action. In Part III, our focus shall be shifted to the problem of modelling group behaviours. When there are interactions between objects in a group, automatic modelling and interpretation of their behaviour become more challenging, as localised actions and activities of each individual in isolation are no longer the only factors that define the characteristics of the overall group behaviour. This brings about some fundamental changes in the way how visual information about a behaviour is represented and modelled.

References

Ahmad, M., Lee, S.: HMM-based human action recognition using multiview image sequences. In: International Conference on Pattern Recognition, pp. 263–266 (2006)

Andrews, S., Tsochantaridis, I., Hofmann, T.: Support vector machines for multiple-instance learning. In: Advances in Neural Information Processing Systems, pp. 561–568 (2003)

Bach, F., Lanckriet, G., Jordan, M.I.: Multiple kernel learning, conic duality, and the SMO algorithm. In: International Conference on Machine Learning, Banff, Canada (2004)

Bishop, C.M.: Pattern Recognition and Machine Learning. Springer, Berlin (2006)

Blank, M., Gorelick, L., Shechtman, E., Irani, M., Basri, R.: Actions as space-time shapes. In: IEEE International Conference on Computer Vision, pp. 1395–1402 (2005)

Bregonzio, M., Gong, S., Xiang, T.: Recognising action as clouds of space-time interest points. In: IEEE Conference on Computer Vision and Pattern Recognition, Miami, USA, June 2009, pp. 1948–1955 (2009)

Brzezinski, J.R.: Logistic regression modeling for context-based classification. In: International Workshop on Database and Expert Systems Applications, Florence, Italy, September 1999, pp. 755–759 (1999)

Byers, S., Raftery, A.: Nearest-neighbour clutter removal for estimating features in spatial point processes. J. Am. Stat. Assoc. **93**(442), 577–584 (1998)

Chapelle, O., Haffner, P., Vapnik, V.: Support vector machines for histogram-based image classification. IEEE Trans. Neural Netw. **10**(5), 1055–1064 (1999)

Dalal, N., Triggs, B.: Histograms of oriented gradients for human detection. In: IEEE Conference on Computer Vision and Pattern Recognition, pp. 886–893 (2005)

Decety, J., Grezes, J., Costes, N., Perani, D., Jeannerod, M., Procyk, E., Grassi, F., Fazio, F.: Brain activity during observation of actions. Brain **120**, 1763–1777 (1997)

Dollár, P., Rabaud, V., Cottrell, G., Belongie, S.: Behavior recognition via sparse spatio-temporal features. In: IEEE International Workshop on Visual Surveillance and Performance Evaluation of Tracking, pp. 65–72 (2005)

Felzenszwalb, P.F., Girshick, R.B., McAllester, D., Ramanan, D.: Object detection with discriminatively trained part-based models. IEEE Trans. Pattern Anal. Mach. Intell. **32**(9), 1627–1645 (2010)

Figueiredo, M.A.T., Jain, A.K.: Unsupervised learning of finite mixture models. IEEE Trans. Pattern Anal. Mach. Intell. **24**(3), 381–396 (2002)

Freeman, H.: On encoding arbitrary geometric configurations. IRE Trans. Electron. Comput. **10**(2), 260–268 (1961)

Fukunaga, K.: Introduction to Statistical Pattern Recognition. Academic Press, San Diego (1990)

Gehler, P., Nowozin, S.: On feature combination for multiclass object classification. In: IEEE International Conference on Computer Vision, Kyoto, Japan, October 2009, pp. 221–228 (2009)

Gong, S., Walter, M., Psarrou, A.: Recognition of temporal structures: Learning prior and propagating observation augmented densities via hidden Markov states. In: IEEE International Conference on Computer Vision, Corfu, Greece, pp. 157–162 (1999)

Gunawardana, A., Mahajan, M., Acero, A., Platt, J.C.: Hidden conditional random fields for phone classification. In: International Conference on Speech Communication and Technology, pp. 1117–1120 (2005)

HOSDB: Imagery library for intelligent detection systems (i-LIDS). In: The Institution of Engineering and Technology Conference on Crime and Security, pp. 445–448 (2006)

Ke, Y., Sukthankar, R., Hebert, M.: Event detection in crowded videos. In: IEEE International Conference on Computer Vision, Rio de Janeiro, Brasil, October 2007, pp. 1–8 (2007)

Lafferty, J., McCallum, A., Pereira, F.: Conditional random fields: probabilistic models for segmenting and labeling sequence data. In: International Conference on Machine Learning, pp. 282–289 (2001)

Laptev, I., Marszalek, M., Schmid, C., Rozenfeld, B.: Learning realistic human actions from movies. In: IEEE Conference on Computer Vision and Pattern Recognition, pp. 1–8 (2008)

Lee, H., Kim, J.H.: An HMM-based threshold model approach for gesture recognition. IEEE Trans. Pattern Anal. Mach. Intell. **21**(10), 961–973 (1999)

Liu, J.G., Luo, J.B., Shah, M.: Recognizing realistic actions from videos 'in the wild'. In: IEEE Conference on Computer Vision and Pattern Recognition, pp. 1996–2003 (2009)

Lowe, D.: Distinctive image features from scale-invariant keypoints. Int. J. Comput. Vis. **2**(60), 91–110 (2004)

Niebles, J., Wang, H., Fei-Fei, L.: Unsupervised learning of human action categories using spatial-temporal words. Int. J. Comput. Vis. **79**(3), 299–318 (2008)

Parker, J.: Algorithms for Image Processing and Computer Vision. Wiley Computer, New York (1997)

Quattoni, A., Wang, S., Morency, L.P., Collins, M., Darrell, T.: Hidden-state conditional random fields. Technical report, MIT (2006)

Rissanen, J.: A universal prior for integers and estimation by minimum description length. Ann. Stat. **11**(2), 417–431 (1983)

Savarese, S., Pozo, A.D., Niebles, J., Fei-Fei, L.: Spatial-temporal correlations for unsupervised action classification. In: IEEE Workshop on Motion and Video Computing, Copper Mountain, USA, January 2008, pp. 1–8 (2008)

Schüldt, C., Laptev, I., Caputo, B.: Recognizing human actions: a local SVM approach. In: International Conference on Pattern Recognition, pp. 32–36 (2004)

Siva, P., Xiang, T.: Action detection in crowd. In: British Machine Vision Conference, Aberystwyth, UK, September 2010

Sminchisescu, C., Kanaujia, A., Metaxas, D.: Conditional models for contextual human motion recognition. Comput. Vis. Image Underst. **104**(2–3), 210–220 (2006)

Sonnenburg, S., Rätsch, G., Schäfer, C., Schölkopf, B.: Large scale multiple kernel learning. J. Mach. Learn. Res., **7**, 1531–1565 (2006)

Sun, J., Wu, X., Yan, S., Cheong, L., Chua, T., Li, J.: Hierarchical spatio-temporal context modeling for action recognition. In: IEEE Conference on Computer Vision and Pattern Recognition, Miami, USA, June 2009, pp. 2004–2011 (2009)

Sutton, C., McCallum, A.: An introduction to conditional random fields for relational learning. In: Getoor, L., Taskar, B. (eds.) Introduction to Statistical Relational Learning, pp. 134–141. MIT Press, Cambridge (2007)

Viola, P., Jones, M.: Robust real-time face detection. Int. J. Comput. Vis. **57**(2), 137–154 (2004)

Viola, P., Jones, M., Snow, D.: Detecting pedestrians using patterns of motion and appearance. In: IEEE International Conference on Computer Vision, Nice, France, October 2003, pp. 734–741 (2003)

Wang, S., Quattoni, A., Morency, L., Demirdjian, D., Darrell, T.: Hidden conditional random fields for gesture recognition. In: IEEE Conference on Computer Vision and Pattern Recognition, pp. 1521–1527 (2006)

Zhang, J., Gong, S.: Action categorisation with modified hidden conditional random field. Pattern Recognit. **43**(1), 197–203 (2010)

Part III
Group Behaviour

Chapter 7
Supervised Learning of Group Activity

In Part II, different computational models for visual analysis of behaviours of a single object are considered. In a public space, actions of individuals are more commonly observed as elements of group activities, and are likely to involve multiple objects interacting or co-existing in a shared common space. Examples of group activities include 'people playing football' and 'shoppers checking out at a supermarket'. Group activity modelling is thus concerned with modelling not only the actions of individual objects in isolation, but also the interactions and causal relationships among individual actions. The goal of automatic group activity modelling for interpretation is to infer the semantic meaning of an activity by extracting and computing higher level information from visual observation.

We consider a continuous video stream input from visual observation of a wide-area scene, where activities of multiple objects occur over time. A supervised learning model for activity recognition aims to first automatically segment temporally the video stream into plausible activity elements, followed by constructing a model from the observed visual data so far for describing different categories of activities, and recognising a new instance of activity by classifying it into one of the known categories. To that end, the following three problems need be addressed:

1. How to select visual features that best represent activities of multiple interacting objects with occlusions between each other. Such a representation should facilitate both activity modelling and activity based segmentation of video.
2. How to perform automatic video segmentation according to the activities occurring in the scene so that each video segment ideally contains a single category of activities.
3. How to model the temporal and causal correlations among objects whose actions are considered to form meaningful group activities. Both the structure and parameters of such a model should be learned from data. With such a model, a semantic description of activities can be inferred. Interpretation of the behaviours of each individual object can also be made more accurate.

A computational process for activity segmentation and recognition relies heavily on a suitable representation of activities. We consider a contextual event-based activity representation for this purpose. Given this representation, two different methods

S. Gong, T. Xiang, *Visual Analysis of Behaviour*,
DOI 10.1007/978-0-85729-670-2_7, © Springer-Verlag London Limited 2011

for activity based video segmentation are studied. Suppose that a sufficient number of labelled activity training samples are available, supervised learning of an activity model for each category of activity is considered. In particular, a dynamic Bayesian network (DBN) is exploited, with its model structure automatically learned from visual observations.

7.1 Contextual Events

Contextual events are defined as significant scene changes characterised by the location, shape and direction of these changes (Xiang and Gong 2006). They are object-independent and location specific in a visual environment. Contextual events are also autonomous, in that both the number of these events and their whereabouts in a scene are determined automatically from learning without pre-defined top-down rules and specifications. It should be noted that although events are detected in each image frame, each event is represented by and estimated based on accumulated visual changes over time, measured by pixel-change-history (Gong and Xiang 2003a).

The advantage of using a contextual event-based representation is that it enables a model to represent activity without the need for object tracking and trajectory matching. Events are detected and classified by unsupervised clustering using mixture of Gaussian with automatic model order selection. A group activity is thus represented as a group of co-occurring events and modelled through the interpretation of the temporal and causal correlations among different classes of events. This event-based representation is easier and more robust to compute as compared to a trajectory-based representation. In particular, it is more suitable for modelling complex group activities when object tracking is unreliable.

7.1.1 Seeding Event: Measuring Pixel-Change-History

The first step towards computing a contextual event-based representation is to detect visual changes in a scene. Adaptive mixture background models are commonly used to memorise and maintain the background pixel distribution of a dynamic scene in order to detect visual changes (McKenna et al. 2000; Ng and Gong 2003; Stauffer and Grimson 1999). The strength of such a model is its potential to cope with persistent movements of background objects, such as waving tree leaves, given an appropriate model parameter setting. However, an adaptive mixture background model cannot differentiate, although may still be able to detect, the presence of pixel-level changes of different temporal scales. In general, a pixel-level change of different temporal scales can have different levels of significance in its semantics:

1. A short-term change is most likely to be caused by instant moving objects, such as passing-by people or vehicles.
2. A medium-term change is most likely to be caused by localised moving objects, such as a group of people standing and talking to each other.

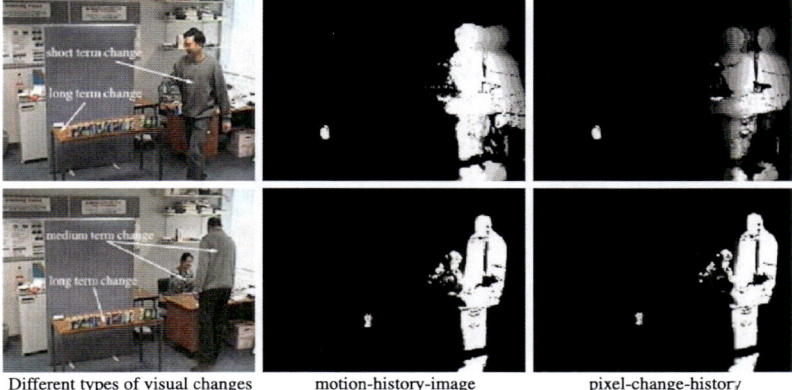

Different types of visual changes   motion-history-image   pixel-change-history
(a) Examples of motion-history-image and pixel-change-history images.

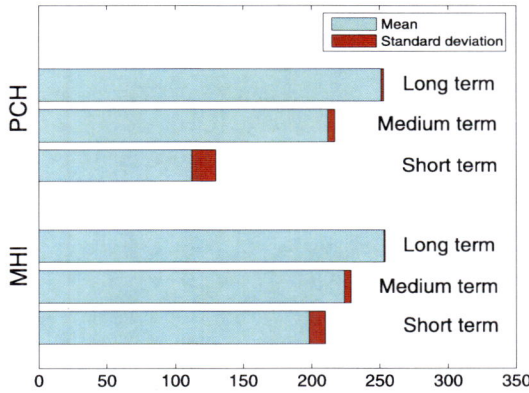

(b) The mean and standard deviation values of motion-history-image
and pixel-change-history for different types of visual changes.

Fig. 7.1 Comparing motion-history-image (Bobick and Davis 2001) with pixel-change-history (Xiang and Gong 2006). (**a**) Examples of motion-history-image and pixel-change-history-images from an indoor scene. (**b**) The mean and standard deviation of motion-history-image and pixel-change-history are shown for different types of visual change including short-, medium- and long-term change. Pixel-change-history exhibits greater discriminative power for differentiating short- from medium- and long-term changes

3. A long-term change is most likely to be caused by either the introduction of novel static objects into a scene, or the removal of existing objects from the scene, for example, a piece of furniture is moved in the background or a car is parked in a car park.

Examples of visual changes of different temporal scales are shown in Fig. 7.1(a). Ideally, a representation of pixel-level visual change should be a single, unified multi-scale temporal representation capable of differentiating changes of different scale and rate at the pixel level. An example of such a representation is the pixel-change-history for measuring multi-scale temporal changes at each pixel (Gong and Xiang 2003a; Xiang and Gong 2006). More precisely, the pixel-change-history of a pixel is defined as

$$P_{\varsigma,\tau}(x,y,t) = \begin{cases} \min(P_{\varsigma,\tau}(x,y,t-1) + \frac{255}{\varsigma}, 255), & \text{if } D(x,y,t) = 1 \\ \max(P_{\varsigma,\tau}(x,y,t-1) - \frac{255}{\tau}, 0), & \text{otherwise} \end{cases} \qquad (7.1)$$

where $P_{\varsigma,\tau}(x,y,t)$ is the pixel-change-history for a pixel at (x,y), $D(x,y,t)$ is a binary image of foreground pixels computed by subtracting the current image frame from a background model,[1] ς is an accumulation factor and τ is a decay factor. When $D(x,y,t) = 1$, instead of jumping to the maximum value, the value of a pixel-change-history increases gradually according to the accumulation factor ς. When no significant pixel-level visual change is detected at a particular location (x,y) in the current frame, pixel (x,y) will be treated as part of the background scene. The corresponding pixel-change-history starts to decay, and the speed of decay is controlled by a decay factor τ. The accumulation factor and the decay factor give a model the flexibility of characterising pixel-level changes over time. In particular, large values of ς and τ indicate that the history of visual change at (x,y) is considered over a longer backward temporal window. In the meantime, the ratio between ς and τ determines how much bias is put on the recent change.

There are a number of alternative techniques for representing pixel-level visual changes. One of them is the motion-history-image (Bobick and Davis 2001). A motion-history-image is less expensive to compute by keeping a history of temporal changes at each pixel location which then decays over time. Motion-history-images have been used to build holistic motion templates for the recognition of human actions (Bobick and Davis 2001) and for object tracking (Piater and Crowley 2001). An advantage of motion-history-image is that although it is a representation of the history of pixel-level change, only one image frame needs be stored. However, at each pixel location, explicit information about its past is also lost in a motion-history-image. The current changes are updated to the model with their corresponding motion-history-image values jumping to the maximum value. A motion-history-image can be considered as a special case of a pixel-change-history-image in that these two are identical when ς in (7.1) is set to 1 for a pixel-change-history-image. Figure 7.1 shows an example of how pixel-change-history can be used for representing visual changes of different temporal scales. Figure 7.1(b) shows that the mean value of pixel-change-history of pixels corresponding to short-term changes caused by, for example, a person passing by, is significantly lower than those for medium- and long-term changes introduced by, for instance, a shopper paying or a drink can being removed. This provides a good measurement for discriminating different types of visual changes. Its advantage over motion-history-image is also illustrated in Fig. 7.1(b).

---

[1]A time-varying background model can be either maintained adaptively using a mixture of Gaussian functions (Stauffer and Grimson 1999), or estimated from combinatorial optimisation labelling (Russell and Gong 2006).

7.1.2 *Classification of Contextual Events*

Given a binary map from detected pixel changes, the connected component procedure (Ballard and Brown 1982) can be utilised to group those changed pixels into
connected groups. Small groups are then removed by a size filter. For each group
a mean pixel-change-history value is computed. Those groups with a mean change
value greater than a threshold T_B are referred to as 'salient pixel groups' and are
considered for constructing contextual events. An event is represented by a seven-
dimensional feature vector extracted from each salient pixel group as follows:

$$\mathbf{v} = [\bar{x}, \bar{y}, w, h, R_m, M_p x, M_p y] \qquad (7.2)$$

where $(\bar{x}, \bar{y})$ is the centroid of the salient pixel group, (w, h) are the width and
height of the salient pixel group, R_m represents the percentage of foreground pixels $D(x, y, t)$ in the group (7.1), and $(M_p x, M_p y)$ are a pair of first order moments of the pixel-change-history-image within the salient pixel group. Among
these features, $(\bar{x}, \bar{y})$ are location features, (w, h) are shape features, R_m is a visual change type feature and $(M_p x, M_p y)$ are motion features capturing the direction of object motion direction.[2] Despite that these salient pixel groups are defined
in 2D image frames, they are essentially local space-time clusters grouped by a co-
occurrence criterion through thresholding the mean pixel-change-history value from
each group, whilst each group is computed from pixel-wise spatio-temporal change
information. Salient pixel groups can be also defined in a space-time volume. For
instance, one could adopt a method considered by Greenspan et al. (2004), or compute salient events over a spatio-temporal volume using multi-scale entropy ratio
over space and time (Hung and Gong 2004; Kadir and Brady 2001). However, a
spatio-temporal volume based event representation is always computationally more
expansive, and may not be tractable given the complexity of the activities captured
in video footages.

In essence, it can be considered that a salient pixel group is similar to a space-
time interest point local patch or volume used for gesture and action representation.
Furthermore, a contextual event computed from each salient pixel group as a local
feature vector defined by (7.2) is also similar to a descriptor of local features extracted from a local space-time volume as defined by (6.13). The difference is that
they use different criteria for measuring what is spatio-temporally 'interesting' at
each pixel location (Bregonzio et al. 2009; Dollár et al. 2005).

Salient pixel groups are clustered unsupervised into different contextual events in
the seven-dimensional feature space using a mixture of Gaussian model (Sect. 5.1).
The model parameters of a mixture of Gaussian model is commonly learned using
the EM algorithm, whilst the number of Gaussian components and their mixture
probabilities, known as the model order, can be determined using the Bayesian information criterion (BIC) (Schwarz 1978).

---

[2]Similar to the motion-history-image (Bobick and Davis 2001), a pixel-change-history implicitly
represents the direction of movement. The first order moment from the pixel-change-history value
distribution within the bounding box estimates the direction of movement quantitatively.

Specifically, given a training data set $\mathbf{O}$ of salient pixel groups from some training image sequences, the optimal model order $\hat{k}$ is defined as the most likely number of different event classes. This number is selected automatically by the BIC from a set of K competing models $\mathbf{m}_k$ parameterised by $\boldsymbol{\theta}_{\mathbf{m}_k}$ where $k \in \{1, \ldots, K\}$. This BIC based model order selection is formulated as

$$\hat{\mathbf{m}}_k = \arg\min_{\mathbf{m}_k}\left\{ -\log P(\mathbf{O}|\mathbf{m}_k, \hat{\boldsymbol{\theta}}_{\mathbf{m}_k}) + \frac{D_k}{2}\log N \right\} \tag{7.3}$$

where $\hat{\boldsymbol{\theta}}_{\mathbf{m}_k} = \arg\max_{\boldsymbol{\theta}_{\mathbf{m}_k}}\{P(\mathbf{O}|\mathbf{m}_k, \boldsymbol{\theta}_{\mathbf{m}_k})\}$ is the maximum likelihood estimation of $\boldsymbol{\theta}_{\mathbf{m}_k}$, D_k is the dimensionality of $\boldsymbol{\theta}_{\mathbf{m}_k}$, and N is the size of the training data set. The model order is the number of mixture components k. If k ranges from 1 to K for the candidate mixture models, the optimal model order $\hat{k}$ estimated by the BIC is given by

$$\hat{k} = \arg\min_k\left\{ -\sum_{i=1}^{N}\log f\left(\mathbf{y}_i|k, \hat{\boldsymbol{\theta}}(k)\right) + \frac{D_k}{2}\log N \right\} \tag{7.4}$$

where $f(\mathbf{y}_i|k, \hat{\boldsymbol{\theta}}(k))$ is the class-conditional Gaussian density function, $\mathbf{y}_i$ is the feature vector representing one data simple, $\hat{\boldsymbol{\theta}}(k)$ are the mixture parameters estimated using the EM algorithm, and D_k is the number of parameters needed for a k-component mixture of Gaussian model. If a full covariance matrix is used, (7.4) can be re-written as

$$\hat{k} = \arg\min_k\left\{ -\sum_{i=1}^{N}\log f\left(\mathbf{y}_i|k, \hat{\boldsymbol{\theta}}(k)\right) + \frac{k-1}{2}\log N + \frac{q^2+3q}{4}k\log N \right\} \tag{7.5}$$

where q is the dimensionality of the feature space. To summarise, $\hat{k}$ estimated by (7.5) yields the most likely number of event classes given the training data set $\mathbf{O}$. Salient pixel groups detected in unseen image frames independent from the training data set are classified as one of the $\hat{k}$ event classes in the seven-dimensional feature space.

Figure 7.2(a) shows an example of computing contextual events from group activities. This scene shows aircraft frontal cargo loading and unloading activities, in which trucks, cargo-lift and cargo container boxes are moving or being moved to transfer cargoes to and from a docked aircraft on the ground. Four different classes of events are automatically detected (Fig. 7.2(b)). They are labelled as 'moving truck', 'moving cargo', 'moving cargo-lift' and 'moving truck/cargo', and shown colour-coded in Fig. 7.2. It is evident from Figs. 7.2(a) and (c) that these events correspond correctly to the four key constituents of a cargo service activity. The first three event classes correspond, respectively, to a truck, a cargo container and a cargo-lift moving into a specific location with a particular direction of motion in the image space. The last event class corresponds to any occurrence of simultaneous movements of trucks and cargo containers when they are overlapped. It is noted that different classes of events occurred simultaneously, and the model makes errors in event discovery and classification. Some of the errors are caused by unstable lighting, occlusion, and visual similarity among different events. Activity interpretation

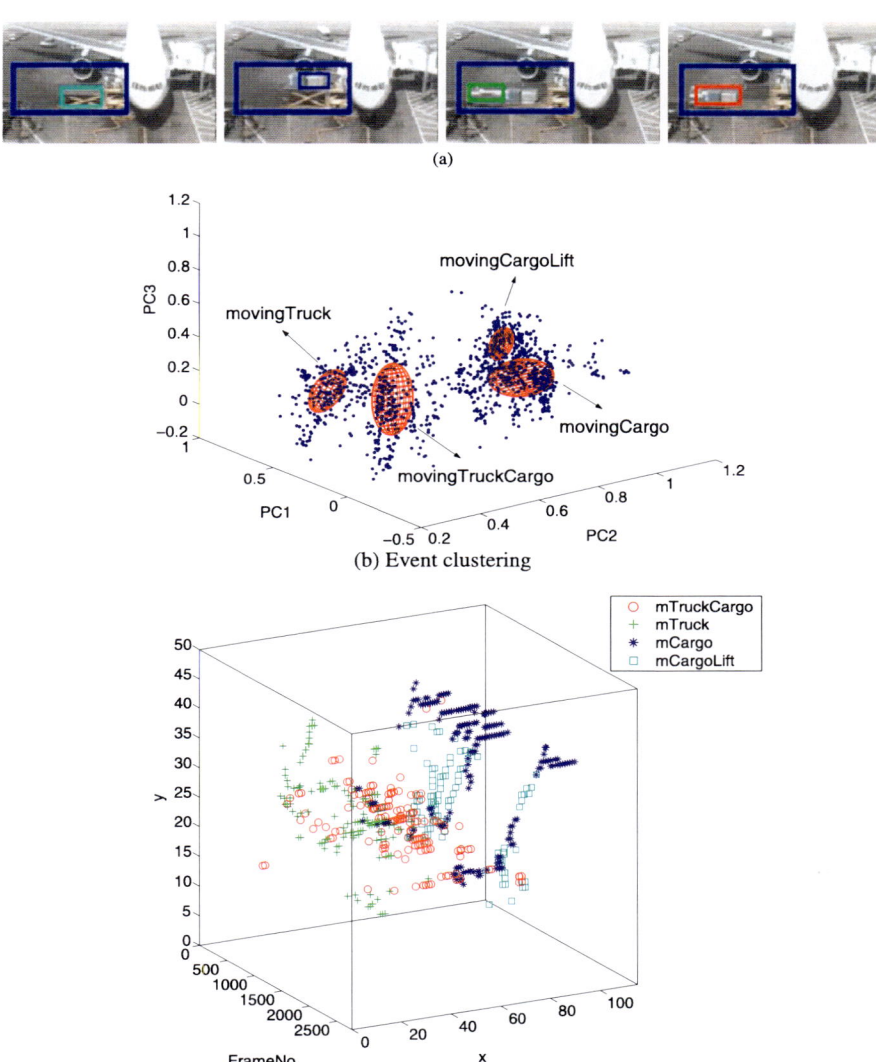

Fig. 7.2 Contextual event discovery from aircraft cargo loading and unloading activities. (**a**) Examples of detected and classified events in the image frames. Different event classes are shown by *colour-coded bounding boxes*. The cargo loading/unloading area is also highlighted using a *thick blue box* in each frame. (**b**) Unsupervised event clustering of the training set in a seven-dimensional feature space, where the first three principal components are shown for visualisation. (**c**) The whereabouts and temporal order of four event classes are auto-detected. Centroids of different event classes are depicted by different symbols

based on local event classification alone does not take into account the temporal and causal correlations among different events. Exploring such temporal and causal context is important for modelling group activity.

7.2 Activity Segmentation

In order to classify a group activity by some known categories, what constitutes a group activity needs be segmented first from a continuous video input. In general, the problems of object segmentation and classification are closely related, often posing a 'chicken-and-egg' challenge. That is, to classify an object, a model needs to segment the object from its surroundings. However, without knowing what the object is, segmentation is ill-posed. To overcome a similar problem in segmenting and locating a 2D object, a common technique is to apply a sliding 2D window that searches exhaustively in images through all the possible positions and scales of the object of interest (Dalal and Triggs 2005; Felzenszwalb et al. 2010; Viola and Jones 2004; Viola et al. 2003). This Catch-22 problem becomes harder still for segmenting groups of activities from continuous video feed when a grouping is not necessarily well-defined visually either in space or over time. The aim is to divide a video temporally into variable-length segments by activity content so that each segment may plausibly contain a complete activity circle. The problem is similar to segmenting a person's gesture (Sect. 5.2) or action from continuous image sequences, although this is now a harder problem due to the more uncertain nature of group activities involving multiple objects.

To illustrate the problem, we consider object activities involved in an aircraft docking operation. In each operation, there are typically fifteen different kinds of activities taking place in the same scene, most of which are not visually detectable (Xiang and Gong 2008). Figure 7.3 shows a typical sequence of activities from a complete aircraft docking operation. One activity can be followed by either another activity immediately, or a period of inactivity when no meaningful visual changes can be detected. The durations of these activities and inactivity gaps vary significantly during a single aircraft docking operation, and across different operations. There is no standard activity circle. There are also variations in the temporal order of activities and inactivity gaps. For example, the temporal order of 'frontal cargo service' and 'frontal catering service' appears to be arbitrary. It is noted that some activities such as 'aircraft arrival' involve a single object, whilst other activities such as 'frontal cargo service' and 'frontal catering service' consist of multiple objects, which may leave and re-appear in the scene. For the latter case, there often exist a number of short inactivity break-ups within an activity.

7.2.1 Semantic Content Extraction

For video segmentation by activity as semantic content, a video input stream is represented by contextual events detected automatically over time. The semantic content of the video is defined as both the distribution of the detected contextual events and the temporal correlations among them during different activities.

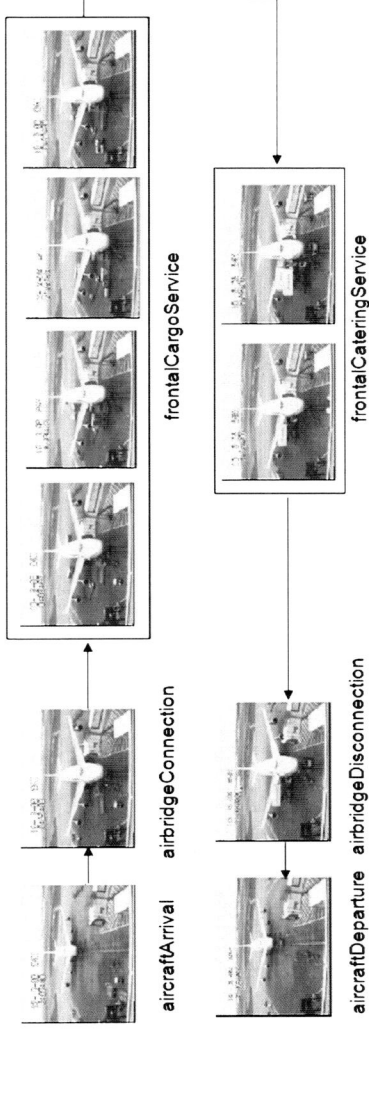

Fig. 7.3 A typical sequence of object group activities from an aircraft docking operation

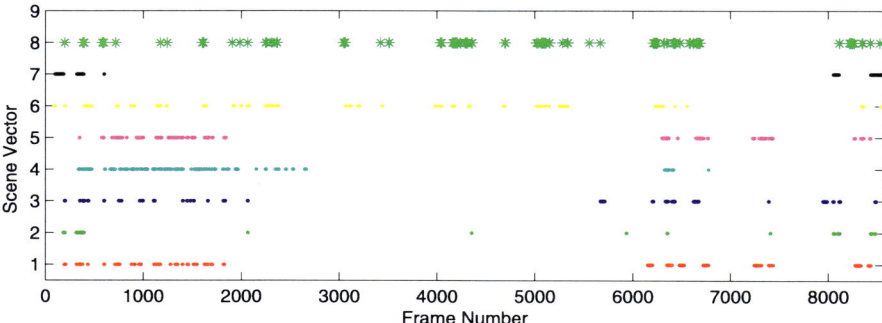

Fig. 7.4   An example of scene vectors evolving over time. There are eight different classes of contextual events detected in this video. There are frequent short inactivity break-ups within activities (frames 1 to 2000), and a long inactivity gaps between activities (frames 3000 to 6000)

Constructing a Scene Vector

In each video frame, contextual events are detected and classified into different classes using the method described in Sect. 7.1. These events can be considered as 'snapshots' of activities captured in a scene. A scene vector $\mathbf{sv}_t$ is constructed for each video frame t, and defined as

$$\mathbf{sv}_t = \left[s_t^1, \ldots, s_t^k, \ldots, s_t^{K_e} \right] \tag{7.6}$$

where K_e is the number of event classes automatically determined using BIC. The value of s_t^k is the number of events of the kth event class detected in frame t.

A scene vector gives a description of 'what is happening' in the scene through the class labels of the detected events. It is thus a concise representation of the video content at a more semantic level than simple pixel colour or motion block characteristics. However, using directly this scene vector to detect changes in content for video segmentation is not reliable. Specifically, the value of a scene vector $\mathbf{sv}_t$ can become $\mathbf{0}$ when there is absence of any event at a given frame (Fig. 7.4). This can be caused by either frequent but short inactivity break-ups within activities or long inactivities between activities. Each 'coming to zero' is reflected as a dramatic change in the value of $\mathbf{sv}_t$ due to the discrete nature of s_t^k. Genuine changes in video semantic content can be overwhelmed by these inactivity break-ups.

Cumulative Scene Vector over Time

To overcome this problem, one solution is to represent video content using a cumulative scene vector computed at frame t using $\mathbf{sv}_t$ from frame 1 to frame t. More specifically, the kth element of the cumulative scene vector (denoted as $\widetilde{\mathbf{sv}}_t$) is computed as

$$\widetilde{s}_t^k = \sum_{i=1}^t s_i^k \tag{7.7}$$

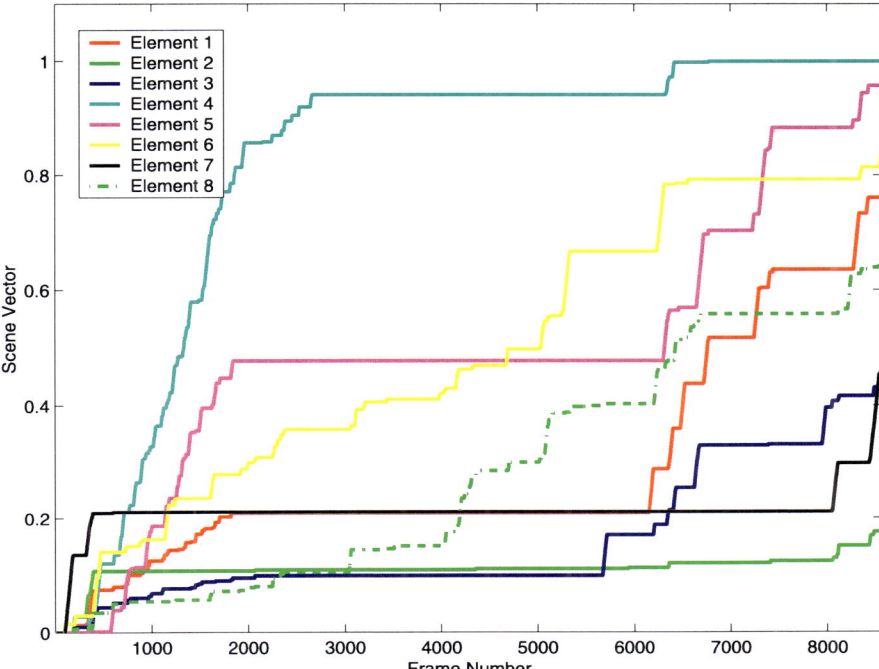

Fig. 7.5 An example of a cumulative scene vector involving over time for a video containing a complete aircraft docking operation. The normalised distributions of the eight event elements in the scene vector are shown separately

The value of each element of $\widetilde{\mathbf{sv}}_t$ increases monotonically with time (Fig. 7.5). Compared to frame-wise scene vector $\mathbf{sv}_t$ (Fig. 7.4), the short inactivity break-ups at individual frames have little impact on the values of the cumulative scene vector evolving over time. It enables more reliable detection of breakpoints that correspond to more significant changes in video content.

To illustrate the usefulness of computing cumulative scene vectors from video contextual events over time, Fig. 7.6(a) shows the evolution of the first two principal components of $\widetilde{\mathbf{sv}}_t$ of seven different video sequences, each of which contains a different and complete aircraft docking operation. Among the seven videos, Video-1 to Video-6 follow a typical sequence of activities with cargo services performed before catering services. The seventh video is different in that the cargo services were carried out after catering services, which is abnormal. It can be seen that the video content trajectory of Video-7 is distinctively different from those of the other six videos. For comparison, the same video sequences were also represented using video frame colour histograms. This image feature characteristic based representation captures 'what is present' rather is 'what is happening' in the scene. Figure 7.6(b) shows the evolution of the first two principal components of a colour histogram representation over time from the same seven videos as in Fig. 7.6(a). It is evident that the colour histogram representation is more sensitive to changes in individual frames therefore

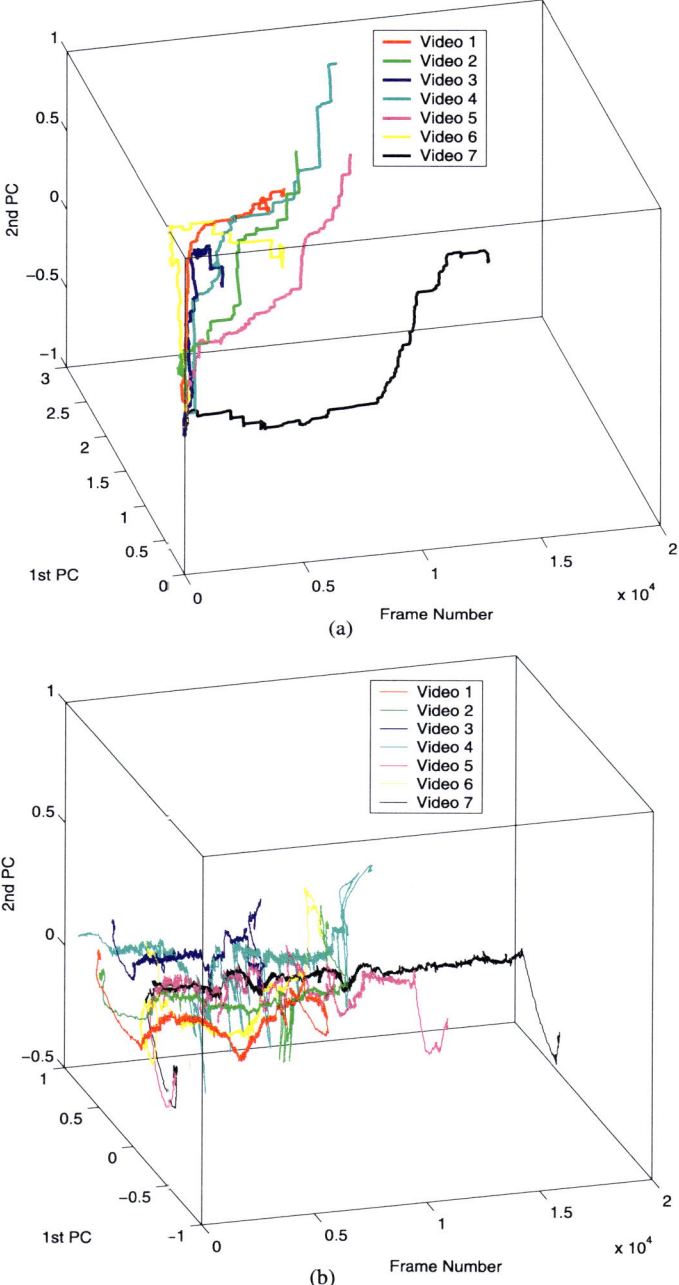

Fig. 7.6 (**a**) and (**b**) show the first two principal components of cumulative scene vectors and colour histograms evolving over time for seven different videos of aircraft docking operations. The principal component analysis is only used for visualisation

less robust compared to the cumulative scene vector representation. Only the arrival and departure of an aircraft are captured clearly by colour histogram changes at the beginning and ending of those trajectories in Fig. 7.6(b). Moreover, Video-7 is no longer distinct from other videos with the colour histogram representation.

7.2.2 Semantic Video Segmentation

A cumulative scene vector trajectory[3] is able to represent reliably video semantic content over time, where the breakpoints on a trajectory correspond to video frames of significant content change, despite variations in activity duration and occurrence of inactivity break-ups within activities. We consider the video segmentation problem as a video semantic content trajectory breakpoint detection problem. To solve this problem, we consider both off-line processing, as in the case of processing recorded videos, and on-line processing, as in the case of processing live video feeds.

Detecting Video Content Breakpoint by Dynamic Bayesian Networks

Let us first consider detecting video content breakpoints using a dynamic Bayesian network (DBN). When a DBN is adopted for segmenting video by content, the network observation variables are the cumulative scene vectors and the network hidden states correspond to the video content breakpoints which the model aims to discover. After model parameter estimation and hidden state inference given some training videos, a model is used to detect changes between hidden states when a new video is analysed by the model. These detected state changes indicate video content breakpoints that can be used for video segmentation.

Different DBNs of different topologies can be considered to model video content trajectories as statistical temporal processes. One may consider a hidden Markov model (HMM) (Fig. 7.7(a)). In an HMM, the observation variable at each time instance is the cumulative scene vector $\tilde{\mathbf{s}}\mathbf{v}_t$, with K dimensions corresponding to the number of event classes discovered from the training videos. However, if the number of event classes K is large giving a high dimensional observation variable, an HMM model requires a large number of parameters to be estimated. This will result in poor model learning unless a very large training data set is made available. To overcome this problem, a common technique is to factorise the observational space by using multiple observation variables, with each observation variable assumed to be independent from the others. Such an HMM is known as a multi-observation HMM, as shown in Fig. 7.7(b). Specifically, $O_t^{(j)}$ ($j \in \{1, \ldots, K\}$) in Fig. 7.7(b) is an one-dimensional (1D) random variable corresponding to one of the K elements of $\tilde{\mathbf{s}}\mathbf{v}_t$. Now, instead of modelling a single seven-dimensional random variable, the model only needs to deal with seven 1D random variables. The observational space

---

[3]More precisely, it is a video polyline due to the discrete nature of video frames.

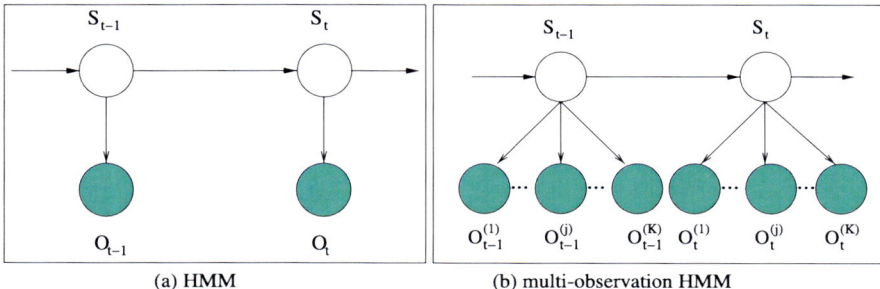

(a) HMM   (b) multi-observation HMM

Fig. 7.7 An HMM and a multi-observation HMM, with the observation variables shown as *shaded circles* and hidden states as *clear circles*

is therefore factorised based on the assumption that the occurrence of instances of different event classes is independent from each other. This assumption is mostly valid in practice.

More precisely, to automatically determine N_h, the number of hidden states corresponding to the number of plausible video content breakpoints, Schwarz's Bayesian information criterion (BIC) is applied (Schwarz 1978). This is similar to the problem in event classification (Sect. 7.1). For a model $\mathbf{m}_i$, parameterised by a C_i-dimensional vector $\boldsymbol{\theta}_{\mathbf{m}_i}$, the BIC is defined as

$$\text{BIC} = -2 \log L(\boldsymbol{\theta}_{\mathbf{m}_i}) + C_i \log N \tag{7.8}$$

where $L(\boldsymbol{\theta}_{\mathbf{m}_i})$ is the maximum likelihood under $\mathbf{m}_i$, C_i is the dimensionality of the parameters of $\mathbf{m}_i$ and N is the number of videos in a training data set. For a multi-observation HMM, $L(\boldsymbol{\theta}_{\mathbf{m}_i})$ can be written as

$$-2 \log \left\{ \sum_{S_t} \left\{ P(S_1) \prod_{t=2}^{T} P\big(S_t | \text{Pa}(S_t)\big) \prod_{t=1}^{T} \prod_{j=1}^{K} P\big(O_t^{(j)} | \text{Pa}(O_t^{(j)})\big) \right\} \right\} \tag{7.9}$$

where $\text{Pa}(S_t)$ are the parents of S_t and similarly, $\text{Pa}(O_t^{(i)})$ for observations.

The search for the optimal N_h that produces the minimal BIC value involves parameter learning. More specifically, for each candidate state number, the corresponding parameters are learned iteratively using the EM algorithm (Dempster et al. 1977). The EM algorithm consists of an expectation step (E-step) and a maximisation step (M-step):

- E-step: inferring the hidden states given the parameters estimated in the last M-step. The E-step can be implemented using an exact inference algorithm such as the junction tree algorithm (Huang and Darwiche 1996).
- M-step: updating the model parameters using the inferred hidden states in the last E-step.

The algorithm alternates between the two steps until a stopping criterion is met. After parameter learning, the BIC value is computed using (7.8) where $L(\boldsymbol{\theta}_{\mathbf{m}_i})$ has been obtained from the final M-step of EM in parameter learning. Alternatively, both parameter and structure learning can be performed within a single EM process using a structured EM algorithm (Friedman et al. 1998).

An example of automatic video segmentation using a multi-observation HMM model is shown in Fig. 7.8. A multi-observation HMM is trained to segment a videos of aircraft docking operation. The number of observation variables is eight, automatically determined by the number of event classes discovered from some training videos. The number of hidden states is also automatically discovered to be eight by the BIC. The learned state transition probabilities are shown in Fig. 7.8(a). The inferred states of the learned multi-observation HMM are shown in Fig. 7.8(b). The video is segmented by the inferred content breakpoints from the model (Fig. 7.8(c)). To validate the segmentation, a manually labelled ground-truth description of the video activities is also shown in Fig. 7.8(d).

Video Content Trajectory Segmentation by Forward–Backward Relevance

The inference and learning of a multi-observation HMM is computational expensive, making it mostly only suitable for off-line processing. For on-line processing of live video feeds, we consider a forward–backward-relevance model (Xiang and Gong 2008). At each time instance, the model estimates the relevance of one vertex on a video content trajectory with respect to both the starting point of the current video segment along the content trajectory (looking backward) and a short distance ahead on the trajectory (looking forward). The relevance of this vertex is minimal when these three vertices are collinear. When the video content trajectory turns significantly at a vertex, a high relevance value is expected. This model is motivated by the discrete-curve-evolution segmentation algorithm (Latecki and Lakamper 1999). It aims to overcome the over-segment tendency of a conventional on-line sliding-window model.

Discrete-curve-evolution is designed for 2D shape decomposition and can be extended for detecting video content trajectory breakpoints. The algorithm starts with the finest possible segmentation, namely $T - 1$ segments for a trajectory of length T considering every time frame is a breakpoint, and progresses with two segments being merged at each step. Specifically, the cost of merging all adjacent pairs of segments is computed and the pair with the minimal cost is merged. This process is performed iteratively until a stopping criterion is met. Discrete-curve-evolution computes the merging cost as a relevance measure R. This is computed for every vertex v, given three points including v and its two neighbour vertices u and w, as follows:

$$R(u, v, w) = d(v, u) + d(v, w) - d(u, w) \tag{7.10}$$

where function $d(\cdot)$ can be any distance metric or similarity measure, such as the Euclidean distance. In each iteration, the vertex with minimum R is deleted until the minimum R exceeds a threshold, or when a user-specified number of vertices are left in the trajectory. The remaining vertices correspond to the most likely video content breakpoints. Discrete-curve-evolution is similar in spirit to bottom-up segmentation using piece-wise linear approximation for time series data analysis (Keogh 2001). However, it is much faster to compute compared to bottom-up segmentation by

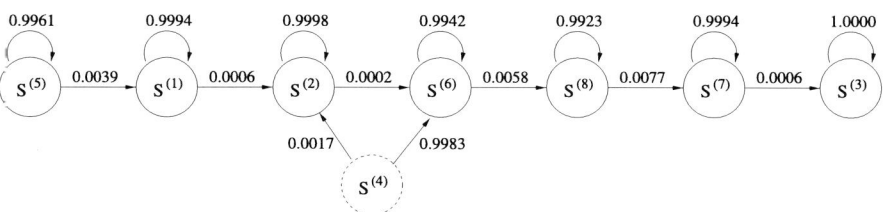

(a) Learned transition probabilities of the hidden states of a multi-observation HMM

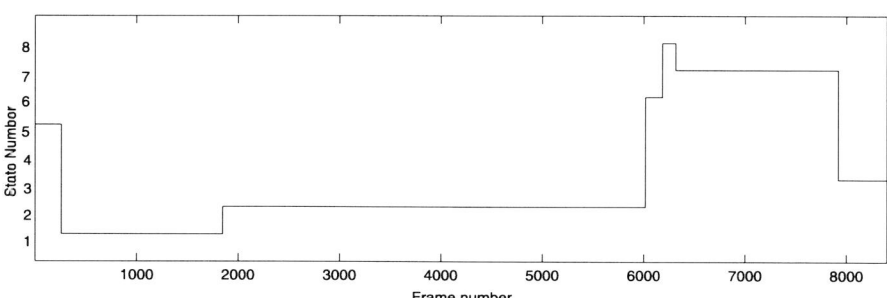

(b) Inferred states of the multi-observation HMM

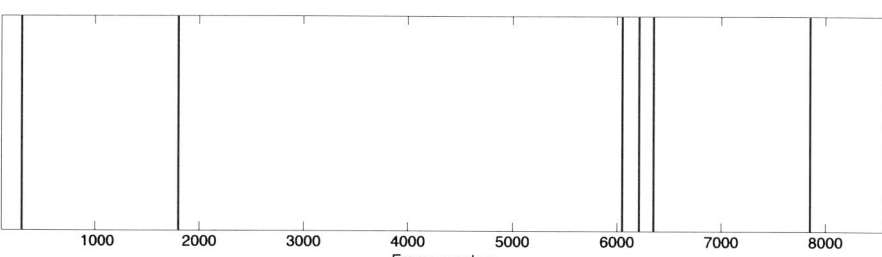

(c) Automatic video segmentation by the multi-observation HMM model

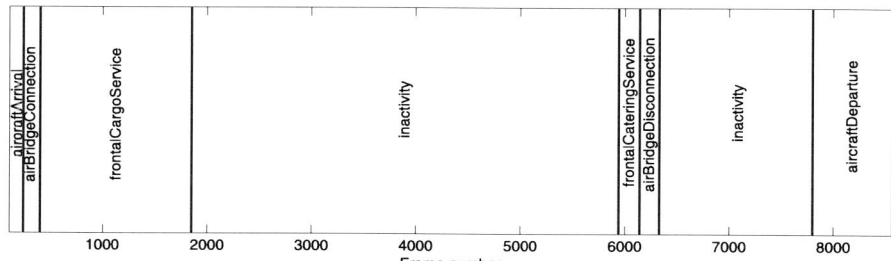

(d) Manually labelled ground-truth for video content breakpoints

Fig. 7.8 An example of video segmentation by automatic video content breakpoint detection using a multi-observation HMM (Xiang and Gong 2008)

Algorithm 7.1: A forward–backward-relevance algorithm for on-line video segmentation by semantic content, where $R(t - L_{\min}/2)$ is computed by (7.10) with the function $d(\cdot)$ defined as the Euclidean distance

---

input: A K dimensional trajectory of length T: $\widetilde{sv}_t$
output: Detected breakpoints on the trajectory: breakPoint[ ]

1 breakpointNo $= 0$;
2 **while** $t < T$ **do**
3   **if** $t > ($breakPoint[breakpointNo]$+ L_{\min})$ **then**
4   compute the relevance of the $t - L_{\min}/2$th vertex as
           $R(t - L_{\min}/2) = R($breakPoint[breakpointNo]$, t - L_{\min}/2, t)$;
5   **if** $R(t - L_{\min}/2) >$ Th **then**
6   breakpointNo $=$ breakpointNo $+ 1$;
7   breakPoint[breakpointNo] $= t - L_{\min}/2$;
8   **end**
9   **end**
10 **end**

---

either linear interpolation or linear regression, especially when the dimensionality of a trajectory space is high.

For on-line video segmentation, instead of computing the relevance of each vertex iteratively, the relevance of only one vertex is computed at each time instance when a new video frame has arrived. Two parameters, $L_{\min}$ and Th need be learned from a training data set. To that end, discrete-curve-evolution is performed on the training data. $L_{\min}$ is set to half of the length of the shortest segment in the training data. Th is computed as the average relevance value of the breakpoints detected by discrete-curve-evolution. An on-line model is outlined in Algorithm 7.1.

This on-line algorithm for video segmentation by semantic content has the following characteristics:

1. The computation complexity is independent of a video segment length and much lower than a sliding-window algorithm and its variations (Keogh 2001). For each new data point, instead of fitting the trajectory to a straight line, forward–backward-relevance computes a relevance measure of a single vertex. This requires computing only three Euclidean distances.
2. The representation is very compact. At each time instance, only the coordinates of $L_{\min}/2 + 1$ vertices need be stored for computation. This makes the algorithm suitable for analysing open-ended long video streams from 24/7 surveillance videos,[4] where at times very infrequent changes in video content occur.
3. Compared to a sliding-window algorithm and its variations, the forward–backward-relevance algorithm is more robust to noise. This is because the model

---

[4]24/7 stands for 24 hours a day, 7 days a week round-the-clock without interruption.

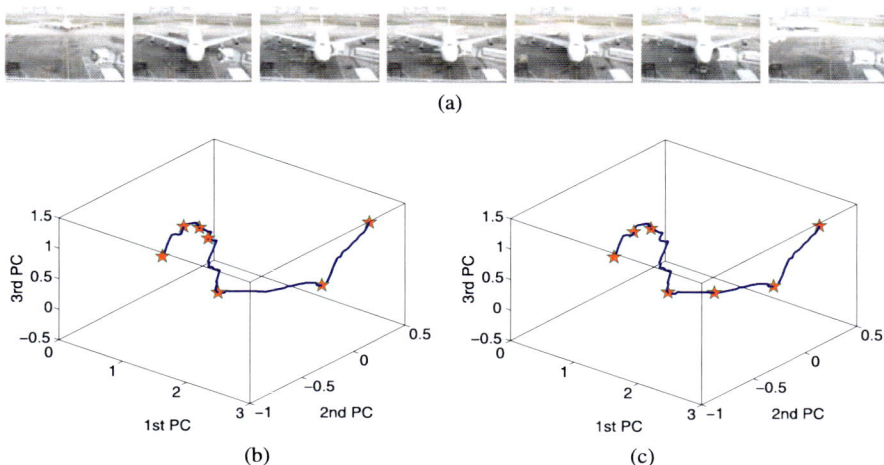

Fig. 7.9 An example of video content breakpoint detection using the forward–backward-relevance algorithm (Xiang and Gong 2008). (**a**) The video frames corresponding to the detected video content breakpoints. (**b**) Detected breakpoints on the video content trajectory. (**c**) The ground-truth on video content breakpoints from manual labelling for comparison

takes a more holistic view of a scene trajectory with relevance measure both forward and backward.

4. This more holistic approach to measuring forward–backward-relevance has a $L_{min}/2$ delay in detecting video content breakpoints. In practice, this is insignificant compared of the length of a typical video segment between content breakpoints.

Figure 7.9 shows an example of breakpoint detection using the forward–backward-relevance algorithm. Figure 7.9(b) shows the video content trajectory in a 3D principal component analysis space for a complete aircraft docking operation. Video content is computed by cumulative scene vectors (see examples in Fig 7.6). There are seven detected breakpoints on this video content trajectory, with the trajectory evolving from left to right. Figure 7.9(a) shows those video frames corresponding to the detected content breakpoints. Xiang and Gong (2008) show that a multi-observation HMM based model achieves higher accuracy on video segmentation by semantic content compared to the forward–backward-relevance algorithm. However, the former is computationally costly and only suitable for off-line analysis. On the other hand, the forward–backward-relevance model is fast to compute whilst improves the discrete-curve-evolution algorithm when applied to video content trajectory segmentation.

Video segmentation by detecting object activity patterns as semantic content addresses the more general problem of 'bridging the semantic gap' (Enser and Sandom 2003; Grosky and Zhao 2001; Naphade et al. 2002). To represent and segment a video by semantic content, there is a computational gap between measurable low level image features, such as colour and motion characteristics, and quantifiable higher level concepts describing more holistically the underlying nature of the scene captured in the video, such as object behaviour or spatio-temporal event characteristics. A contextual event-based scene content representation provides a plausible model towards bridging the semantic gap between video pixels and video content representation and analysis.

7.3 Dynamic Bayesian Networks

Let us now consider the problem of modelling group activities for behavioural interpretation of multiple objects captured by dynamic scene changes in an unconstrained visual environment. A group activity is considered as a group of inter-related probabilistic temporal processes. These correlated temporal processes are described by a dynamic graph of a set of discrete states correlated in space and over time in a state space that can be modelled by a dynamic Bayesian network (DBN). Such a probabilistic dynamic graph describes the spatio-temporal inter-relationships among different events at the overall scene level, therefore provides a mechanism for dynamic scene analysis over time.

7.3.1 Correlations of Temporal Processes

The states of a DBN are linked by a set of probabilistic causal and temporal connections referred to as the structure of the model. Without a priori top–down knowledge, the model requires a computational process to both learn the number of states through unsupervised clustering of some labelled training data, and discover the model structure by factorising the underlying state space of these labelled training data. To solve this problem, we consider a dynamically multi-linked hidden Markov Model (HMM) (Gong and Xiang 2003b; Xiang and Gong 2006). This model aims to discover and connect dynamically only a subset of relevant hidden state variables across multiple probabilistic temporal processes representing different events in a dynamic scene.

Learning Model Structure

Learning a dynamically multi-linked HMM involves estimating both its structure and parameters from data. The structure of a DBN refers primarily to (1) the number of hidden states of each hidden variables of a model, and (2) the conditional independence structure of a model, that is, factorisation of the state space for determining the topology of a probabilistic graph. The factorisation reduces the number of unnecessary parameters and caters for better network structure discovery. Using an event-based activity representation, the number of hidden temporal processes in a dynamically multi-linked HMM corresponds to the number of event classes which is automatically discovered through unsupervised clustering (Sect. 7.1).

In each temporal process, there are two states for each hidden state variable, signalling whether the corresponding class of event is visually detected at each time instance. This binary variable switches between the status of `True` and `False`. The observation variables are continuous, given by a seven-dimensional feature vector representing different event classes. Their distributions are mixtures of Gaussian with respect to the states of their discrete parent nodes. The structure learning problem for a dynamically multi-linked HMM is to determine directed probabilistic links among the temporal processes, known as the model topology. By learning the model topology, the temporal and causal correlations among different event classes are discovered.

More precisely, the aim is to learn a DBN model $\mathbf{B}$ parameterised by $(\lambda, \theta(\lambda))$ that can best explain the observed events $\mathbf{O}$. Such a best explanation, as the most probable explanation given the labelled training data, is quantified by the minimisation of a cost function. For a maximum likelihood estimation, the cost function is

$$-\log P(\mathbf{O}|\lambda, \hat{\theta}(\lambda)) \tag{7.11}$$

The negative logarithm of the probability of observing $\mathbf{O}$ by model $\mathbf{B}$ where $\hat{\theta}(\lambda)$ are the parameters for the candidate structure λ that maximise the likelihood of observing data $\mathbf{O}$ and give the most probable explanation of the data.

A maximum likelihood estimation (MLE) of the structure of $\mathbf{B}$ in the most general case results in a fully connected DBN, which implies that any class of events would possibly cause all classes of events in the future. This is computationally neither tractable nor plausible. Therefore, adding a penalty factor in the cost function to count for the complexity of a dynamic graphical model is essential for extracting meaningful and computationally tractable causal relationships. To address this problem, Bayesian information criterion (BIC) is exploited as a constrained MLE cost function for measuring the goodness of one hypothesised graphical model against that of another in describing the labelled training data.

Let us consider K competing DBN models of an activity consisting of a group of events. These models are denoted as $\mathbf{B}_k$ and parameterised by $(\lambda_k, \hat{\theta}(\lambda_k))$, where $k \in \{1, \ldots, K\}$ and $\hat{\theta}(\lambda_k)$ is a D_k-dimensional vector. Let $\mathbf{O} = (O_1, \ldots, O_T)$ be an observation sequence, where $O_t = (O_t^{(1)}, \ldots, O_t^{(N_o)})$ are the observation variables at time t, the BIC model selection criterion is computed as

$$\hat{\mathbf{B}}_k = \arg\min_{\mathbf{B}_k} \left\{ -P(\mathbf{O}|\lambda_k, \hat{\theta}(\lambda_k)) + \frac{D_k}{2} \log T \right\} \tag{7.12}$$

For learning model structure λ_k, the distributions of the detected events are used to initialise the distributions of the observation vectors. The priors and transition matrices of states are initialised randomly. The joint probability value in the BIC cost function of (7.12) is

$$P(\mathbf{O}|\lambda_k, \hat{\theta}(\lambda_k))$$
$$= \sum_{S_t^{(i)}} \left\{ \prod_{i=1}^{N_h} P(S_1^{(i)}) \prod_{t=2}^{T} \prod_{i=1}^{N_h} P(S_t^{(i)}|\mathrm{Pa}(S_t^{(i)})) \prod_{t=1}^{T} \prod_{j=1}^{N_o} P(O_t^{(j)}|\mathrm{Pa}(O_t^{(j)})) \right\} \tag{7.13}$$

where N_h is the overall number of hidden state variables and N_o is the overall number of observation variables describing all the temporal processes in a DBN model. To effectively evaluate $P(\mathbf{O}|\lambda_k, \hat{\boldsymbol{\theta}}(\lambda_k))$, an extended forward–backward algorithm, also known as Baum–Welch algorithm (Baum and Petrie 1966), is formulated in the following for a dynamically multi-linked HMM with a specific topology.

Learning Model Parameter

Consider a dynamically multi-linked HMM with C temporal processes. Each temporal process has one hidden variable and one observation variable at a time instance, giving $N_h = N_o = C$. It is assumed that all the hidden state variables are discrete and all the observation variables are continuous whose probability density functions are Gaussian with respect to each state of their parent hidden state variables. The model parameter space is given by the following.

1. The initial state distribution $\pi = \{\pi_{i^{(c)}}\}$ where $\pi_{i^{(c)}} = P(S_1^{(c)} = q_{i^{(c)}})$, $1 \leq i^{(c)} \leq N^{(c)}$ and $1 \leq c \leq C$.
2. The state transition probability distribution $A = \{a_{\text{Pa}(j^{(c)})j^{(c)}}\}$ where $a_{\text{Pa}(j^{(c)})j^{(c)}} = P(S_{t+1}^{(c)} = q_{j^{(c)}}|\text{Pa}(S_{t+1}^{(c)}) = q_{\text{Pa}(j^{(c)})})$, $\text{Pa}(S_{t+1}^{(c)})$ are the hidden variables at time t on which $S_{t+1}^{(c)}$ is conditionally dependent, $\text{Pa}(j^{(c)})$ are subscripts of those discrete values that $\text{Pa}(S_{t+1}^{(c)})$ can assume, $1 \leq j^{(c)} \leq N^{(c)}$ and $1 \leq c \leq C$.
3. The observation probability distribution $B = \{b_{i^{(c)}}(O_t^{(c)})\}$ where $b_{i^{(c)}}(O_t^{(c)}) = \mathcal{N}(O_t^{(c)}; \boldsymbol{\mu}_{i^{(c)}}, \mathbf{U}_{i^{(c)}})$, $\boldsymbol{\mu}_{i^{(c)}}$ and $\mathbf{U}_{i^{(c)}}$ are the mean vector and covariance matrix of the Gaussian distribution with respect to $S_t^{(c)} = q_{i^{(c)}}$, $1 \leq i^{(c)} \leq N^{(c)}$ and $1 \leq c \leq C$.

Given an observation sequence $\mathbf{O}$ and a model structure λ, one needs to determine the model parameters $\boldsymbol{\theta}(\lambda) = \{A, B, \pi\}$ that maximise the probability of the observation sequence given the model structure $P(\mathbf{O}|\lambda, \boldsymbol{\theta}(\lambda))$. There is no analytical solution to determine the optimal parameters given a finite observation sequence. However, the parameters can be estimated iteratively using an extended forward–backward Baum–Welch algorithm (Baum and Petrie 1966).

Let us first define the following variables:

- The forward variable, $\alpha_t(i^{(1)}, \ldots, i^{(C)}) = P(O_1, O_2, \ldots, O_t, S_t^{(1)} = q_{i^{(1)}}, \ldots, S_t^{(C)} = q_{i^{(C)}}|\lambda, \boldsymbol{\theta}(\lambda))$, is the probability of a partial observation sequence until time t with states for $S_t^{(1)}, \ldots, S_t^{(C)}$, given the model λ and $\boldsymbol{\theta}(\lambda)$. The forward variable is computed as

$$\begin{cases} \prod_{c=1}^C \pi_{i^{(c)}} b_{i^{(c)}}(O_t^{(c)}), & \text{if } t = 1 \\ \prod_{c=1}^C \left(\left(\sum_{j^{(1)}, \ldots, j^{(C)}} \alpha_{t-1}(j^{(1)}, \ldots, j^{(C)}) a_{\text{Pa}(i^{(c)})i^{(c)}} \right) b_{i^{(c)}}(O_t^{(c)}) \right), & \text{if } 1 < t \leq T \end{cases}$$

- The backward variable, $\beta_t(i^{(1)}, \ldots, i^{(C)}) = P(O_t, \ldots, O_T, S_t^{(1)} = q_{i^{(1)}}, \ldots, S_t^{(C)} = q_{i^{(C)}}|\lambda, \boldsymbol{\theta}(\lambda))$, is the probability of a partial observation sequence from

$t + 1$ to T, given the states for $S_t^{(1)}, \ldots, S_t^{(C)}$ and the model λ and $\theta(\lambda)$. The backward variable is computed as

$$
\begin{cases}
1, & \text{if } t = T \\
\sum_{j^{(1)},\ldots,j^{(C)}} \left(\prod_{c=1}^{C} (a_{\mathrm{Pa}(j^{(c)})j^{(c)}} b_{j^{(c)}}(O_t^{(c)})) \right) \beta_{t+1}(j^{(1)},\ldots,j^{(C)}), & \text{if } 1 \leq t < T
\end{cases}
$$

- $\xi_t(i^{(1)},\ldots,i^{(C)}, j^{(1)},\ldots,j^{(C)})$ is the probability of being at certain states at time t and $t + 1$, given the model and observation sequence. It is defined as

$$
P\big(S_t^{(1)} = q_{i^{(1)}}, \ldots, S_t^{(C)} = q_{i^{(C)}}, S_{t+1}^{(1)} = q_{j^{(1)}}, \ldots, S_{t+1}^{(C)} = q_{j^{(C)}} | \lambda, \theta(\lambda)\big)
$$

This can be re-written as

$$
\frac{\beta_{t+1}(j^{(1)},\ldots,j^{(C)}) \prod_{c=1}^{C} \alpha_t(i^{(1)},\ldots,i^{(C)}) a_{\mathrm{Pa}(j^{(c)})j^{(c)}} b_{j^{(c)}}(O_{t+1}^{(c)})}{P(\mathbf{O}|\lambda, \theta(\lambda))}
$$

where $P(\mathbf{O}|\lambda, \theta(\lambda))$ is computed using the forward and backward variables:

$$
P\big(\mathbf{O}|\lambda, \theta(\lambda)\big) = \sum_{i^{(1)},\ldots,i^{(C)}} \alpha_t\big(i^{(1)},\ldots,i^{(C)}\big) \beta_t\big(i^{(1)},\ldots,i^{(C)}\big) \tag{7.14}
$$

- $\gamma_t(i^{(1)},\ldots,i^{(C)}) = P(S_t^{(1)} = q_{i^{(1)}}, \ldots, S_t^{(C)} = q_{i^{(C)}}|\mathbf{O},\lambda, \theta(\lambda))$ is the probability of being at certain states at time t, given the model and observation sequence:

$$
\gamma_t\big(i^{(1)},\ldots,i^{(C)}\big) = \frac{\alpha_t(i^{(1)},\ldots,i^{(C)}) \beta_t(i^{(1)},\ldots,i^{(C)})}{\sum_{i^{(1)},\ldots,i^{(C)}} \alpha_t(i^{(1)},\ldots,i^{(C)}) \beta_t(i^{(1)},\ldots,i^{(C)})}
$$

Denote the current parameter estimates as $\theta(\lambda) = \{A, B, \pi\}$, the parameters $\bar{\theta}(\lambda) = \{\bar{A}, \bar{B}, \bar{\pi}\}$ can be re-estimated as follows:

$$
\bar{\pi}_{i^{(c)}} = \sum_{i^{(1)},\ldots,i^{(c-1)},i^{(c+1)},\ldots,i^{(C)}} \gamma_1\big(i^{(1)},\ldots,i^{(C)}\big) \tag{7.15}
$$

$$
\bar{a}_{\mathrm{Pa}(j^{(c)})j^{(c)}} = \frac{\sum_{t=1}^{T-1} \sum_{j^{(1)},\ldots,j^{(c-1)},j^{(c+1)},\ldots,j^{(C)},i^{(c')} \neq \mathrm{Pa}(j^{(c)})} \xi_t(i^{(1)},\ldots,i^{(C)}, j^{(1)},\ldots,j^{(C)})}{\sum_{t=1}^{T} \sum_{i^{(c')} \neq \mathrm{Pa}(j^{(c)})} \gamma_t(i^{(1)},\ldots,i^{(C)})} \tag{7.16}
$$

$$
\bar{\mu}_{i^{(c)}} = \frac{\sum_{t=1}^{T} (\sum_{i^{(1)},\ldots,i^{(c-1)},i^{(c+1)},\ldots,i^{(C)}} \gamma_t(i^{(1)},\ldots,i^{(C)})) O_t^{(c)}}{\sum_{t=1}^{T-1} \sum_{i^{(1)},\ldots,i^{(c-1)},i^{(c+1)},\ldots,i^{(C)}} \gamma_t(i^{(1)},\ldots,i^{(C)})} \tag{7.17}
$$

$$
\bar{\mathbf{U}}_{i^{(c)}} = \frac{\sum_{t=1}^{T} (\sum_{i^{(1)},\ldots,i^{(c-1)},i^{(c+1)},\ldots,i^{(C)}} \gamma_t(i^{(1)},\ldots,i^{(C)}))(O_t^{(c)} - \mu_{i^{(c)}})(O_t^{(c)} - \mu_{i^{(c)}})^T}{\sum_{t=1}^{T} \sum_{i^{(1)},\ldots,i^{(c-1)},i^{(c+1)},\ldots,i^{(C)}} \gamma_t(i^{(1)},\ldots,i^{(C)})} \tag{7.18}
$$

If a model uses $\bar{\theta}(\lambda)$ to replace iteratively $\theta(\lambda)$ and repeats this re-estimation process until some stopping criterion is met, the final result is a MLE estimate of parameters $\hat{\theta}(\lambda)$. In search for the optimal model $\hat{\mathbf{B}}_k$ with a minimum BIC value given by (7.12), for each candidate model structure λ_k, the corresponding MLE estimate of the parameters $\hat{\theta}(\lambda_k)$ are estimated iteratively using this extended forward–backward algorithm. An alternative approach to dynamic probabilistic network parameter and structure learning is to use a structured EM algorithm (Friedman et al. 1998).

Since the forward–backward algorithm is essentially an EM algorithm (Rabiner 1989), $P(\mathbf{O}|\lambda, \theta(\lambda))$ is only locally maximised by the estimated parameters. In a

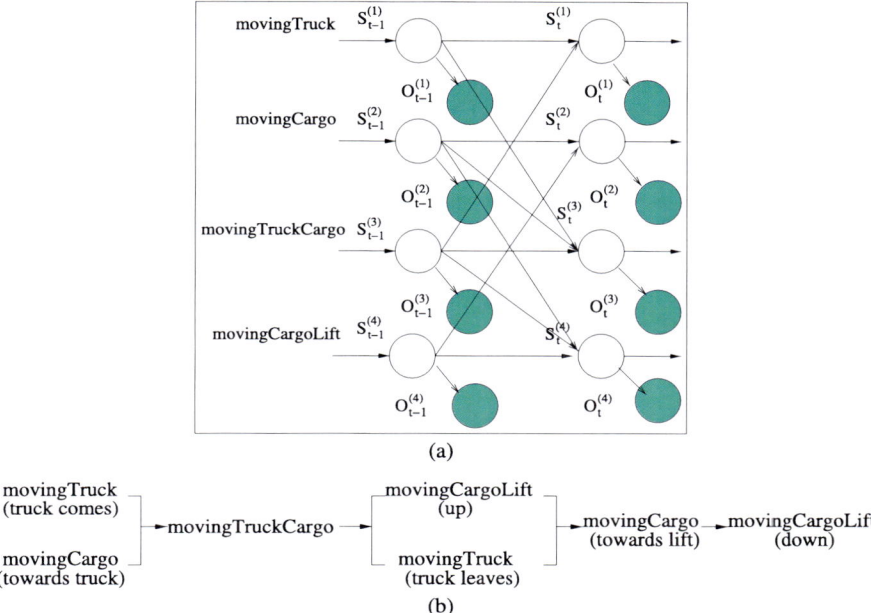

Fig. 7.10 (**a**) Automatically learned topology of a dynamically multi-linked HMM that describes the aircraft cargo loading and unloading activities as shown in Fig. 7.2. (**b**) Expected causal and temporal structure of these activities defined by human operational knowledge

general case, the cost function error surface has multiple local maxima and optimisation using the forward–backward algorithm is sensitive to initialisation. A strategy to overcome this problem is to start with multiple initialisations and choose the model by which the training data can be best explained.

Figure 7.10 shows an example of temporal dependency modelling based on correlation learning. The aircraft frontal cargo loading and unloading activities are modelled, as shown in Fig. 7.2, where four different event classes are detected automatically. They correspond to four basic constituents of the cargo loading and unloading activities: 'moving truck', 'moving cargo', 'moving cargo-lift' and 'moving truck/cargo'. The topology of the dynamically multi-linked HMM is learned from some training data using BIC (Fig. 7.10(a)). Compared to the expected structure of these activities defined by human operational knowledge (Fig. 7.10(b)), the automatically discovered causal relationships among different event classes and estimated temporal structure of the activities are mostly correct.

7.3.2 Behavioural Interpretation of Activities

After supervised learning a group activity model from some labelled training data, the model can be deployed to interpret behavioural patterns of multiple objects in new unseen data. The model operates by identifying different spatio-temporal events at key phases of a group activity and characterising the temporal and causal correlations among those events over time. In essence, the model provides higher level semantic interpretation of object activities based on validating probabilistically current visual observations against learned parameters from previous repeated observations given by the training data. Moreover, the learned model can be used to improve the accuracy of event detection and classification through inference of the hidden states. This in effect leads to a better understanding of the behaviours of each individual object involved in the group activity. By doing so, it can be considered that such a model provides both a bottom-up and top-down mechanism for understanding correlated behaviours of multiple objects. In the following, we consider in more details how such a dynamic graphical model of activities can be exploited for behavioural interpretation.

Activity Graphs

The temporal and causal correlations among events are quantified by the model structure and parameters of a dynamically multi-linked HMM learned from some labelled training data. To describe explicitly at a more semantic level the overall structure of events detected from a group activity, such a model can be used to generate automatically an activity transition matrix. Holistic activity phases can then be identified. At a higher level description, correlations among different holistic activity phases are also encoded in this matrix. This suggests a layered, hierarchical representation of activities. To that end, we consider a process to construct activity graphs as follows:

1. Compute an activity transition matrix $AT = \{at_{ij}\}$ where $1 \le i, j \le 2^C$. This is a $2^C \times 2^C$ matrix. Each entry of the matrix at_{ij} is computed as

$$at_{ij} = \prod_{c=1}^{C} a_{\text{Pa}(j^{(c)})j^{(c)}} \qquad (7.19)$$

where $j = \sum_{c=1}^{C} 2^{(c-1)}(j^{(c)} - 1)$ and $i = \sum_{c=1}^{C} 2^{(c-1)}(i^{(c)} - 1)$ for $i^{(c)}$ that satisfies: $i^{(c)} \in \{\text{Pa}(j^{(c)})\}$; at_{ij} represents the probability of transferring from activity phase i at time instance t to activity phase j at time instance $t + 1$.

2. Extract a simplified transition matrix $AT' = \{at'_{ij}\}$:

$$at'_{ij} = \begin{cases} at_{ij}, & \text{if } at_{ij} > \text{Th}_{\text{tr}} \\ 0, & \text{otherwise} \end{cases}$$

   The aim is to remove the activity phase transition that is unlikely to happen caused by errors in event recognition.

3. Generate an activity graph automatically from AT'. Each state of the activity transition matrix is represented as a node in the graph. The temporal and causal correlations with non-zero entries of AT' are assigned to directed arcs pointing from one node to another.

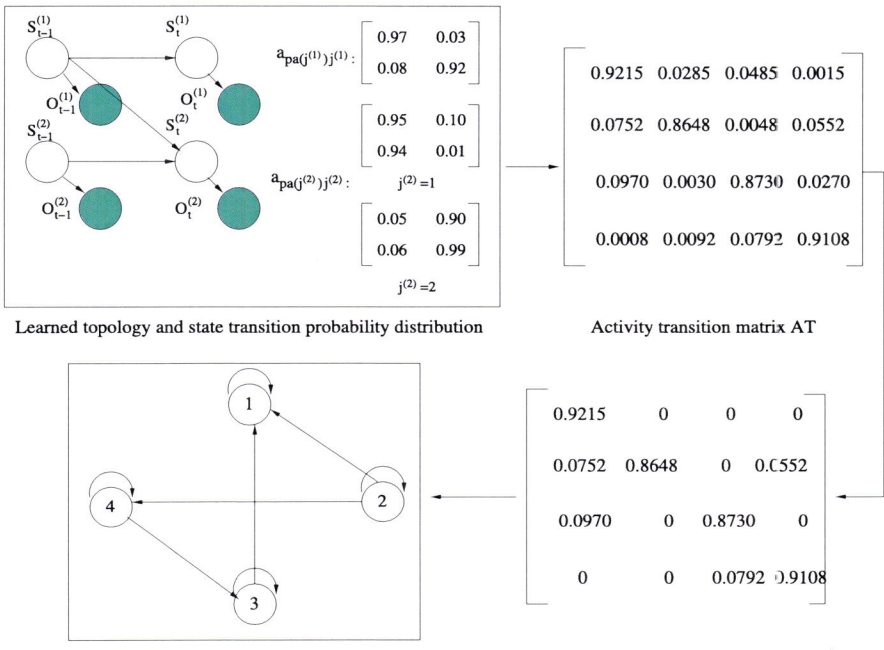

Learned topology and state transition probability distribution

Activity transition matrix AT

Activity graph generated from AT'

Simplified activity transition matrix AT'

Fig. 7.11 An example of automatically generating an activity graph from a learned dynamically multi-linked HMM. Two classes of events, denoted as $e1$ and $e2$, are modelled using temporal processes 1 & 2, respectively. The learned topology indicates that $e2$ can be caused by $e1$, reflected by the arc pointing from temporal processes 1 to 2. A simplified activity transition matrix AT' was obtained by thresholding the activity transition matrix AT. The number of non-zero diagonal elements of AT' corresponds to the number of key activity phases interpreted as the 'co-occurrence' of the two classes of events. In this case, four key activity phases were detected. The causal relationships among these four phases are reflected by the directed arcs connecting the nodes in the activity graph

The process of generating an activity graph is illustrated using a simple dynamically multi-linked HMM with two temporal processes (Fig. 7.11). An activity graph for the aircraft frontal cargo loading and unloading activities is shown in Fig. 7.12. In this graph, it is evident that important activity phases are discovered by the model. It also shows that the learned activity transition matrix has sparse structure, indicating effective state space factorisation.

Activity Classification

Suppose K different DBNs $\mathbf{B}_k$ are learned from some training data, respectively, for K different activity patterns involving multiple objects, where $1 \leq k \leq K$. The problem of classifying an unknown activity pattern by K known activity classes can be considered as a Bayesian model selection problem (Kass and Raftery 1995).

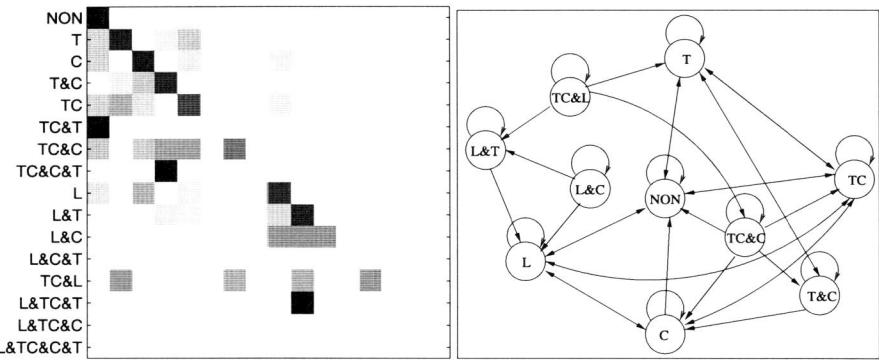

Fig. 7.12 An activity graph of the aircraft frontal cargo loading and unloading activities. *Left*: Activity transition matrix. Each entry corresponds to the transition probabilities of two states (*black* for true and *white* for false). Each state corresponds to the occurrence of one single event class or multiple co-occurrence event classes. States 'T', 'C', 'TC', 'L' and 'NON' correspond to 'moving truck', 'moving cargo', 'moving truck/cargo', 'moving cargo-lift' and 'no activity' events, respectively. State 'T&C' refers to 'moving truck' and 'moving cargo' occurring events. *Right*: Activity graph automatically generated from the activity transition matrix

More precisely, the model $\hat{\mathbf{B}}_k$ associated with the most likely activity is determined as

$$\hat{\mathbf{B}}_k = \arg\max_{\mathbf{B}_k} P(\mathbf{O}|\mathbf{B}_k)P(\mathbf{B}_k) \tag{7.20}$$

where $P(\mathbf{O}|\mathbf{B}_k)$ is the probability of observing the new data given the learned model structure and parameters, computed by (7.14). $P(\mathbf{B}_k)$ is the a priori probability of observing the kth candidate activity, and $P(\mathbf{B}_k)$ represents a priori knowledge about the unknown activity. If this a priori knowledge is unavailable, the unknown activity is recognised as one of the candidate activities whose learned model can best explain the observation. Xiang and Gong (2006) show that a dynamically multi-linked HMM activity model is robust for classifying both indoor and outdoor group activities, especially when the training data size is small and visual observations are corrupted by noise.

Improving Event Recognition Through State Inference

Since each hidden state variable of a dynamically multi-linked HMM represents a probability on whether a particular event class is likely to be detected at a particular time instance, it can be utilised as conditional probability to improve event detection accuracy. More precisely, given a sequence of detected events, an extended Viterbi algorithm (Forney 1973) can be formulated to infer the hidden states using the learned model. Let us first define the following variables:

- $\delta_t(i^{(1)}, \ldots, i^{(C)})$ is the highest probability of a partial observation sequence until time t with a sequence of hidden states from time 1 to time t, given the model λ

and $\theta(\lambda)$. It is computed as

$$\delta_t\left(i^{(1)}, \ldots, i^{(C)}\right)$$

$$= \max_{\{S_1^c\}, \ldots, \{S_{t-1}^c\}} P\left(\left\{S_1^{(c)}\right\}, \ldots, \left\{S_{t-1}^{(c)}\right\},\right.$$

$$\left. S_t^{(1)} = q_{i^{(1)}}, \ldots, S_t^{(C)} = q_{i^{(C)}}, O_1, \ldots, O_t \mid \lambda, \theta(\lambda)\right)$$

where $\{S_t^c\} = \{S_t^{(1)}, \ldots, S_t^{(C)}\}$.

- $\varphi_t(i^{(1)}, \ldots, i^{(C)})$ is an array representing the most probable state sequence. The most probable hidden states at time t for the C hidden state variables are denoted as $\{S_t^{*(c)}\}$.

An extended Viterbi algorithm is as follows:

1. Initialisation:

$$\delta_1\left(i^{(1)}, \ldots, i^{(C)}\right) = \prod_{c=1}^{C} \pi_{i^{(c)}} b_{i^{(c)}}\left(O_t^{(c)}\right)$$

$$\varphi_1\left(i^{(1)}, \ldots, i^{(C)}\right) = \mathbf{0}$$

2. Recursion:

$$\delta_t\left(i^{(1)}, \ldots, i^{(C)}\right) = \max_{j^{(1)}, \ldots, j^{(C)}} \left\{ \prod_{c=1}^{C} \delta_{t-1}\left(j^{(1)}, \ldots, j^{(C)}\right) a_{\mathrm{Pa}(i^{(c)})i^{(c)}} b_{i^{(c)}}\left(O_t^{(c)}\right) \right\}$$

$$\varphi_t\left(i^{(1)}, \ldots, i^{(C)}\right) = \arg \max_{j^{(1)}, \ldots, j^{(C)}} \left\{ \prod_{c=1}^{C} \delta_{t-1}\left(j^{(1)}, \ldots, j^{(C)}\right) a_{\mathrm{Pa}(i^{(c)})i^{(c)}} b_{\cdot(c)}\left(O_t^{(c)}\right) \right\}$$

where $1 < t \le T$.

3. Termination:

$$\left\{S_T^{*(c)}\right\} = \arg \max_{i^{(1)}, \ldots, i^{(C)}} \delta_t\left(i^{(1)}, \ldots, i^{(C)}\right)$$

4. The most probable state sequence backtracking:

$$\left\{S_t^{*(c)}\right\} = \varphi_{t+1}\left(\left\{S_{t+1}^{*(c)}\right\}\right)$$

where $t = T - 1, T - 2, \ldots, 1$.

An inferred most probable hidden state sequence represents the correct explanations of currently detected events conditioned by the model expectation learned from the training data. The learned temporal and causal correlations among different events from the training data are utilised to explain away the errors in current event detection and classification. The inferred hidden states given the new observations override detected events. Such a dynamically multi-linked HMM activity model provides a top-down probabilistic mechanism for explaining away inconsistent bottom-up event detection due to image noise or incomplete visual observations. Figure 7.13 shows an example of utilising top-down model inference to improve the accuracy of bottom-up autonomous event detection in an aircraft cargo unloading activity.

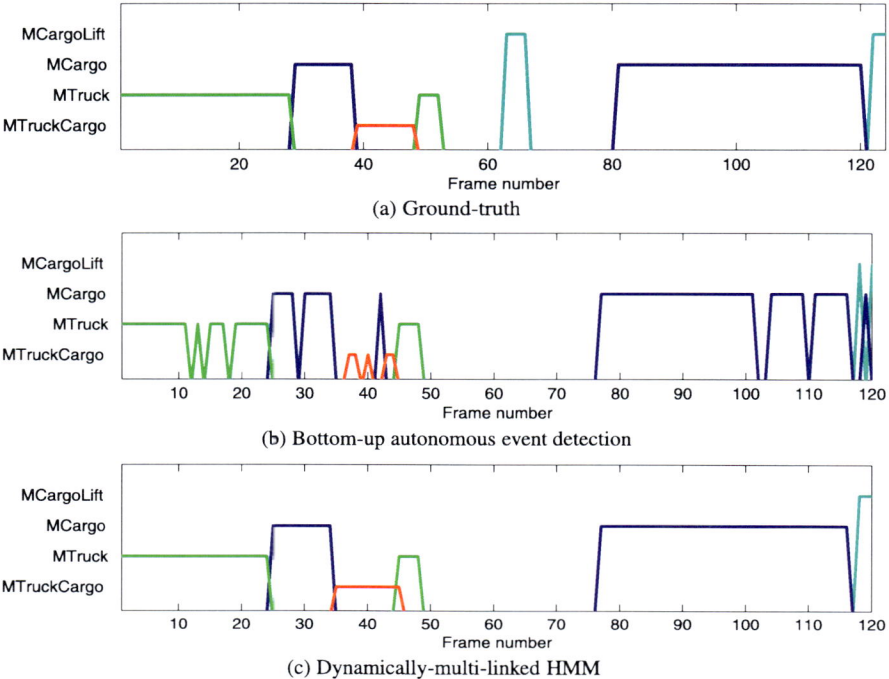

Fig. 7.13 Improving bottom-up autonomous event detection in the aircraft cargo activities by top-down probabilistic inference using a dynamically multi-linked HMM activity model

7.4 Discussion

For modelling group activities involving more than a single object in isolation, we consider the notion of contextual event-based representation defined in a visual scene regardless object identity. This approach to activity representation does not rely on object segmentation and tracking, so is particularly suitable for modelling complex group activities featured with severe occlusions between objects.

We also consider the problem of segmenting temporally a continuous video stream into different activity segments. Activities are considered as multiple probabilistic temporal processes of different event classes. A special case of a dynamic Bayesian network in the form of a dynamically multi-linked HMM is described in details as a suitable model for measuring the causal and temporal correlations among multiple temporal processes of an activity. Such correlations are discovered by learning a model topology from some training data.

The models described in this chapter are predominantly based on supervised learning that requires labelled training data. This covers all stages of a model building process, from event detection to activity classification. Even though the temporal segmentation method is unsupervised, the process makes an assumption that within each segment, only one type of group activity is present. This assumption is invalid in general, especially in a crowded wide-area scene where multiple group activities often occur simultaneously. For modelling multiple activities concurrently in a complex dynamic scene, we shall return to investigate other plausible techniques in Chaps. 9, 10, and 11. Until then, we first focus on the need for behaviour profiling.

In order to learn activity models of known classes for interpretation, sufficient and accurate labelling of training data play an important role. However, supervised model learning may not be plausible for describing complex activity patterns when activity classes are not well-defined. Different activities may share common characteristics with their unique features only subtly defined visually, resulting in manual labelling of activities that is not only difficult but also highly inconsistent.

A plausible solution to this problem is automatic behaviour profiling by unsupervised learning. Behaviour profiling by analysing visual observation of object activities aims to directly discover automatically from data hidden trends and patterns of regularity associated with visually indistinct observations without a priori knowledge. It is essentially unsupervised video data mining. Behaviour profiling exploits two types of information: (1) information on what is normal about all the activities captured in massive video data from observing a visual environment, and (2) the quantified distributions of their common components. The goal is to discover what is unique about different types of activities and how to measure quantitatively their uniqueness. This problem will be studied in more details in the next chapter, where on-line abnormal behaviour detection is addressed.

References

Ballard, D., Brown, C.: Computer Vision. Prentice Hall, New York (1982)

Baum, L.E., Petrie, T.: Statistical inference for probabilistic functions of finite state Markov chains. Ann. Math. Stat. **37**, 1554–1563 (1966)

Bobick, A.F., Davis, J.: The recognition of human movement using temporal templates. IEEE Trans. Pattern Anal. Mach. Intell. **23**(3), 257–267 (2001)

Bregonzio, M., Gong, S., Xiang, T.: Recognising action as clouds of space-time interest points. In: IEEE Conference on Computer Vision and Pattern Recognition, Miami, USA, June 2009, pp. 1948–1955 (2009)

Dalal, N., Triggs, B.: Histograms of oriented gradients for human detection. In: IEEE Conference on Computer Vision and Pattern Recognition, pp. 886–893 (2005)

Dempster, A.P., Laird, N.M., Rubin, D.B.: Maximum likelihood from incomplete data via the EM algorithm. J. R. Stat. Soc. **39**(1), 1–38 (1977)

Dollár, P., Rabaud, V., Cottrell, G., Belongie, S.: Behavior recognition via sparse spatio-temporal features. In: IEEE International Workshop on Visual Surveillance and Performance Evaluation of Tracking, pp. 65–72 (2005)

Enser, P., Sandom, C.: Towards a comprehensive survey of the semantic gap in visual image retrieval. In: ACM International Conference on Image and Video Retrieval, pp. 291–299 (2003)

Felzenszwalb, P.F., Girshick, R.B., McAllester, D., Ramanan, D.: Object detection with discriminatively trained part based models. IEEE Trans. Pattern Anal. Mach. Intell. **32**(9), 1627–1645 (2010)

Forney, G.D.: The Viterbi algorithm. Proc. IEEE **61**, 268–278 (1973)

Friedman, N., Murphy, K.P., Russell, S.: Learning the structure of dynamic probabilistic networks. In: Uncertainty in Artificial Intelligence, pp. 139–147 (1998)

Gong, S., Xiang, T.: Scene event recognition without tracking. Acta Autom. Sin. **29**(3), 321–331 (2003a)

Gong, S., Xiang, T.: Recognition of group activities using dynamic probabilistic networks. In: IEEE International Conference on Computer Vision, Nice, France, October 2003, pp. 742–749 (2003b)

Greenspan, H., Goldberger, J., Mayer, A.: Probabilistic space-time video modelling via piecewise GMM. IEEE Trans. Pattern Anal. Mach. Intell. **26**(3), 384–396 (2004)

Grosky, W.I., Zhao, R.: Negotiating the semantic gap: from feature maps to semantic landscapes. In: Lecture Notes in Computer Science, vol. 2234, pp. 33–42 (2001)

Huang, C., Darwiche, A.: Inference in belief networks: a procedural guide. Int. J. Approx. Reason. **15**(3), 225–263 (1996)

Hung, H., Gong, S.: Quantifying temporal saliency. In: British Machine Vision Conference, Kingston-upon-Thames, UK, September 2004, pp. 727–736 (2004)

Kadir, T., Brady, M.: Scale, saliency and image description. Int. J. Comput. Vis. **45**(2), 83–105 (2001)

Kass, R., Raftery, A.: Bayes factors. J. Am. Stat. Assoc. **90**, 377–395 (1995)

Keogh, E.: An online algorithm for segmenting time series. In: IEEE International Conference on Data Mining, pp. 289–296 (2001)

Latecki, L., Lakamper, R.: Convexity rule for shape decomposition based on discrete contour evolution. Comput. Vis. Image Underst. **73**, 441–454 (1999)

McKenna, S., Jabri, S., Duric, Z., Rosenfeld, A., Wechsler, H.: Tracking group of people. Comput. Vis. Image Underst. **80**, 42–56 (2000)

Naphade, M.R., Kozintsev, I., Huang, T.S.: A factor graph framework for semantic indexing and retrieval in video. IEEE Trans. Veh. Technol. **12**(1), 40–52 (2002)

Ng, J., Gong, S.: Learning pixel-wise signal energy for understanding semantics. Image Vis. Comput. **21**(13–14), 1171–1182 (2003)

Piater, J.H., Crowley, J.: Multi-modal tracking of interacting targets using Gaussian approximation. In: IEEE Workshop on Performance Evaluation of Tracking and Surveillance, pp. 141–147 (2001)

Rabiner, L.R.: A tutorial on hidden Markov models and selected applications in speech recognition. Proc. IEEE **77**(2), 257–286 (1989)

Russell, D., Gong, S.: Minimum cuts of a time-varying background. In: British Machine Vision Conference, Edinburgh, UK, September 2006, pp. 809–818 (2006)

Schwarz, G.: Estimating the dimension of a model. Ann. Stat. **6**, 461–464 (1978)

Stauffer, C., Grimson, W.E.L.: Adaptive background mixture models for real-time tracking. In: IEEE Conference on Computer Vision and Pattern Recognition, Fort Collins, USA, June 1999, pp. 246–252 (1999)

Viola, P., Jones, M.: Robust real-time face detection. Int. J. Comput. Vis. **57**(2), 137–154 (2004)

Viola, P., Jones, M., Snow. D.: Detecting pedestrians using patterns of motion and appearance. In: IEEE International Conference on Computer Vision, Nice, France, October 2003, pp. 734–741 (2003)

Xiang, T., Gong, S.: Beyond tracking: modelling activity and understanding behaviour. Int. J. Comput. Vis. **67**(1), 21–51 (2006)

Xiang, T., Gong, S.: Activity based surveillance video modelling. Pattern Recognit. **41**(7), 2309–2326 (2008)

Chapter 8
Unsupervised Behaviour Profiling

Given a large quantity of either on-line video stream input or recorded videos, the goal of automatic behaviour profiling is to learn a model that is capable of detecting unseen abnormal behaviour patterns whilst recognising novel instances of expected normal behaviour patterns. Each behaviour pattern may correspond to a video segment of group activity. In this context, an anomaly is defined as an atypical behaviour pattern that is not represented by sufficient examples in previous observations captured by a training data set. Critically, an anomaly should also satisfy a specificity constraint for a true abnormal pattern (Bernstein et al. 2010). Anomaly detection is thus treated as a binary classification problem. One of the main challenges for a binary classification model is to differentiate true anomaly from outliers that give false positives (Altman and Bland 1994; Zweig and Campbell 1993). This can be caused by noisy visual features which blur the boundary between normal and abnormal patterns. Therefore, the effectiveness of a behaviour profiling algorithm needs be measured by (1) how well anomalies can be detected by measuring specificity to expected patterns of behaviour, namely low false positives, and (2) how accurately and robustly different classes of normal behaviour patterns can be recognised by maximising between-class discrimination and classification sensitivity.

For unsupervised behaviour profiling, we consider a clustering model that discovers the intrinsic grouping of behaviour patterns. The method is fully unsupervised since it does not require manual data labelling for either feature extraction or discovery of grouping. There are two good reasons for behaviour profiling using unlabelled data:

1. Manual labelling of behaviour patterns is laborious, often rendered impractical given the vast amount of video data to be processed. More critically, manual labelling of behaviour patterns is subject to inconsistency and error prone. This is because a human tends to interpret behaviour based on a priori cognitive knowledge of what should be present in a scene rather than solely based on what is visually detectable in the scene. This introduces bias due to differences in experience and one's mental state.

S. Gong, T. Xiang, *Visual Analysis of Behaviour*,
DOI 10.1007/978-0-85729-670-2_8, © Springer-Verlag London Limited 2011

2. The method also performs on-line anomaly detection. That is, a decision on
   whether a behaviour pattern is normal is made as soon as sufficient visual evidence is collected without needing to wait for the completion of the pattern.

Automatic behaviour profiling for anomaly detection is challenging not only because of the complexity and variety of behaviours in an unconstrained visual environment, but also due to the ambiguity in defining normality and abnormality given
the visual context that can change over time. In particular, a behaviour can be considered as either being normal or abnormal depending on when and where it takes
place. This causes problems for a behaviour profiling model that is learned off-line.
Such a model assumes what is normal and abnormal in the learning stage would
continue to hold true regardless of any circumstantial changes over time. To overcome this problem, an incremental unsupervised learning algorithm is considered to
enable a model to be adaptive to context change and effective for processing large
volume of unlabelled video data on-the-fly.

8.1 Off-line Behaviour Profile Discovery

To learn a model for automatic behaviour profiling given some training data, each
training video needs be processed into a series of smaller chunks. Given a continu-
ous video footage $\mathbf{V}$, it is divided into N video segments $\mathbf{V} = \{\mathbf{v}_1, \ldots, \mathbf{v}_n, \ldots, \mathbf{v}_N\}$,
such that each segment contains at least one plausible behaviour pattern (Sect. 7.2).
The nth video segment $\mathbf{v}_n$ consisting of T_n image frames is represented as $\mathbf{v}_n =
[\mathbf{I}_{n1}, \ldots, \mathbf{I}_{nt}, \ldots, \mathbf{I}_{nT_n}]$, where $\mathbf{I}_{nt}$ is the tth image frame. When there is no clearly
separable behaviour patterns in a crowded scene such as a shopping mall, a video
can be divided evenly into overlapping segments with each segment having a constant time duration.

8.1.1 Behaviour Pattern

Suppose that K_e classes of contextual event are automatically discovered from a
training data set (Sect. 7.1), the nth video segment $\mathbf{v}_n$ as a behaviour pattern is then
represented as a feature vector $\mathbf{P}_n$ of event classes, given by

$$\mathbf{P}_n = [\mathbf{p}_{n1}, \ldots, \mathbf{p}_{nt}, \ldots, \mathbf{p}_{nT_n}] \tag{8.1}$$

where T_n is the number of frames in the nth video segment as a behaviour pattern.
The tth element of $\mathbf{P}_n$ as a video frame is represented by a K_e dimensional variable:

$$\mathbf{p}_{nt} = \left[p_{nt}^1, \ldots, p_{nt}^k, \ldots, p_{nt}^{K_e} \right] \tag{8.2}$$

In essence, $\mathbf{p}_{nt}$ represents the tth image frame of video segment $\mathbf{v}_n$ as a contextual
event feature vector, where p_{nt}^k is the posterior probability for an event of the kth
event class being detected in that frame given the prior learned from the training

data. If an event of the kth class is detected in the tth frame of segment $\mathbf{v}_n$, $0 <$ $p_{nt}^k \leq 1$; otherwise, $p_{nt}^k = 0$.

Using this representation, a video segment as a behaviour pattern is decomposed into and represented by temporally ordered contextual events. Both the membership of events and their temporal order in a behaviour pattern describe this behaviour pattern. Two behaviour patterns can differ by either their constituent event classes, or the temporal order in the occurrence of their respective member events. For instance, behaviour patterns A and B are deemed as being different if (1) A is composed of events of classes a, b, and d, whilst B is composed of events of classes a, c and e; or (2) both A and B are composed of events of classes a, c and d. However, in A, event a is followed by c, while in B, event a is followed by d.

8.1.2 Behaviour Profiling by Data Mining

The unsupervised behaviour profiling problem is defined formally as follows. Consider a training data set $\mathbf{D}$ consisting of N feature vectors:

$$\mathbf{D} = \{\mathbf{P}_1, \ldots, \mathbf{P}_n, \ldots, \mathbf{P}_N\} \tag{8.3}$$

where $\mathbf{P}_n$ is given by (8.1) representing a behaviour pattern in the nth video segment $\mathbf{v}_n$. The problem to be addressed is to discover any intrinsic groupings among the training behaviour patterns from which a model for describing normal behaviours can be found. This is essentially a data mining by unsupervised clustering problem with the number of clusters unknown. The following characteristics of the feature space make this data mining by clustering problem challenging:

1. Each feature vector $\mathbf{P}_n$ can be of different dimensions, because the number of detected event classes in each video frame can vary. Conventional clustering techniques such as k-means and mixture models require that each data example is represented as a fixed dimension feature vector.
2. Different behaviour patterns can have different durations resulting in variable length feature vectors. To compute a distance metric for measuring affinity among variable length feature vectors is not straightforward, as it often requires dynamic time warping (Kruskal and Liberman 1983). One can consider a string distance based approach, such as the Levenshtein distance, to dynamic time warping. Such an approach attempts to treat a feature vector $\mathbf{P}_n$ as a K_e dimensional trajectory, and measure the distance between two trajectories of different lengths by finding correspondences between discrete vertices on the two trajectories (Ng and Gong 2002). Alternatively, a model can be built to match key frames in two image sequences of different lengths based on salient object detection and matching. Consequently, the affinity of the most similar pair of images from two sequences can be used to estimate the overall sequence affinity (Shan et al. 2004). This is less applicable to behaviour patterns involving multiple objects where no visually dominant object may be detected and matched. Fundamentally, we consider that a behaviour pattern is a sequence of temporally correlated

events. A stochastic process based model which considers a distance metric for not only time warping but also temporal ordering would provide a more accurate behaviour pattern affinity measure.

3. Automatic model order selection is required to determine the number of clusters of behaviour patterns in the training data set, whilst both the behaviour pattern feature vector dimension and feature vector length are variable.

8.1.3 Behaviour Affinity Matrix

A dynamic Bayesian network (DBN) provides a plausible stochastic process based solution for measuring the affinity between different behaviour patterns. More specifically, each behaviour pattern in a training data set is modelled using a multi-observation hidden Markov model (HMM) (Sect. 7.2). To measure the affinity between two behaviour patterns represented as $\mathbf{P}_i$ and $\mathbf{P}_j$, two multi-observation HMMs denoted as $\mathbf{B}_i$ and $\mathbf{B}_j$ are constructed from $\mathbf{P}_i$ and $\mathbf{P}_j$, respectively, using the EM algorithm. The affinity between $\mathbf{P}_i$ and $\mathbf{P}_j$ is then computed as

$$S_{ij} = \frac{1}{2}\left\{ \frac{1}{T_j} \log P(\mathbf{P}_j|\mathbf{B}_i) + \frac{1}{T_i} \log P(\mathbf{P}_i|\mathbf{B}_j) \right\} \tag{8.4}$$

where $P(\mathbf{P}_j|\mathbf{B}_i)$ is the likelihood of observing $\mathbf{P}_j$ given $\mathbf{B}_i$, and T_i and T_j are the lengths of $\mathbf{P}_i$ and $\mathbf{P}_j$ respectively.

An $N \times N$ affinity matrix $\mathbf{S} = [S_{ij}]$, where $1 \leq i, j \leq N$ and N is the number of behaviour patterns in the entire training data set, provides a new representation for the training data set, denoted as $\mathbf{D_s}$. In this representation, a behaviour pattern is represented by its affinity to each of the other behaviour patterns in the training set. Specifically, the nth behaviour pattern is represented as the nth row of $\mathbf{S}$, denoted as $\mathbf{s}_n$. This gives

$$\mathbf{D_s} = \{\mathbf{s}_1, \ldots, \mathbf{s}_n, \ldots, \mathbf{s}_N\} \tag{8.5}$$

Each behaviour pattern is now represented by an affinity feature vector of a fixed length N, measuring N relative distances to all other behaviour patterns in the training data set. As the number of data examples is equal to the dimension of the affinity feature vector space, dimensionality reduction is necessary before clustering is attempted in this affinity feature vector space. This is to avoid the 'curse of dimensionality' problem (Bishop 2006). We consider a spectral-clustering model to solve this problem, which both reduces the data dimensionality and performs unsupervised clustering with automatic model order selection determined by the eigenvectors of a data affinity matrix.

8.1.4 Eigendecomposition

Dimensionality reduction on the N dimensional feature space given by (8.5) can be achieved through eigen-decomposition of the affinity matrix $\mathbf{S}$. The eigenvectors of the affinity matrix can then used for data clustering. It is shown that clustering by the eigenvectors of a normalised affinity matrix is more desirable (Shi and Malik 2000; Weiss 1999). A normalised affinity matrix is defined as

$$\bar{\mathbf{S}} = \mathbf{L}^{-\frac{1}{2}} \mathbf{S} \mathbf{L}^{-\frac{1}{2}} \tag{8.6}$$

where $\mathbf{L} = [L_{ij}]$ is an $N \times N$ diagonal matrix with $L_{ii} = \sum_j S_{ij}$. Weiss (1999) shows that under certain constraints, the largest K eigenvectors[1] of $\bar{\mathbf{S}}$ are sufficient to partition the data set into K clusters. Representing behaviour patterns by the K largest eigenvectors reduces the dimensionality of the feature space from N, the number of behaviour patterns in the training set, to K, the number of statistically most dominant classes of behaviour pattern. For a given K, a standard clustering technique such as k-means or a mixture model can be adopted. The remaining problem is to determine what is the optimal K, which is unknown. This number is not necessarily the same as the largest K eigenvectors of the affinity matrix. The problem needs be solved through automatic model order selection.

8.1.5 Model Order Selection

It is assumed that the optimal number of clusters K is between 1 and K_m, whilst K_m is a number sufficiently larger than the true value of K. The training data set is now represented by projecting all the behaviour patterns to the K_m largest eigenvectors, denoted as $\mathbf{D_e}$, where:

$$\mathbf{D_e} = \{\mathbf{y}_1, \ldots, \mathbf{y}_n, \ldots, \mathbf{y}_N\} \tag{8.7}$$

Each element of $\mathbf{D_e}$ is a K_m-dimensional coefficient vector representing a behaviour pattern by its projection weights to the K_m eigenvectors:

$$\mathbf{y}_n = [e_{1n}, \ldots, e_{kn}, \ldots, e_{K_m n}] \tag{8.8}$$

where e_{kn} is the weight coefficient from projecting the nth behaviour affinity feature vector to the kth largest eigenvector $\mathbf{e_k}$. Since $K < K_m$, it is guaranteed that all the information needed for grouping K clusters is preserved in this K_m dimensional space.

   To help determine the optimal cluster numbers as the model order for the distribution of $\mathbf{D_e}$, a mixture of Gaussian model can be considered (Sect. 5.2). To that end, the Bayesian information criterion (BIC) is a plausible technique for selecting the optimal number of mixture components K that can describe the most probable intrinsic groupings of behaviour classes in the training data set. However, Xiang and Gong (2008b) show experimental results to indicate that the BIC tends to underestimate the optimal number of clusters. The BIC is known to have a tendency of under-fitting a model when training data is sparse (Roberts et al. 1998). Due to a

---

[1]The largest eigenvectors are eigenvectors with largest eigenvalues in magnitude.

wide variety of different object behaviour patterns exhibited in a behaviour training data set, examples for many classes of behaviours are likely to be sparse, with either uneven biased examples or a lack of examples. A solution to this problem is to reduce the dimensionality of a data space by unsupervised feature selection. This process aims to select only relevant and informative eigenvectors of a normalised affinity matrix in order to minimise under-fitting in model order selection given sparse training data.

8.1.6 Quantifying Eigenvector Relevance

Let us consider a suitable criterion for measuring the relevance of each eigenvector from a normalised affinity matrix $\bar{\mathbf{S}}$. We suggest that only a subset of the K_m largest eigenvectors are sufficiently relevant in contributing to the K unknown intrinsic behaviour clusters. An eigenvector is deemed relevant if it can be used to separate at least one cluster of data from the others. A technique is needed to identify and remove those irrelevant and uninformative eigenvectors so to reduce the data space dimensionality and improve the accuracy of model learning.

An intuitive measure of an eigenvector's relevance is its eigenvalue. Ng et al. (2001) show that in an ideal case where intrinsic data clusters are infinitely far apart, the top K_{true} relevant eigenvectors have a corresponding eigenvalue of magnitude 1 whilst others do not. In situations where data are clearly separated, close to this ideal case, the problem of determining the optimal number of clusters is reduced to a trivial case such that the optimal cluster number is equal to the number of eigenvalues of magnitude 1. However, the eigenvalues for eigenvectors of a more realistic data set are no longer good indicators of relevance. We consider addressing this problem by a relevance learning algorithm.

The algorithm is based on measuring the relevance of an eigenvector according to how well it can separate a data set into different clusters. This is achieved through analysing the distributions of the elements of each eigenvector with the following considerations: (1) The distribution of the elements of a relevant eigenvector must enable it to be used for separating at least one cluster from others. More specifically, the distribution of the elements of an eigenvector needs be multimodal if the eigenvector is relevant and unimodal otherwise. (2) An eigenvector with large eigenvalue magnitude is more likely to be relevant than that of a small one.

Let us denote the likelihood of the kth largest eigenvector $\mathbf{e_k}$ being relevant as $R_{\mathbf{e_k}}$, where $0 \leq R_{\mathbf{e_k}} \leq 1$. It is assumed that the elements of $\mathbf{e_k}$, e_{kn}, follow two different distributions, unimodal or multimodal, depending on whether $\mathbf{e_k}$ is relevant. The probability density function of e_{kn} is a mixture model of two components:

$$p(e_{kn}|\theta_{e_{kn}}) = (1 - R_{\mathbf{e_k}})p(e_{kn}|\theta_{e_{kn}}^1) + R_{\mathbf{e_k}} p(e_{kn}|\theta_{e_{kn}}^2) \tag{8.9}$$

where $\theta_{e_{kn}}$ are the parameters describing the distribution; $p(e_{kn}|\theta_{e_{kn}}^1)$ is the probability density function of e_{kn} when $\mathbf{e_k}$ is irrelevant, and $p(e_{kn}|\theta_{e_{kn}}^2)$ otherwise; $R_{\mathbf{e_k}}$ acts as the weight, or mixing probability, of the second mixture component. The distribution of e_{kn} is assumed to be a unimodal single Gaussian to reflect the fact that $\mathbf{e_k}$ cannot be used for data clustering when it is irrelevant:

$$p\left(e_{kn}|\theta_{e_{kn}}^1\right) = \mathcal{N}(e_{kn}|\mu_{k1}, \sigma_{k1}) \tag{8.10}$$

where $\mathcal{N}(\cdot|\mu, \sigma)$ denotes a Gaussian, or a Normal distribution, of mean μ and covariance σ. We assume that the second component of $P(\mathbf{e_k}|\boldsymbol{\theta_{e_k}})$ is a mixture of two Gaussian components, so to reflect the fact that $\mathbf{e_k}$ can separate one cluster of data from the others when it is relevant:

$$p\left(e_{kn}|\theta_{e_{kn}}^2\right) = w_k \mathcal{N}(e_{kn}|\mu_{k2}, \sigma_{k2}) + (1 - w_k)\mathcal{N}(e_{kn}|\mu_{k3}, \sigma_{k3}) \tag{8.11}$$

where w_k is the weight of the first Gaussian in $p(e_{kn}|\theta_{e_{kn}}^2)$.

There are two reasons for using a mixture of two Gaussian components even when e_{kn} forms more than two clusters, or the distribution of each cluster is not Gaussian: (1) In these cases, a mixture of two Gaussian components ($p(e_{kn}|\theta_{e_{kn}}^2)$) still fits better to the data compared to a single Gaussian ($p(e_{kn}|\theta_{e_{kn}}^1)$). (2) Its simple form means that only small number of parameters are needed to describe $p(e_{kn}|\theta_{e_{kn}}^2)$. This makes model learning possible even given sparse data.

There are eight parameters required for describing the distribution of e_{kn}:

$$\theta_{e_{kn}} = \{R_{\mathbf{e_k}}, \mu_{k1}, \mu_{k2}, \mu_{k3}, \sigma_{k1}, \sigma_{k2}, \sigma_{k3}, w_k\} \tag{8.12}$$

The maximum likelihood estimation (MLE) of $\theta_{e_{kn}}$ can be obtained using the following procedure:

1. The parameters of the first mixture component $\theta_{e_{kn}}^1$ are estimated as $\mu_{k1} = \frac{1}{N}\sum_{n=1}^{N} e_{kn}$ and $\sigma_{k1} = \frac{1}{N}\sum_{n=1}^{N}(e_{kn} - \mu_{k1})^2$.
2. The rest six parameters are estimated iteratively using the EM algorithm. It is important to note that $\theta_{e_{kn}}$ as a whole is not estimated iteratively using a standard EM algorithm, although the EM is employed for estimating a part of $\theta_{e_{kn}}$, namely $\theta_{e_{kn}}^1$. This subtle difference is critical for this eigenvector selection algorithm because if all the eight parameters are re-estimated in each iteration, the distribution of e_{kn} is essentially modelled as a mixture of three Gaussian components. Then, the estimated $R_{\mathbf{e_k}}$ would represent the weight of two of the three Gaussian components. This is very different from what $R_{\mathbf{e_k}}$ is designed to represent, namely, the likelihood of $\mathbf{e_k}$ being relevant for data clustering.
3. This relevance learning process is essentially a local (greedy) searching method. The algorithm is sensitive to parameter initialisation, especially given noisy and sparse data (Dempster et al. 1977). To overcome this problem, the a priori knowledge on the relationship between the relevance of each eigenvector and its corresponding eigenvalue is utilised to set the initial value of $R_{\mathbf{e_k}}$. Specifically, we set $\tilde{R}_{\mathbf{e_k}} = \bar{\lambda}_k$, where $\tilde{R}_{\mathbf{e_k}}$ is the initial value of $R_{\mathbf{e_k}}$ and $\bar{\lambda}_k \in [0, 1]$ is the normalised eigenvalue for $\mathbf{e_k}$ with $\bar{\lambda}_1 = 1$ and $\bar{\lambda}_{K_m} = 0$. We then randomly initialise the values of the other five parameters, $\mu_{k2}, \mu_{k3}, \sigma_{k2}, \sigma_{k3}$ and w_k. The solution that yields the largest $p(e_{kn}|\theta_{e_{kn}}^2)$ over different initialisations is chosen.

Although this relevance learning algorithm is based on estimating the distribution of the elements of each eigenvector, the model only seeks to learn whether the distribution is unimodal or multimodal, reflected by the value of $R_{\mathbf{e_k}}$. In other words, among the eight free parameters of the eigenvector distribution (8.12), $R_{\mathbf{e_k}}$ is the only parameter that matters. This also explains why the algorithm is able to estimate the relevance accurately when there are more than 2 clusters, and when the distribution of each cluster is not Gaussian.

The MLE of $\hat{R}_{\mathbf{e_k}}$ provides a real-value measurement of the relevance of $\mathbf{e_k}$. Since a 'hard-decision' is needed for dimensionality reduction, the kth eigenvector $\mathbf{e_k}$ is eliminated among the K_m candidate eigenvectors if

$$\hat{R}_{\mathbf{e_k}} < 0.5 \tag{8.13}$$

After eliminating those irrelevant eigenvectors, the selected relevant eigenvectors are used to determine the number of clusters K. Clustering is carried out by a mixture of Gaussian and the BIC is used for model order selection. Each behaviour pattern in the training data set is labelled by one of the K behaviour classes as a result.

Figure 8.1 shows an example of automatic behaviour profiling in a doorway entrance scene, where a camera was mounted on the ceiling of an entrance corridor, monitoring people entering and leaving an office area. The office area was secured by an entrance-door which can only be opened by scanning an entry card on the wall next to the door (see middle frame in row (b) of Fig. 8.1). Two side-doors were also located at the right hand side of the corridor. People from both inside and outside the office area have access to those two side-doors. Typical behaviour occurring in the scene would be people entering or leaving either the office area or the side-doors, and walking towards the camera. Each behaviour pattern would normally last a few seconds. There are six different classes of behaviours typically taking place in this scene (Table 8.1). Figure 8.2 shows that six eigenvectors are automatically selected as being relevant for behaviour pattern clustering, whilst six behaviour clusters are also discovered from the training data. Each discovered cluster consists of behaviour patterns from one of the six behaviour classes listed in Table 8.1.

8.2 On-line Anomaly Detection

Given automatically discovered typical behaviour patterns from a training data set, we consider the problem of how these profiled behaviour groupings can be utilised to build a model capable of detecting behavioural anomalies.

8.2.1 A Composite Behaviour Model

Suppose that N behaviour patterns in a training data set are classified into K_o behaviour classes, we consider the problem of building a model that describes the expected, normal behaviours as follows. First, a model is built for the kth behaviour class using a multi-observation HMM $\mathbf{B}_k$, with the model parameters $\theta_{\mathbf{B}_k}$ estimated

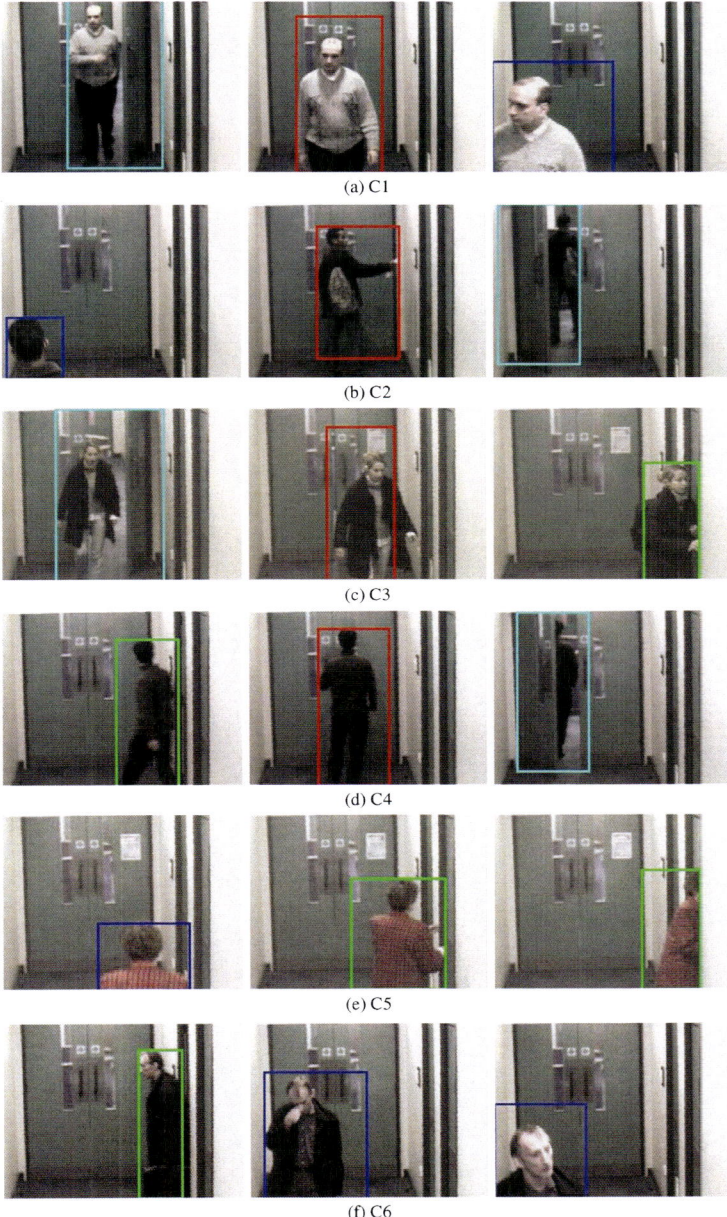

(a) C1

(b) C2

(c) C3

(d) C4

(e) C5

(f) C6

Fig. 8.1 Examples of typical behaviour patterns at a doorway entrance scene. (**a**)–(**f**) show video frames of typical behaviour patterns from six behaviour classes (see Table 8.1). Four different classes of events are detected in these examples, and shown in each frame by bounding boxes in *blue*, *cyan*, *green*, and *red*, respectively, representing: 'entering/leaving the corridor', 'entering/leaving the entry-door', 'entering/leaving the side-doors', and 'in corridor with the entry door closed'

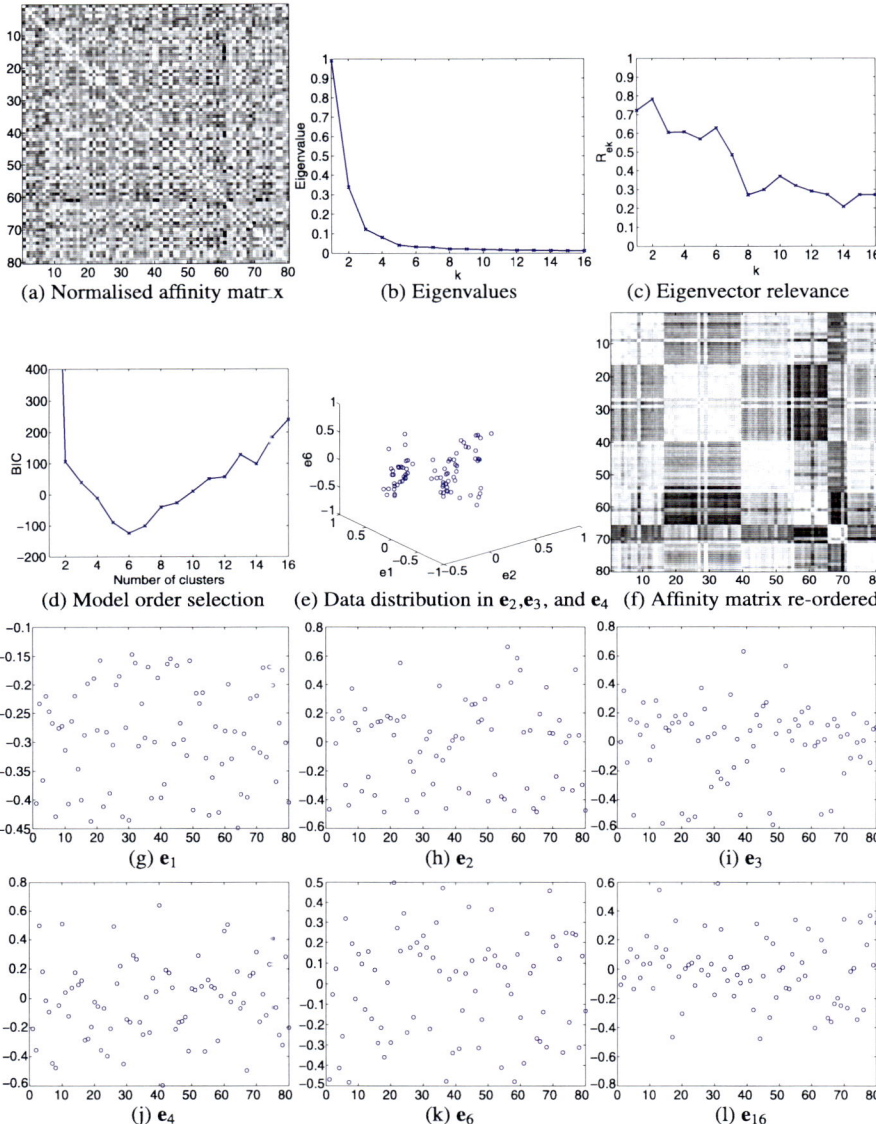

Fig. 8.2 An example of behaviour profiling in a doorway entrance scene, where (**d**) shows that the number of behaviour classes is determined as six with relevant eigenvector selection; (**b**) shows that eigenvalues cannot be used to determine the number of clusters; (**e**) suggests the existence of distinctive groups in a feature space defined by three relevant eigenvectors ($\mathbf{e}_2$, $\mathbf{e}_3$, and $\mathbf{e}_4$); (**g**)–(**l**) indicate that eigenvectors with high relevance values (see (**c**)) tend to have a multimodal distribution whilst those with low relevant values, such as $\mathbf{e}_{16}$, has a unimodal distribution

Table 8.1 Six classes of typical behaviour patterns occur in a doorway entrance scene

C1	From the office area to the near end of the corridor
C2	From the near end of the corridor to the office area
C3	From the office area to the side-doors
C4	From the side-doors to the office area
C5	From the near end of the corridor to the side-doors
C6	From the side-doors to the near end of the corridor

from using all the patterns of the kth class in the training data set. Second, a composite behaviour model $\mathbf{M}$ is constructed as a mixture of K multi-observation HMM behaviour class models. Third, given an unseen behaviour pattern, represented by a behaviour pattern feature vector $\mathbf{P}$, the likelihood of observing $\mathbf{P}$ given $\mathbf{M}$ is computed as

$$P(\mathbf{P}|\mathbf{M}) = \sum_{k=1}^{K} \frac{N_k}{N} P(\mathbf{P}|\mathbf{B}_k) \tag{8.14}$$

where N is the total number of training behaviour patterns and N_k is the number of patterns that belong to the kth behaviour class.

Once constructed, this composite behaviour model $\mathbf{M}$ is employed to detect whether an unseen behaviour pattern, considered as a probe pattern, is normal by a run-time anomaly measure. If detected as being normal, the probe pattern is then classified as one of the K classes of normal behaviour patterns using an on-line likelihood ratio test.

An unseen behaviour pattern of length T is given by $\mathbf{P} = [\mathbf{p}_1, \ldots, \mathbf{p}_i, \ldots, \mathbf{p}_T]$. At the tth frame, the accumulated visual information for the behaviour pattern, $\mathbf{P}_t = [\mathbf{p}_1, \ldots, \mathbf{p}_t]$, is used for on-line anomaly detection. To that end, a normalised log-likelihood of observing $\mathbf{P}$ at the tth frame given the behaviour model $\mathbf{M}$ is computed as

$$l_t = \frac{1}{t} \log P(\mathbf{P}_t|\mathbf{M}) \tag{8.15}$$

l_t is computed on-line using the forward-backward procedure (Rabiner 1989). Specifically, the K_e forward probabilities at time t are computed using the K_e forward probabilities computed at time $t-1$ together with the observation at time t. It should be noted that l_t is computed using l_{t-1} and $\mathbf{p}_t$ without the need to infer the hidden states in the multi-observation HMM. The complexity of computing l_t is $\mathcal{O}(K_e^2)$ and does not increase with t.

8.2.2 Run-Time Anomaly Measure

The decision on whether $\mathbf{P}_t$ is an anomaly is made based on a run-time anomaly measure Q_t, defined as

$$Q_t = \begin{cases} l_1, & \text{if } t = 1 \\ (1 - \alpha)Q_{t-1} + \alpha(l_t - l_{t-1}), & \text{otherwise} \end{cases} \tag{8.16}$$

where α is a cumulative factor determining how important the visual information extracted from the current frame is for anomaly detection, and $0 < \alpha \leq 1$. Compared to using l_t as an indicator of normality, Q_t places more weight on more recent observations. Anomaly is detected at frame t if

$$Q_t < \text{Th}_A \tag{8.17}$$

where Th_A is an anomaly detection threshold. The value of Th_A should be set according to a joint requirement on both the detection rate, namely the model sensitivity, and the false alarm rate, namely the model specificity. It takes a time delay for Q_t to stabilise at the beginning of evaluating a probe behaviour pattern due to the nature of the forward-backward procedure.

8.2.3 On-line Likelihood Ratio Test

If $Q_t > \text{Th}_A$, a probe behaviour pattern is detected as being normal at time frame t. It is then considered to be classified as one of the K profiled normal behaviour classes. This is determined by an on-line likelihood ratio test. More specifically, a hypothesis test is considered between:

H_k: $\mathbf{P}_t$ is from the hypothesised model $\mathbf{B}_k$ and belongs to the kth normal behaviour class

H_0: $\mathbf{P}_t$ is from a model other than $\mathbf{B}_k$ and does not belong to the kth normal behaviour class

where H_0 is called the alternative hypothesis. Using this likelihood ratio test, the likelihood ratio of accepting the two hypotheses is computed as

$$r_k = \frac{P(\mathbf{P}_t; H_k)}{P(\mathbf{P}_t; H_0)} \tag{8.18}$$

The hypothesis H_k is represented by model $\mathbf{B}_k$ learned from profiled and clustered behaviour patterns in class k. The key to likelihood ratio test is to construct the alternative model which represents H_0. In principle, the number of possible alternatives is unlimited. In practice, $P(\mathbf{P}_t; H_0)$ is computed by approximation (Higgins et al. 1991; Wilpon et al. 1990). Fortunately in this case, it has already been determined at the tth time frame that $\mathbf{P}_t$ is normal so it should only be generated by one of the K normal behaviour classes. A reasonable candidate for this alternative model is a mixture of the rest $K - 1$ normal behaviour class models. The problem of classifying the probe behaviour patterns becomes an one-versus-rest binary classification problem. More precisely, (8.18) is re-written as

$$r_k = \frac{P(\mathbf{P}_t | \mathbf{B}_k)}{\sum_{i \neq k} \frac{N_i}{N - N_k} P(\mathbf{P}_t | \mathbf{B}_i)} \tag{8.19}$$

For behaviour classification, r_k is a function of t and computed over time. $\mathbf{P}_t$ is only reliably classified as the kth behaviour class when $1 \ll \text{Th}_r < r_k$ where Th_r

is a threshold. When there are more than one r_k being greater than Th_r, the probe behaviour pattern is recognised as the class with the largest r_k.

For detecting anomalies from an unseen video, Xiang and Gong (2008b) show that a behaviour model trained using an unlabelled data set is superior to a model trained using the same but labelled data set. The former also outperforms the latter in differentiating abnormal behaviour patterns from normal behaviours contaminated by errors in event detection. In contrast, a model trained using manually labelled data is more accurate in explaining data that are well-defined. However, training a model using labelled data does not necessarily help the model with identifying novel instances of abnormal behaviour patterns. The model tends to be brittle and less robust in dealing with unclear behaviour instances in an open environment when the number of expected normal and abnormal behaviour cannot be pre-defined exhaustively.

Compared with a likelihood ratio test based on-line behaviour recognition and anomaly detection, a commonly used maximum likelihood estimation (MLE) method recognises $\mathbf{P}_t$ as the kth behaviour class when $k = \arg\max_k\{P(\mathbf{P}_t|\mathbf{B}_k)\}$. Using a MLE based model, classification is only performed at every single time frame without taking into consideration how reliable and sufficient the accumulated visual evidence is. This can cause errors, especially when there are ambiguities between different behaviour classes. For example, a behaviour pattern can be explained away equally well by multiple plausible behaviour models at its early stage. In contrast to maximum likelihood estimation, the on-line likelihood ratio test holds the decision on classification until sufficient visual evidence has been accumulated to overcome ambiguities, giving more reliable recognition than that of maximum likelihood estimation. An example of on-line anomaly detection using the run-time anomaly measure and behaviour classification using the on-line likelihood ratio test is shown in Fig. 8.3. It is evident from this example that likelihood ratio test is superior to maximum likelihood estimation for on-line classification.

8.3 On-line Incremental Behaviour Modelling

The behaviour model of normality introduced so far does not change once learned from a given training data set. It is thus unable to cope with changes in behaviour context. To overcome this problem, Xiang and Gong (2008a) suggest a method for incremental and adaptive behaviour model learning. This algorithm initialises a behaviour model using a small bootstrapping data set. The model then undertakes both on-line abnormal behaviour detection and incremental model parameter updating simultaneously when each new behaviour pattern is presented to the model. More importantly, the model is designed to detect changes in both visual context and definitions of anomaly. It carries out model adaptation to reflect these changes.

In essence, given a newly observed behaviour pattern, it is detected as either normal or abnormal by the model learned using the patterns observed so far. The model parameters are then updated incrementally using only the new data based on an incremental expectation-maximisation (EM) algorithm.

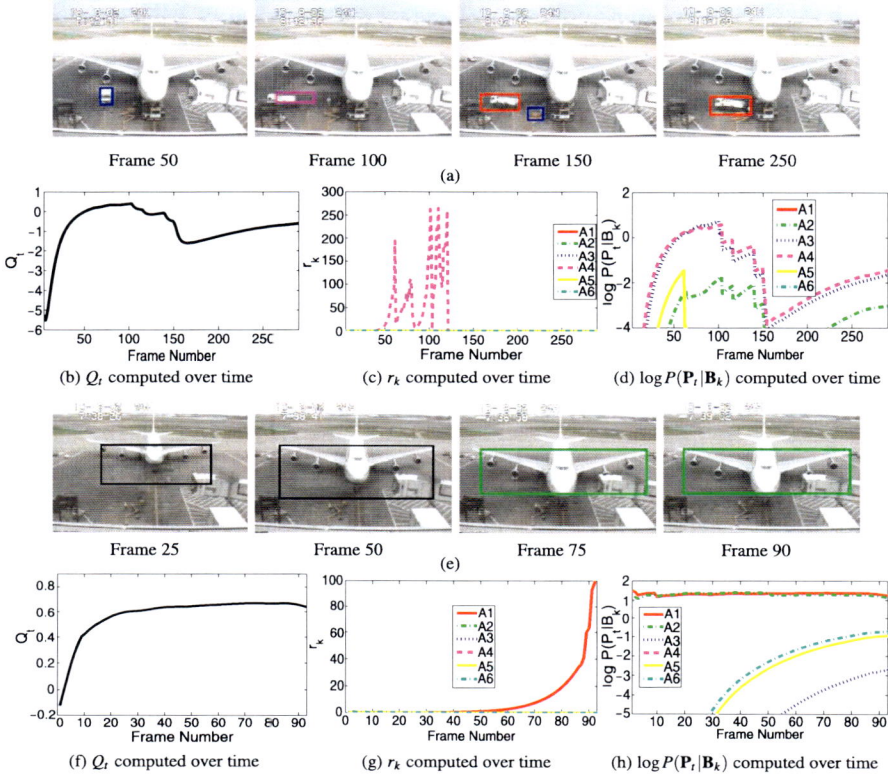

Fig. 8.3 Compare the run-time anomaly measure (**b**), the likelihood ratio test (**c**) and the MLE (**d**) for on-line normal behaviour classification in an aircraft docking scene, where six behaviour pattern classes are automatically discovered, denoted as A1–A6, respectively. (**a**) An abnormal behaviour pattern where a truck brought engineers to fix a ground power cable problem. (**b**) It is detected as anomaly from Frame 147 till the end based on Q_t. (**c**) The behaviour pattern between Frame 53 to 145 is recognised reliably as A4 (a normal class) using likelihood ratio test before becoming abnormal and being detected by Q_t. (**d**) The behaviour pattern was wrongly recognised as A3 between Frame 80 to 98 using the maximum likelihood estimation. (**e**) A normal A1 behaviour pattern where an aircraft moved into a docking position. (**f**) The behaviour pattern was detected as normal throughout based on Q_t. (**g**) It was recognised reliably as A1 from Frame 73 till the end using the likelihood ratio test (**h**) It was wrongly recognised as A2 between Frame 12 to 49 using MLE

8.3.1 Model Bootstrapping

Given a small bootstrapping data set **D** consisting of N behaviour patterns, an initial behaviour model is built. This initial model is constructed as a mixture of multi-observation HMMs (Sec. 8.2.1). However, a key difference from the earlier single model of normal behaviours is that two different mixture models are learned for normal and abnormal behaviours respectively. Specifically, model bootstrapping is proceeded as follows:

1. By using the spectral-clustering technique (Sect. 8.1), each of the N behaviour patterns in the bootstrapping data set $\mathbf{D}$ is labelled as one of the K_c behaviour classes discovered automatically by clustering.
2. The kth ($1 \leq k \leq K_c$) behaviour class is modelled using a multi-observation HMM denoted as $\mathbf{B}_k$. The parameters of $\mathbf{B}_k$, $\boldsymbol{\theta}_{\mathbf{B}_k}$ are estimated using all the behaviour patterns that belong to the kth class in $\mathbf{D}$.
3. Critically, each of the K_c behaviour classes is labelled as being either normal and abnormal according to the number of patterns within the class. This implies that the model initialises anomaly by rarity. More specifically, we have the following.
   a. The K_c classes are ordered in descending order according to the number of class members. The first K_n classes are labelled as being normal, whilst K_n is computed as

$$K_n = \arg\min_{K_n} \left(\sum_{k=1}^{K_n} \frac{N_k}{N} > Q \right) \tag{8.20}$$

   where N_k is the number of members in the kth class and Q corresponds to the minimum portion of the behaviour patterns in the bootstrapping training set to be deemed as being normal, and $0 < Q \leq 1$.
   b. A normal behaviour model $\mathbf{M}_n$ is constructed as a mixture of K_n multi-observation HMMs for the K_n normal behaviour classes.
   c. Let $\mathbf{P}$ be an example of $\mathbf{M}_n$. The probability density function of $\mathbf{M}_n$ can be written as

$$P(\mathbf{P}|\mathbf{M}_n) = \sum_{k=1}^{K_n} w_{nk} P(\mathbf{P}|\mathbf{B}_{nk}) \tag{8.21}$$

   where w_{nk} is the mixing probability, or weight, of the kth mixture component with $\sum_{k=1}^{K_n} w_{nk} = 1$. $\mathbf{B}_{nk}$ is the kth multi-observation HMM corresponding to the kth normal behaviour class.
   d. The parameters of the normal behaviour model $\mathbf{M}_n$ are

$$\boldsymbol{\theta}_{\mathbf{M}_n} = \{K_n, w_{n1}, \ldots, w_{ni}, \ldots, w_{nK_n}, \boldsymbol{\theta}_{\mathbf{B}_{n1}}, \ldots, \boldsymbol{\theta}_{\mathbf{B}_{ni}}, \ldots, \boldsymbol{\theta}_{\mathbf{B}_{nK_n}}\}$$

4. An approximate abnormal model $\mathbf{M}_a$ is also initialised using the $K_a = K_c - K_n$ abnormal behaviour classes in the bootstrapping data set. More precisely,
   a. The probability density function of $\mathbf{M}_a$ is written as

$$P(\mathbf{P}|\mathbf{M}_a) = \sum_{k=1}^{K_a} w_{ak} P(\mathbf{P}|\mathbf{B}_{ak}) \tag{8.22}$$

   b. The parameters of the approximate abnormal behaviour model $\mathbf{M}_a$ are

$$\boldsymbol{\theta}_{\mathbf{M}_a} = \{K_a, w_{a1}, \ldots, w_{aj}, \ldots, w_{aK_a}, \boldsymbol{\theta}_{\mathbf{B}_{a1}}, \ldots, \boldsymbol{\theta}_{\mathbf{B}_{aj}}, \ldots, \boldsymbol{\theta}_{\mathbf{B}_{aK_a}}\}$$

During model initialisation, given a very small bootstrapping training set with poor statistics, abnormal behaviour detection is performed based on learning from the initial training set simply according to the rarity of behaviours. This is a reasonable starting position as there is no other meaningful discriminative information available in the small bootstrapping training set. For further abnormality detection as more data becomes available on-line, a more sophisticated incremental learning mechanism is formulated. This incremental learning model takes into consideration the generalisation capability of mixture models learned using an incremental EM algorithm.

8.3.2 Incremental Parameter Update

Given a pair of bootstrapped initial normal and abnormal behaviour models, the models are deployed to detect anomaly in new on-line video stream input using a likelihood ratio test (Sect. 8.2). Specifically, given a newly observed behaviour pattern $\mathbf{P}_{new}$ and the current models $\mathbf{M}_n$ and $\mathbf{M}_a$, anomaly detection is based on a likelihood ratio test given by

$$\Lambda(\mathbf{P}_{new}) = \frac{P(\mathbf{P}_{new}; H_0)}{P(\mathbf{P}_{new}; H_i)} \begin{cases} \geq \text{Th}_\Lambda, & \text{accept } H_0 \\ < \text{Th}_\Lambda, & \text{accept } H_i \end{cases} \tag{8.23}$$

where $P(\mathbf{P}_{new}; H_0)$ and $P(\mathbf{P}_{new}; H_i)$ are the likelihood functions of a normal (H_0) and abnormal (H_i) hypotheses respectively. Th_Λ is a rejection threshold.

The key to this likelihood ratio test is constructing accurately an abnormal behaviour model, which is much more difficult compared to the construction of a normal behaviour model. This is because abnormal behaviours are generally rare and unpredictable. There are potentially infinite ways of being abnormal and they do not repeat often. To overcome this problem, a mixture of multi-observation HMMs ($\mathbf{M}_a$), constructed from a small number of abnormal behaviour patterns observed so far, is employed to approximate the abnormal behaviour model. The likelihood ratio test is then re-written as

$$\Lambda(\mathbf{P}_{new}) = \frac{P(\mathbf{P}_{new}|\mathbf{M}_n)}{P(\mathbf{P}_{new}|\mathbf{M}_a)} \begin{cases} \geq \text{Th}_\Lambda, & \text{accept } H_0 \\ < \text{Th}_\Lambda, & \text{accept } H_i \end{cases} \tag{8.24}$$

where $P(\mathbf{P}_{new}|\mathbf{M}_n)$ and $P(\mathbf{P}_{new}|\mathbf{M}_a)$ are computed by (8.21) and (8.22) respectively. As the newly observed behaviour pattern $\mathbf{P}_{new}$ is classified as either normal or abnormal, the parameters of $\mathbf{M}_n$ and $\mathbf{M}_a$ are updated, respectively, as follows.

Normal Behaviour Parameter Update

If $\mathbf{P}_{new}$ is normal, $\mathbf{P}_{new}$ is expected to be matched with a mixture component of $\mathbf{M}_n$ that has the maximum posterior probability, the probability that $\mathbf{P}_{new}$ is generated by

---

Algorithm 8.1: An on-line incremental EM algorithm for updating the parameters of a mixture component of $\mathbf{M}_n$ by a newly observed behaviour pattern $\mathbf{P}_{\text{new}}$. The forward and backward procedure is given by Rabiner (1989) and computing sufficient statistics is by Berger (1995). Convergence of the algorithm is reached when $P(\mathbf{P}_{\text{new}}|\theta_{\mathbf{B}_{ni}}^{[p+1]}) - P(\mathbf{P}_{\text{new}}|\theta_{\mathbf{B}_{ni}}^{[p]}) < \text{Th}_p$ where Th_p is a threshold.

---

1 Set iteration counter $p = 0$;

2 Set $\theta_{\mathbf{B}_{ni}}^{[0]} = \theta_{\mathbf{B}_{ni}}^{[\text{old}]}$, the parameters of $\mathbf{B}_{ni}$ before seeing $\mathbf{P}_{\text{new}}$;

3 **while** no convergence **do**

4   **E-step**;

5   Given $\mathbf{P}_{\text{new}}$ and $\theta_{\mathbf{B}_{ni}}^{[p]}$, compute the sufficient statistics of $\mathbf{P}_{\text{new}}$, $S_{\mathbf{P}_{\text{new}}}^{[p+1]}$ using the forward and backward procedure over $\mathbf{P}_{\text{new}}$;

6   Compute the sufficient statistics for the complete data (all the behaviour patterns observed so far that belong to $\mathbf{B}_{ni}$) as
$$S^{[p+1]} = S^{[p]} + S_{\mathbf{P}_{\text{new}}}^{[p+1]} - S_{\mathbf{P}_{\text{new}}}^{[p]};$$

7   **M-step**;

8   Set $\theta_{\mathbf{B}_{ni}}^{[p+1]}$ to the $\theta_{\mathbf{B}_{ni}}$ that yields the maximum likelihood given $S^{[p+1]}$;

9   Set $p = p + 1$;

10 **end**

---

that component. This best matched component is denoted as $\mathbf{B}_{ni}$ and the posterior probability for $\mathbf{B}_{ni}$ is computed as

$$P(\mathbf{B}_{ni}|\mathbf{P}_{\text{new}}) = \frac{w_{ni} P(\mathbf{P}_{\text{new}}|\mathbf{B}_{ni})}{P(\mathbf{P}_{\text{new}}|\mathbf{M}_n)} \tag{8.25}$$

where $P(\mathbf{P}_{\text{new}}|\mathbf{M}_n)$ is given by (8.21).

The parameters of $\mathbf{B}_{ni}$, $\theta_{\mathbf{B}_{ni}}$, are updated using an incremental EM algorithm (Algorithm 8.1). After $\theta_{\mathbf{B}_{ni}}$ are updated, the weight of the matched mixture component is also updated as

$$w_{ni}^{[\text{new}]} = w_{ni}^{[\text{old}]} + \alpha\left(1 - w_{ni}^{[\text{old}]}\right) \tag{8.26}$$

where $w_{ni}^{[\text{old}]}$ is the weight before seeing $\mathbf{P}_{\text{new}}$ and α is a learning rate, $0 \leq \alpha \leq 1$. The weights for all the components of $\theta_{\mathbf{M}_n}$ are re-normalised so that they satisfy $\sum_{k=1}^{K_n} w_{nk} = 1$. Consequently, the weight of the matched component is increased whilst the weights for the other components of $\mathbf{M}_n$ are decreased. The learning rate α determines the speed at which the weights are updated.

The general principle of incremental EM was originally introduced by Neal and Hinton (1998). We use a specific algorithm for on-line incremental learning of $\mathbf{B}_{ni}$ given a detected normal behaviour pattern $\mathbf{P}_{\text{new}}$, as outlined in Algorithm 8.1. It is proven that stable convergence is guaranteed for such an incremental EM algorithm (Neal and Hinton 1998). Both the conventional, batch EM and an incremental EM have the identical M-step. The difference lies in where sufficient statistics $S^{[p+1]}$ is computed in the E-step. Specifically, the batch EM algorithm computes sufficient statistics on the whole data set at each iteration. In contrast, an incremental EM updates $S^{[p+1]}$ incrementally using a subset of the data set, in this case a single data item.

The main rationale for an incremental EM in an off-line learning process is that faster convergence can be achieved. This is because that the information from the new data contributes to the parameter estimation more quickly than the batch EM algorithm. In other words, the parameter estimation efficiency is the main concern. For on-line behaviour learning, the main motivation for an incremental EM is that data only become available sequentially, as the whole data set never exists. The E-step of Algorithm 8.1 only considers a single data item $\mathbf{P}_{new}$. Furthermore, both the E-step and the M-step take constant time, regardless of the number of behaviour patterns observed so far. These characteristics make the algorithm both computation and memory efficient, particularly suitable for real-time processing.

Abnormal Behaviour Parameter Update

If $\mathbf{P}_{new}$ is detected as being abnormal, a model needs to further establish whether $\mathbf{P}_{new}$ belongs to one of the existing abnormal behaviour classes. Specifically, the best matched mixture component of $\mathbf{M}_a$ is estimated using a posterior probability $\mathbf{B}_{aj}$. The similarity distance between $\mathbf{P}_{new}$ and $\mathbf{B}_{aj}$ is measured as the normalised log-likelihood of observing $\mathbf{P}_{new}$ given $\mathbf{B}_{aj}$:

$$d(\mathbf{P}_{new}, \mathbf{B}_{aj}) = \frac{1}{T_{\mathbf{P}_{new}}} \log P(\mathbf{P}_{new} | \boldsymbol{\theta}_{\mathbf{B}_{aj}})$$

where $T_{\mathbf{P}_{new}}$ is the length of $\mathbf{P}_{new}$ (total number of frames). Now, consider the following:

1. If

$$d(\mathbf{P}_{new}, \mathbf{B}_{aj}) > \text{Th}_d \qquad (8.27)$$

then $\mathbf{P}_{new}$ is considered to be best matched with mixture component $\mathbf{B}_{aj}$. Both $\boldsymbol{\theta}_{\mathbf{B}_{aj}}$ and w_{aj} are updated in the same way as $\boldsymbol{\theta}_{\mathbf{B}_{ni}}$ and w_{ni} (Algorithm 8.1 and (8.26)).

2. Otherwise, $\mathbf{P}_{new}$ is detected as being abnormal but Condition (8.27) is not satisfied. Then, a new mixture component corresponding to a new abnormal behaviour class is created and added to $\mathbf{M}_a$, with its parameters estimated using $\mathbf{P}_{new}$ and its weight set to the smallest weight of the existing components of $\mathbf{M}_a$. Weight re-normalisation is performed to ensure that $\sum_{k=1}^{K_a} w_{ak} = 1$.

8.3.3 Model Structure Adaptation

Model structure adaptation is achieved through a mixture component trimming process. Unlike model parameter updating which is carried out whenever new data are available, model structure adaptation is performed only when changes in visual context are detected.

More specifically, when a normal behaviour class, represented as one of the mixture component of $\mathbf{M}_n$, has not been supported by any new observations, its weight is decreased gradually following the model parameter updating procedure described above. When its weight is smaller than a threshold Th_{w1}, it can be assumed that this behaviour class has become 'abnormal'. The corresponding mixture component would be regrouped into the approximate abnormal behaviour model $\mathbf{M}_a$.

In the meantime, when an abnormal behaviour class is triggered repeatedly by new observations, with Condition (8.27) being satisfied, the weight of the corresponding mixture component will increase gradually. When this weight becomes greater than a threshold Th_{w2}, it becomes normal. The corresponding mixture component would be regrouped into the normal behaviour mixture $\mathbf{M}_n$.

The abnormal classes whose weights are smaller than Th_{w1} are discarded in order to impose a limit on the total number of abnormal behaviour classes that a model is designed to cope. We call this process mixture component trimming. After component trimming, the mixture weights of $\mathbf{M}_n$ and $\mathbf{M}_a$ are re-normalised. Mixture component trimming enables a behaviour model adaptive to changes in visual context over time. Consequently the numbers of mixture components representing behaviour classes for both the normal and abnormal models can vary over time. Another reason for component trimming is that there are always limited computing resources whilst the total number of abnormal behaviour classes is potentially unlimited.

Xiang and Gong (2008a) show experimental results to suggest that an incremental learning model is superior to the conventional batch based learning model in terms of both performance on abnormality detection and computational efficiency. An incremental behaviour profiling model is capable of adapting to changes of visual context and can run in real-time. This makes it suitable for a practical application that requires the processing of 24/7 continuous video input streams.

8.4 Discussion

We considered the problem of learning behaviour models without exhaustive manual labelling of the training data. A framework is described for unsupervised behaviour profiling. The key feature of this framework is a spectral-clustering algorithm that can effectively discover the intrinsic groupings of behaviour patterns from a set of unlabelled training data. The method exploits both feature selection and automatic model order selection.

Given learned groupings of behaviour patterns, a composite behaviour model is considered as a mixture of dynamic Bayesian networks, and in particular of multi-observation hidden Markov models. On-line anomaly detection and normal behaviour recognition can be exploited given such an unsupervised model. It is shown that more robust anomaly detection can be achieved using this model, as compared to a supervised model learned from the same data. To make the model adaptive to changes in visual context, an incremental learning algorithm can be exploited to accommodate the need for re-classifying anomaly to being normal as more visual observations become available over time, and vice versa.

The unsupervised clustering based behaviour profiling framework is by no means a general purpose anomaly detection method applicable to any type of visual scene. In particular, the model is limited in coping with a very crowded and unstructured dynamic environment. This is partly due to the limitation of the representation. More specifically, crowded scenes with a non-stationary background can cause problems in contextual event detection. This is because it becomes difficult and ambiguous in such visual environments to define and measure significant visual changes that are associated with useful contextual events.

A mixture of dynamic Bayesian networks also has its limitations in coping with noisy inputs. As a DBN models temporal ordering explicitly, it is susceptible to noise and errors in event detection. Such feature detection errors can be propagated through the model, resulting in behaviours falsely detected as anomalies. To address this problem, a more robust alternative approach based on a different type of graphical model will be examined in the next chapter.

The unsupervised behaviour profiling problem is addressed in this chapter essentially by two separate steps: video data mining by unsupervised clustering to discover intrinsic behaviour pattern groups, and constructing individual stochastic behaviour class models as mixture components for a composite dynamic Bayesian behaviour model. However, it is also possible to address both the clustering and behaviour modelling problems simultaneously using a single framework learned in a single step. In the next chapter, we investigate such a solution in the form of a hierarchical model.

References

Altman, D.G., Bland, J.M.: Diagnostic tests 1: sensitivity and specificity. Br. Med. J. **308**(6943), 1552 (1994)

Berger, J.: Statistical Decision Theory and Bayesian Analysis. Springer, Berlin (1995)

Bernstein, A., Mannor, S., Shimkin, N.: Online classification with specificity constraints. In: Advances in Neural Information Processing Systems, Vancouver, Canada, December 2010

Bishop, C.M.: Pattern Recognition and Machine Learning. Springer, Berlin (2006)

Dempster, A.P., Laird, N.M., Rubin, D.B.: Maximum likelihood from incomplete data via the EM algorithm. J. R. Stat. Soc. **39**(1), 1–38 (1977)

Higgins, A., Bahler, L., Porter, J.: Speaker verification using randomized phrase prompting. Digit. Signal Process. **1**, 89–106 (1991)

Kruskal, J.B., Liberman, M.: The Symmetric Time-warping Problem: From Continuous to Discrete. In: Sankof, D., Kruskal, J. (eds.) Time Warps, String Edits, and Macromolecules: The Theory and Practice of Sequence Comparison, pp. 125–161. CSLI Publications, Standford (1983)

Neal, R., Hinton, G.: A view of the EM algorithm that justifies incremental, sparse, and other variants. In: Jordan, M.I. (ed.) Learning in Graphical Models. Kluwer Academic, Dordrecht (1998)

Ng, A.Y., Jordan, M.I., Weiss, Y.: On spectral clustering: analysis and an algorithm. In: Advances in Neural Information Processing Systems, Vancouver, Canada, December 2001, pp. 849–856 (2001)

Ng, J., Gong, S.: Learning intrinsic video content using Levenshtein distance in graph partition. In: European Conference on Computer Vision, Copenhagen, Denmark, May 2002, pp. 670–684 (2002)

Rabiner, L.R.: A tutorial on hidden Markov models and selected applications in speech recognition. Proc. IEEE **77**(2), 257–286 (1989)

Roberts, S., Husmeier, D., Rezek, I., Penny, W.: Bayesian approaches to Gaussian mixture modelling. IEEE Trans. Pattern Anal. Mach. Intell. **20**(11), 1133–1142 (1998)

Shan, Y., Sawhney, H.S., Pope, A.: Measuring the similarity of two image sequence. In: Asian Conference on Computer Vision, Jeju, Korea, January 2004

Shi, J., Malik, J.: Normalized cuts and image segmentation. IEEE Trans. Pattern Anal. Mach. Intell. **22**(8), 888–905 (2000)

Weiss, Y.: Segmentation using eigenvectors: a unifying view. In: IEEE International Conference on Computer Vision, Corfu, Greece, pp. 975–982 (1999)

Wilpon, J., Rabiner, L.R., Lee, C., Goldman, E.: Automatic recognition of keywords in unconstrained speech using hidden Markov models. IEEE Trans. Acoust. Speech Signal Process. **38**(11), 1870–1878 (1990)

Xiang, T., Gong, S.: Incremental and adaptive abnormal behaviour detection. Comput. Vis. Image Underst. **111**(1), 59–73 (2008a)

Xiang, T., Gong, S.: Video behaviour profiling for anomaly detection. IEEE Trans. Pattern Anal. Mach. Intell. **30**(5), 893–908 (2008b)

Zweig, M.H., Campbell, G.: Receiver-operating characteristic (ROC) plots: a fundamental evaluation tool in clinical medicine. Clin. Chem. **39**(8), 561–577 (1993)

Chapter 9
Hierarchical Behaviour Discovery

Behaviour of groups of objects observed in a crowded public space is typically complex and uncertain with an intrinsically hierarchical structure both in space and over time. Due to the complex and uncertain nature of object behaviour patterns in public spaces, what is considered to be 'subjectively interesting behaviour' to a human observer can be influenced by a wide variety of factors including: (1) the activity of a single object over time; (2) the correlated spatial states of multiple objects, for example, a piece of abandoned luggage is defined by separation from its owner; and (3) higher order spatial and temporal correlations among multiple entities, for instance, traffic flow at a road intersection has a particular spatio-temporal order beyond co-occurrence dictated by traffic lights. In addition, the spatial or temporal range over which correlations might be important can be short or long. Constructing computational models that are both flexible and accurate in representing such complex and uncertain characteristics of behaviour is challenging.

There are three fundamental challenges facing computational modelling of complex behaviours in public spaces. They are robustness, sensitivity and computational tractability. Traditional computational methods to behaviour modelling take a pipelined approach that include processing components such as visual feature extraction, segmentation, classification, identification and tracking. Typical public spaces, however, are crowded and visually incomplete due to occlusion either partially or completely. They may also have extreme lighting variations, and contain a variety of different object classes and poses. All of them create difficulties for the pipelined computational approach. Even with a contextual event based representation, as considered in the preceding two chapters for group activity representation, a behaviour model remains to be sensitive to errors and noise from visual feature extraction and segmentation. Model robustness is important and remains a bottleneck in behaviour modelling.

A model is also required to be sensitive in identifying changes in subtle details in the presence of overwhelming clutter. For instance, an effective behaviour model is required to discover behaviours of interest that are often visually subtle, very short in duration or co-occurring with other less interesting behaviours. Computationally, there is a trade-off between the requirements of robustness and sensitivity. A sensitive model is potentially also sensitive to noise thus lacking robustness, whilst a robust model can lose sensitivity in detecting genuine abnormal behaviour patterns when attempting to filter out noise and errors in visual features. To be useful, a model needs to strike the right balance between robustness and sensitivity. In addition to robustness and sensitivity, computational tractability is critical for modelling complex spatio-temporal correlations among different object behaviours. This is especially true for on-line analysis, designed to alert real-time response and assist instant decision making.

A dynamic topic model possesses unique computational attributes that make it an attractive framework for addressing all three underlying challenges facing modelling behaviours in crowded scenes. One such model is the Markov clustering topic model introduced by Hospedales et al. (2009), and extended by Kuettel et al. (2010). This model aims to address the problem of unsupervised modelling of multi-object spatio-temporal behaviours in crowded and complex public scenes. A Markov clustering topic model draws on machine learning theories on probabilistic topic models and dynamic Bayesian networks to achieve a robust hierarchical model of behaviours and their dynamics.

A Markov clustering topic model of object behaviour has a three-layer hierarchical architecture for extracting visual information and describing behavioural dynamics:

1. A codebook of local motion events is learned, so as to generate discrete input features from video. Co-occurring events are automatically composed into activities, such as a pedestrian crossing a road.
2. Co-occurring activities are automatically composed into complex multi-object behaviours, such as object interactions at a street intersection. These behaviours are correlated over time.
3. By introducing a Markov chain to model behaviour dynamics, a Markov clustering topic model is a dynamic Bayesian network generalisation of a static topic model.

The model is learned off-line from unlabelled training data with Gibbs sampling. A topic codebook based representation in a hierarchical temporal structure addresses the needs for model sensitivity and computational tractability when modelling complex multi-object behaviours. The model is probabilistic in nature, well suited for addressing the need for robustness. In the following, we describe in details the representation and an inference algorithm for learning such a model and deploying the model to on-line processing.

9.1 Local Motion Events

The aim of hierarchical behaviour modelling using a Markov clustering topic model is to construct a generative model capable of automatically mining and screening irregular spatio-temporal patterns as 'salient behaviours' in video data of public spaces with people and vehicles at both far and near-field views. The observational viewpoints typically contain multiple groups of heterogeneous objects, occlusions, and shadows. Such viewing conditions raise difficulties for grouping contextual events into plausible behaviour classes, as discussed in Sect. 7.1. As an alternative, a set of much simpler low-level local motion features are exploited to construct local motion events.

More precisely, a camera view is divided into $C \times C$ pixel cells. An optical flow vector is then computed for each cell. When the magnitude of a flow vector is greater than a threshold Th_o, it is deemed reliable and quantised into one of four cardinal directions. A discrete local motion event is defined by the position of a cell and its motion direction. For a 320×240 video frame with cell size of 10×10, a total of $N_x = 32 \times 24 \times 4 = 3072$ different discrete visual events may therefore occur in combination. For visual scenes where objects may remain static for sustained period of time, for example, people waiting for trains at an underground station, one can also use background subtraction to generate a fifth, stationary foreground pixel, state for each cell, giving a visual event codebook size of 3840. This illustrates the flexibility of this representation: it can easily incorporate other kinds of 'metadata features' that may be relevant in a given scene. The input video is uniformly segmented into unit-length clips typically lasting one second, and the input to the Markov clustering topic model at second t is the bag of all N_t visual events occurring in video clip t, denoted as $\mathbf{x}_t = \{x_{1,t}, ., x_{i,t}, ., x_{N_t,t}\}$.

9.2 Markov Clustering Topic Model

A Markov clustering topic model is closely related to a static topic model latent Dirichlet allocation, known as LDA[1] (Blei et al. 2003). Figure 9.1(a) shows a graph representation of LDA. LDA is originally proposed for text document analysis designed to be a generative model of a set of text documents $\mathbf{x}_m$, $m = 1..M$. Specifically, a document m is represented as a bag of $i = 1, \ldots, N_m$ unordered words $x_{i,m}$. Each word is distributed according to a multinomial distribution, $p(x_{i,m}|\phi_{y_{i,m}})$ indexed by the current topic of discussion $y_{i,m}$. Topics are intrinsically latent and chosen from a per-document Dirichlet distribution θ_m. Inference of latent topics $\mathbf{y}$ with parameters θ and ϕ given data $\mathbf{x}_m$ clusters co-occurring words into topics.

This topic based representation of text documents can facilitate both querying and similarity matching. Querying is a process of searching for documents containing similar topics to the topics of some query words, whilst similarity matching is a process of searching for documents of similar topical content to a query document. Due to this topical representation, similarities between queries and documents can be discovered even with few actual word tokens in common. For modelling object behaviours in video data, we consider that video clips are video documents. Among all video documents, local motion events correspond to video words. Simple actions, such as co-occurring events, correspond to video topics. Behaviours as co-occurring actions within each video document define video document categories.

---

[1]One must not confuse LDA as latent Dirichlet allocation with linear discriminant analysis (Belhumeur et al. 1997; Fisher 1938), also commonly known as LDA.

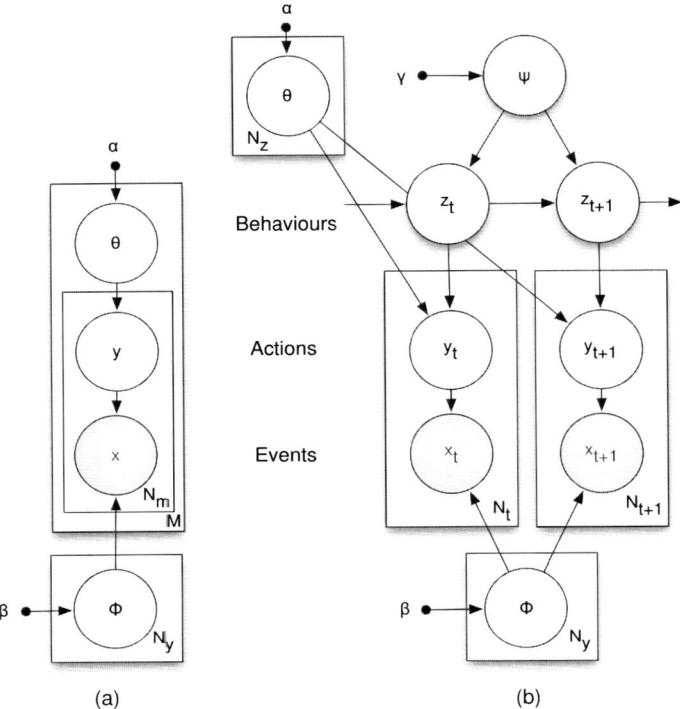

Fig. 9.1 Graphs representing: (**a**) a standard latent Dirichlet allocation model (Blei et al. 2003), (**b**) a Markov clustering topic model (Hospedales et al. 2009)

More precisely, the occurrence of a sequence of video clips as documents, $D = \{\mathbf{x}_t\}$, where $t = 1..T$, can be modelled as having a three-layer latent structure: events, actions and behaviours, as illustrated by the graphical representation in Fig. 9.1(b). The number of possible actions and behaviours in a training data set are assumed to be known and fixed as N_y and N_z respectively, although this assumption will be relaxed later (Sect. 9.4). A generative model is defined as follows:

1. Each clip t exhibits a particular category of behaviour z_t.
2. Behaviour z_t varies systematically over time from clip to clip according to some unknown multinomial distribution $p(z_t|z_{t-1}, \psi)$, denoted as Multi($\cdot$).
3. Within each clip t, a bag of N_t simple actions $\{y_{i,t}\}_{i=1}^{N_t}$ are chosen, with each action chosen independently, based on the clip category $y_{i,t} \sim p(y_{i,t}|z_t, \theta)$.
4. Finally, each observed local motion event $x_{i,t}$ is chosen based on the associated action $y_{i,t}$, $x_{i,t} \sim p(x_{i,t}|y_{i,t}, \phi)$.

All the multinomial parameters $\{\phi, \psi, \theta\}$ are treated as unknowns with Dirichlet priors, denoted as Dir($\cdot$). The complete generative model is then specified by

$$p(\psi_z|\gamma) = \text{Dir}(\psi_z; \gamma)$$

$$p(\theta_z|\alpha) = \text{Dir}(\theta_{z;}\alpha)$$
$$p(\phi_y|\beta) = \text{Dir}(\phi_{y;}\beta)$$
$$p(z_{t+1}|z_t, \psi) = \text{Multi}(z_{t;}\psi_{z_t})$$
$$p(y_{i,t}|z_t, \theta) = \text{Multi}(y_{i,t;}\theta_{z_t})$$
$$p(x_{i,t}|y_{i,t}, \phi) = \text{Multi}(x_{i,t;}\phi_{y_{i,t}})$$

The joint distribution of variables $\{\mathbf{x}_t, \mathbf{y}_t, z_t\}_1^T$ and parameters $\{\theta, \psi, \theta\}$, given some inter-action and inter-document correlation hyper-parameters $\{\alpha, \beta, \gamma\}$, is

$$p\big(\{\mathbf{x}_t, \mathbf{y}_t, z_t\}_1^T, \phi, \psi, \theta|\alpha, \beta, \gamma\big)$$
$$= p(\phi|\beta)p(\psi|\gamma)p(\theta|\alpha)\prod_t\bigg(\prod_i p(x_{i,t}|y_{i,t}, \phi)p(y_{i,t}|z_t, \theta)\bigg)p(z_t|z_{t-1}, \psi)$$

$$(9.1)$$

In essence, the hyper-parameters $\{\alpha, \beta, \gamma\}$ specify an *a priori* belief about how sparsely local motion events, actions and behaviours are distributed (Fig. 9.1(b)). For example, this belief could be about to what extent events should be shared between topics and actions shared between behaviours. Although these hyper-parameters can be optimised by Markov chain Monte Carlo (MCMC) learning (Griffiths et al. 2005; Wallach et al. 2009), they do not strongly affect the performance of a Markov clustering topic model (Hospedales et al. 2009). In some cases, they can be simply fixed (Griffiths and Steyvers 2004; Gruber et al. 2007; Rosen-Zvi et al. 2008; Wang et al. 2009). Later in Sect. 11.1.2, when we consider a different topic model with a different learning strategy, the issue of hyper-parameter learning will be revisited.

Two basic operations of this model concern the learning and inference procedures for the model. To construct a model that learns the typical actions and behaviours from a training data set, a process is required to compute the posterior over all the parameters, $p(\theta, \psi, \phi|\mathbf{x}_{1:t})$. For detecting the occurrence of particular behaviours in a probe video clip using a learned model, the model computes the behaviour posterior $p(z_t|\mathbf{x}_{1:t})$. For detecting salient behaviours in a video clip, the model computes the predictive likelihood of the new clip, $p(\mathbf{x}_t|\mathbf{x}_{1:t-1})$. The model has a number of strengths. Whilst the Bayesian parameter inference increases robustness to over-fitting, the intermediate action layer $\mathbf{y}_t$ increases robustness to occlusion and noise compared to modelling events directly. The variety of discoverable behaviour patterns and sensitivity to saliency is enhanced by the compositional hierarchical structure and Markovian behaviour model.

9.2.1 Off-line Model Learning by Gibbs Sampling

Similar to LDA, exact inference in Markov clustering topic model is intractable. However, it is possible to derive a collapsed Gibbs sampler (Gilks et al. 1995) for approximate Markov chain Monte Carlo learning of the parameters and inference of the latents $p(\mathbf{y}_{1:T}, z_{1:T} | \mathbf{x}_{1:T})$. The Dirichlet–multinomial conjugate structure of the model allows the parameters $\{\phi, \theta, \psi\}$ to be integrated out automatically in a Gibbs sampling procedure. The Gibbs sampling update for action $y_{i,t}$ is derived by integrating out the parameters ϕ and θ in its conditional probability given the other variables:

$$p(y_{i,t} | \mathbf{y}_{\backslash i,t}, z_{1:T}, \mathbf{x}_{1:T}) \propto \frac{n^-_{x,y} + \alpha}{\sum_x n^-_{x,y} + N_x\alpha} \frac{n^-_{y,z} + \beta}{\sum_y n^-_{y,z} + N_y\beta} \tag{9.2}$$

Here $\mathbf{y}_{\backslash i,t}$ denotes all the variables $\mathbf{y}_{1:T}$ excluding $y_{i,t}$; $n^-_{x,y}$ denotes the counts of feature x being associated to action y; $n^-_{y,z}$ denotes the counts of action y being associated to behaviour z. Superscript "$-$" denotes counts excluding item (i, t). N_x is the size of the visual event codebook, and N_y the number of simple actions.

The Gibbs sampling update for the behaviour cluster z_t is derived by integrating out parameters ψ and θ in the conditional $p(z_t | \mathbf{y}, z_{\backslash t}, \mathbf{x})$, and must account for the possible transitions between z_{t-1} and z_{t+1} along the Markov chain of clusters:

$$p(z_t | \mathbf{y}_{1:T}, z_{\backslash t}, \mathbf{x}_{1:T}) \propto \frac{\prod_y \Gamma(\alpha + n_{y,z_t}) \Gamma(N_y\alpha + n^-_{\cdot,z_t})}{\prod_y \Gamma(\alpha + n^-_{y,z_t}) \Gamma(N_y\alpha + n_{\cdot,z_t})} \frac{n^-_{z',z} + \gamma}{n^-_{z',z} + N_z\gamma}$$

$$\times \frac{n_{z_{t+1},z_t} + \mathrm{I}(z_{t-1} = z_t)\mathrm{I}(z_t = z_{t+1}) + \gamma}{n_{\cdot,z_t} + \mathrm{I}(z_{t-1} = z_t) + N_z\gamma} \tag{9.3}$$

Here $n_{z',z}$ represents the counts of behaviour z' following behaviour z, $n_{\cdot,z} \triangleq \sum_{z'} n_{\cdot,z}$, and N_z is the number of clusters. I is the identity function that returns 1 if its argument is true, and Γ is the gamma function.[2]

Iterations of (9.2) and (9.3) entail inference by eventually drawing samples from the posterior $p(\{\mathbf{y}_t, z_t\}_1^T | \{\mathbf{x}\}_1^T, \alpha, \beta, \gamma)$. Parameters $\{\phi, \psi, \theta\}$ are estimated from the expectation of their distribution given a full set of samples (Griffiths and Steyvers 2004; Rosen-Zvi et al. 2008), as follows:

$$\hat{\phi}^s_y = \frac{n_{x,y} + \beta}{n_{\cdot,y} + N_x\beta} \tag{9.4}$$

$$\hat{\theta}^s_z = \frac{n_{y,z} + \alpha}{n_{\cdot,z} + N_y\alpha} \tag{9.5}$$

$$\hat{\psi}^s_z = \frac{n_{z',z} + \gamma}{n_{\cdot,z} + N_z\gamma} \tag{9.6}$$

We consider a street intersection scene to illustrate how a learned Markov clustering topic model is used for implicit hierarchical clustering in order to provide automated dynamic scene understanding given a training data set. The street intersection scene has three traffic flows in different directions regulated by the traffic

---

[2]Here, one does not obtain the simplification of gamma functions as in the case of a standard LDA (Blei et al. 2003) and in (9.2). This is because the inclusive and exclusive counts may differ by more than 1. However, the computation is not prohibitively costly, as (9.3) is computed only once per clip.

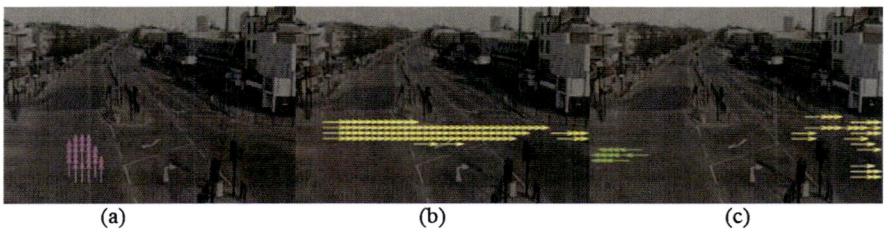

Fig. 9.2 Examples of topics as actions learned in a street intersection scene, illustrated by the most likely events for each topic $\hat{\phi}_y^s$. *Arrow directions* and *colours* represent flow directions of local motion events

lights. These traffic flows are in a certain temporal order. In this example, we assume that the numbers of actions and behaviour clusters are known. In the Markov clustering topic model, the number of actions is set to $N_y = 8$ and the number of behaviour clusters is $N_z = 4$. This assumption is relaxed to a more general case in Sect. 9.4. Clearly, larger N_y and N_z result in a more fine-grained decomposition of scene behaviours. More specifically, we examine two aspects of a model's ability to learn latent behaviours from a training data set:

Clustering visual events into topics as actions: The actions learned by a Markov clustering topic model correspond to co-occurring events. These actions are typically associated with patterns of moving objects. Figure 9.2 shows some example actions y discovered by drawing local motion events $\mathbf{x}$ in the top 50% of the mass of the distribution $p(\mathbf{x}|y, \hat{\phi}_y^s)$ (9.4). Each action has a clear semantic interpretation. In the street intersection scene, Figs. 9.2(a) and (b) represent vertical left lane and horizontal leftwards traffic respectively. Figure 9.2(c) represents the vertical traffic vehicles turning right at a traffic filter.

Discovering behaviours and their dynamics: Co-occurring topics are automatically clustered into behaviours z by matrix θ_z in (9.5). Each behaviour cluster corresponds to a complex behaviour pattern involving multiple interacting objects. Some examples of complex behaviour clusters discovered in a street intersection scene are depicted in Fig. 9.3. Specifically, Figs. 9.3(a) and (b) represent horizontal left and right traffic flows respectively including right turn traffic, as compared to horizontal only traffic in Fig. 9.2(b). Figures 9.3(c) and (d) represent vertical traffic flow with and without interleaved turning traffic. The temporal duration and order of each traffic flow is also discovered accurately. For example, the long duration and exclusiveness of the horizontal traffic flows (a) and (b), and the interleaving of the vertical traffic (c) and vertical turn traffic (d), are clear from the learned transition distribution $\hat{\psi}^s$ (Fig. 9.3(e)).

9.2.2 On-line Video Saliency Inference

One limitation of the model learning and inference method described above is that it is an off-line, batch procedure. For on-line saliency detection in video, we consider

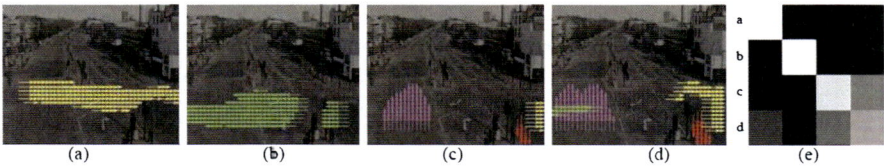

Fig. 9.3 Examples of discovered behaviours in a street intersection scene, illustrated by the most likely local motion events as video words for each behaviour $\hat{\theta}_z^s$, and the transitions between behaviours $\hat{\psi}_z^s$. Brighter shading in (**e**) corresponds to greater probability of behaviour transition

a filtered inference algorithm for the Markov clustering topic model after an off-line batch learning process takes place.

Given a training data set of T_{tr} video clips, N_s samples $\{\{\mathbf{y}_t, z_t\}_{t=1}^{T_{\mathrm{tr}}}, \hat{\phi}^s, \hat{\psi}^s, \hat{\theta}^s\}_{s=1}^{N_s}$ have been generated from the posterior distribution of latent variables in the model $p(\{\mathbf{y}_t, z_t\}_{t=1}^{T_{\mathrm{tr}}} | \{\mathbf{x}\}_1^{T_{\mathrm{tr}}}, \alpha, \beta, \gamma)$. It is assumed that no further adaptation of the parameters is necessary. That is, the training data set is representative, so $p(\phi, \psi, \theta | \mathbf{x}_{t' > T_{\mathrm{tr}}}) = p(\phi, \psi, \theta | \mathbf{x}_{1:T_{\mathrm{tr}}})$. Bayesian filtering is then performed in the Markov chain of clusters to infer the current clip's behaviour $p(z_t | \mathbf{x}_{1:t})$ by approximating the required integral over the parameters with sums over their Gibbs samples (Rosen-Zvi et al. 2008). Conditioned on each set of (sampled) parameters, the other action $y_{i,t}$ and behaviour z_t variables de-correlate, and efficient recursions can be derived to compute the behaviour category for each clip on-line as follows:

$$p(z_{t+1} | \mathbf{x}_{1:t+1}) = \sum_{z_t} \int_{\phi, \theta, \psi} \frac{p(\mathbf{x}_{t+1}, z_{t+1} | z_t, \phi, \theta, \psi, \mathbf{x}_{1:t}) p(z_t, \phi, \theta, \psi | \mathbf{x}_{1:t})}{p(\mathbf{x}_{t+1} | \mathbf{x}_{1:t})}$$
$$\approx \frac{1}{N_s} \sum_s \frac{p(\mathbf{x}_{t+1} | z_{t+1}, \phi^s, \theta^s) p(z_{t+1} | z_t^s, \psi^s)}{p(\mathbf{x}_{t+1} | \mathbf{x}_{1:t})} \qquad (9.7)$$

Bayesian saliency as irregularity is measured by the marginal likelihood of the new observation given all the others, $p(\mathbf{x}_{t+1} | \mathbf{x}_{1:t})$. This can be determined from the normalisation constant of (9.7), as follows:

$$p(\mathbf{x}_{t+1} | \mathbf{x}_{1:t}) = \sum_{z_t} \int_{\phi, \theta, \psi} p(\mathbf{x}_{t+1} | z_t, \psi, \theta, \phi, \mathbf{x}_{1:t}) p(z_t, \phi, \psi, \theta | \mathbf{x}_{1:t})$$
$$\approx \frac{1}{N_s} \sum_{s, z_{t+1}} p(\mathbf{x}_{t+1}, z_{t+1} | \psi^s, \theta^s, \phi^s, z_t^s) \qquad (9.8)$$

Without the iterative sweeps of the Gibbs sampler, even summing over samples s, behaviour inference as clip categorisation and saliency detection can be performed on-line and in real-time by (9.7) and (9.8). In practice, (9.7) may suffer from label switching (Bishop 2006; Gilks et al. 1995). To avoid the problem, a single sample should be used for interpretable results (Griffiths and Steyvers 2004). Equation (9.8) is independent of label switches and should be used with all samples. This on-line model has no direct analogy in a standard LDA (Blei et al. 2003), as the per-document parameter θ requires iterative computation to infer. Finally, clips t

contain varying numbers N_t of events $x_{i,t}$, therefore have varying 'base' probability. To account for that, when searching for behavioural saliency in a test clip[3], a normalised predictive likelihood π is computed for each clip as follows:

$$\log \pi (\mathbf{x}_{t+1}|\mathbf{x}_{1:t}) = \frac{1}{N_t} \log p(\mathbf{x}_{t+1}|\mathbf{x}_{1:t}) \qquad (9.9)$$

The measure of behavioural saliency is given by the predictive likelihood $\pi(\mathbf{x}_t|\mathbf{x}_{1:t-1})$ of test clip $\mathbf{x}_t$ given training video data $\mathbf{x}_{1:T_{tr}}$ and previous test data $\mathbf{x}_{t-1>T_{tr}}$. A low predictive likelihood value indicates irregularity. This can be due to four different reasons, reflecting the following salient aspects of the data:

1. Events: Individual $x_{i,t}$ rarely occurred in training data $\mathbf{x}_{1:T_{tr}}$.
2. Actions: Events in $\mathbf{x}_t$ rarely occurred together in the same activity in $\mathbf{x}_{1:T_{tr}}$.
3. Behaviours: $\mathbf{x}_t$ occurred together in topics, but such topics did not occur together in clusters in $\mathbf{x}_{1:T_{tr}}$.
4. Dynamics: $\mathbf{x}_t$ occurred together in a cluster z_t, but z_t did not occur following the same clusters z_{t-1} in $\mathbf{x}_{1:T_{tr}}$.

Such detections are made possible because the hierarchical structure of the Markov clustering topic model represents behaviour at four different levels including events, actions, behaviours, and behaviour dynamics. In the following, we will refer to clips with rare events as *intrinsically* unlikely, those with rare actions and behaviours as *behaviourally* unlikely, and those with rare behavioural dynamics as *dynamically* unlikely. Algorithm 9.1 depicts the basic off-line learning and on-line inference procedures for the Markov clustering topic model.

9.3 On-line Video Screening

Given the on-line inference algorithm, new video data can be screened on-line. The overall behaviours are identified using (9.7), and visual saliency as irregularity is measured by (9.9). Figure 9.4 shows an example of on-line processing on test data from the street intersection scene. The maximum a posteriori (MAP) estimated behaviour $\hat{z}_t$ at each time is illustrated by the coloured bar in the top graph. This model MAP estimation reports the overall traffic behaviour according to traffic phases in turning, vertical flow, left flow and right flow. The top graph shows the likelihood $\pi(\mathbf{x}_t|\mathbf{x}_{1:t-1})$ of each clip as it is processed on-line over time. Three examples are shown in the bottom of Fig. 9.4 including two typical clips, turning vertical traffic and flowing vertical traffic categories, and one irregular behavioural salient clip where a vehicle drives in the wrong lane. Each is highlighted with the flow vectors (blue arrows) on which computation is based.

Some examples of visual irregularities as 'behavioural surprises' detected by an on-line video screening model are shown in Fig. 9.5. Figures 9.5(a) and (b) show a

---

[3]In this book, we alternate the use of terms 'probe' and 'test' to describe testing data, 'gallery' and 'training' to describe training data.

Algorithm 9.1: Markov Clustering Topic Model Learning and Inference Procedures

---

1 *Learning* (*Off-Line*);

   Input: event detections for every clip, $\{\mathbf{x}_t\}_{t=1}^{T_{tr}}$.

   Output: Parameter estimates $\{\hat{\phi}^s, \hat{\psi}^s, \hat{\theta}^s\}_{s=1}^{N_s}$.

2 Initialise $\{z_t, \mathbf{y}_t\}_{t=1}^{T_{tr}}$ randomly;

3 **for** $j = 1$ *to* 1000 **do**

4   **for** $t = 1$ *to* T_{tr} **do**

5   Resample $p(z_t|\mathbf{y}, \mathbf{z}_{\setminus t}, \mathbf{x})$ by (9.3);

6   For every observation i at t, resample $p(y_{i,t}|\mathbf{y}_{\setminus i,t}, \mathbf{z}, \mathbf{x})$ by (9.2);

7   **if** *100 iterations have passed since the last parameter estimation* **then**

8   Record independent sample $s = \{z_t, \mathbf{y}_t\}_{t=1}^{T_{tr}}$;

9   Estimate model parameters using (9.4), (9.5) and (9.6);

10   **end**

11   **end**

12 **end**

13 *Inference for a new test clip* t (*On-Line*);

   Input: Parameter samples $\{\hat{\phi}^s, \hat{\psi}^s, \hat{\theta}^s\}_{s=1}^{N_s}$, previous posterior $p(z_{t-1}|\mathbf{x}_{1:t-1})$,
             event detections $\mathbf{x}_t$.

   Output: behaviour inference $p(z_t|x_{1:t})$, saliency $\pi(x_t|x_{1:t-1})$.

14 Compute behaviour profile $p(z_t|x_{1:t})$ by (9.7);

15 Compute saliency $\pi(x_t|x_{1:t-1})$ by (9.9);

---

vehicle driving in the wrong lane. This is surprising, because that region of the scene typically only includes down and leftward flows. This clip is *intrinsically* unlikely, as these events were rare in the training data under any circumstances. In Fig. 9.5(c) and (d), a police car jumps a red-light and turns right through opposing traffic. In this region of the scene, the right flow of the other traffic is a typical action, as is the left flow of the police car in isolation. However, their conjunction is not, as it is forbidden by the traffic lights. Moreover, a series of clips alternately suggest left and right flows, but such dynamics are unlikely under the learned temporal model (Fig. 9.3(e)). The model considers this whole series of clips are *behaviourally* and *dynamically* unlikely given global and temporal constraints entailed by $\pi(\mathbf{x}_t|\mathbf{x}_{1:t-1})$.

Figures 9.5(c)–(f) illustrate a feature of the Markov clustering topic model as a dynamic topic model that gives an advantage over a static topic model such as LDA (Li et al. 2008; Wang et al. 2009). That is, the Markov clustering topic model is intrinsically less constrained by the document size for a bag-of-words representation, i.e. less sensitive to the need for determining a suitable video clip size. With a standard LDA, a larger document size would increase the likelihood that vertical and horizontal flows in this traffic scene are captured concurrently and detected as surprising. However, larger document sizes also risk losing model sensitivity and

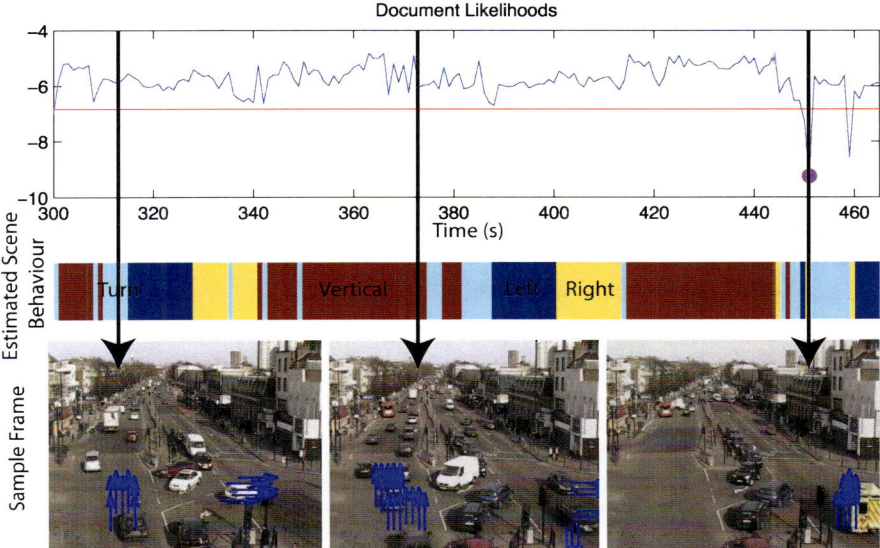

Fig. 9.4 Examples of on-line video screening

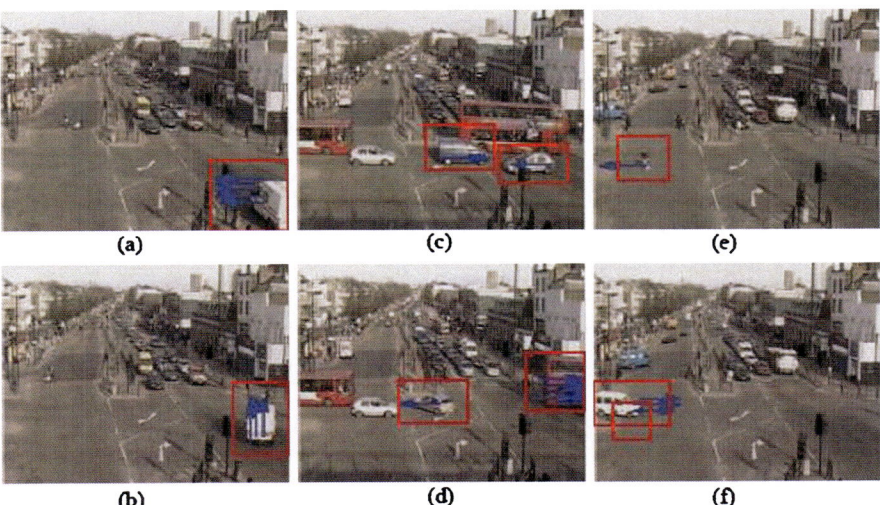

Fig. 9.5 Examples of salient video clips discovered by an on-line video screening model. *Arrows* and *dots* indicate input local motion events and *red boxes* highlight detected regions of 'behavioural surprises'

missing in capturing interesting local motion events in the presence of vast quantity of normal events. In contrast, the Markov clustering topic model facilitates the use of a small document size to increase model sensitivity. This is because such a model represents explicitly temporal information by its Markovian nature. It therefore penalises unlikely behaviour switches. As a result, it can discover not only short events such as those in Figs. 9.5(a) and (b), which are likely to be lost with a representation of large documents, but also long events such as those in Figs. 9.5(c)–(f), which could be lost with a representation of small documents without modelling inter-document temporal correlations.

Hospedales et al. (2009) show that compared with both an HMM and a LDA, the Markov clustering topic model is more sensitive in detecting surprises when screening videos of complex behaviours. In these cases, behavioural saliency is defined by an atypical co-occurrence of actions and sequence of behaviours over time. In other words, a 'behavioural surprise' is measured by hierarchical spatio-temporal correlations of actions rather than the presence or distribution of individual actions. In contrast, conventional LDA can infer actions, but cannot reason about their co-occurrence and temporal correlation simultaneously. On the other hand, an HMM can reason about temporal orders in behaviours. However, with point learning by the EM algorithm and a lack of intermediate action representation, an HMM is prone to model over-fitting.

All three types of models can do fairly well at detecting intrinsically unlikely individual local events as video words that are visually well-defined in isolation, for example, wrong way driving (Fig. 9.5(a)). The Markov clustering topic model differs by being more adept than both LDA and HMM at detecting those more visually subtle salient behaviours. This is due to its simultaneous hierarchical and temporal modelling at four different levels of behaviour, covering from events as video words, actions as visual topics, behaviours as video documents, to behaviour dynamics as inter-document temporal correlations.

9.4   Model Complexity Control

For a probabilistic topic model, model complexity corresponds to how many topics and document clusters are represented in a corpus. For modelling behaviour, this corresponds to how many simple actions N_y and behaviours N_z are represented in a video. In some cases, there may be domain knowledge that can assist in estimating the complexity of a topic model. In other cases, a model needs to discover automatically such information directly from the training data.

From a Bayesian modelling perspective, the marginal likelihood $p(D^{tr}|M)$ of the training data D^{tr}, or the marginal likelihood of a held-out test data set D^{te} given the training data $p(D^{te}|D^{tr}, M)$, should be used for selecting a model M (Bishop 2006; Gilks et al. 1995). Although the marginal likelihood of each model is affected by the chosen hyper-parameters (α, β, γ), these parameters can be fixed in order to determine the most likely model given their fixed values (Griffiths and Steyvers 2004; Wallach et al. 2009).

From a sampling perspective to modelling, the harmonic mean of the sample likelihoods drawn from each model M is an estimator of the model's marginal likelihood $p(D|M)$ (Gilks et al. 1995). This harmonic mean estimation is commonly

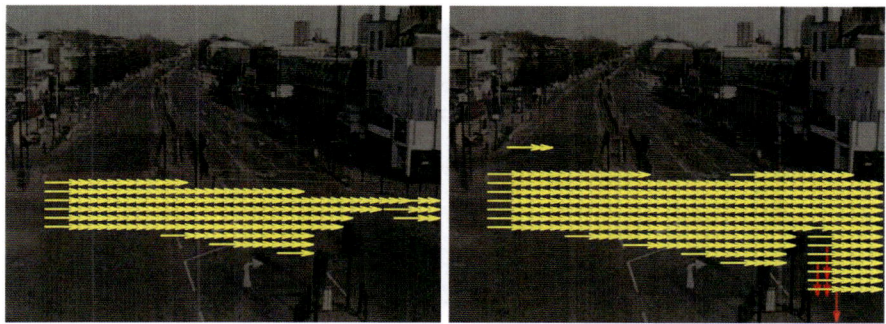

Fig. 9.6 Fine-grained cluster decomposition: horizontal versus right-turning traffic flows

used because of its simplicity and efficiency (Griffiths and Steyvers 2004; Wallach et al. 2009). However, it is also highly unstable (Gilks et al. 1995; Wallach et al. 2009). In particular, its high variance renders it unsuitable for model comparison and model selection.

An alternative to harmonic mean is the predictive likelihood of a test data set, defined as:

$$p\left(D^{\text{te}}|D^{\text{tr}}, M\right) = \prod_{t=1}^{T_{\text{te}}} p\left(\mathbf{x}_t^{\text{te}}|\mathbf{x}_{1:t-1}^{\text{te}}, D^{\text{tr}}, M\right) \tag{9.10}$$

This predictive likelihood can be computed rather efficiently by (9.9). Given a set of Markov clustering topic candidate models M with their complexity depending on the sizes of N_y and N_z, one way to select the best model from them is to learn a two dimensional grid of models across some plausible range of N_y and N_z values. This is computationally expensive, but only needs be done once off-line for a given training data set.

The effectiveness of topic model selection by measuring the predictive likelihood of test data $p(D^{\text{te}}|D^{\text{tr}}, N_y, N_z)$ is examined in more details using the street intersection scene. In this scene, the model size estimated automatically from data is larger than that assumed in Sect. 9.2.1. Notably, the automatically estimated model prefers to break down the data into more clusters, at a finer scale, than previously assumed. That is, the model is more sensitive and tends to separate behaviour clusters into more specific sub-behaviours. For example, Fig. 9.6 shows the prototype flows of two of the 12 automatically learned behaviours from the street intersection scene training data set (Hospedales et al. 2009). In this case, the overall rightward traffic phase is now broken down into horizontal traffic and right-turning traffic. In contrast, these two types of traffics were previously encompassed by a single behaviour cluster, as shown in Fig. 9.3(a). Because the right-turning traffics are sparse, they are only detectable at a finer scale. In practice, there is no clear answer as the choice of model size and sensitivity is dependent on the visual context. Automatically estimated numbers of actions and behaviours should only be taken as a plausible guide in the absence of other domain knowledge, not the absolute best values.

9.5 Semi-supervised Learning of Behavioural Saliency

In unsupervised behaviour discovery, a model defines behavioural saliency by statistical rarity or irregularity in the observed visual data. This is reasonable for a fully automated video screening system geared towards profiling typical behaviours and discovering atypical surprises from data directly. However, there are situations in which behavioural saliency is not defined by statistical rarity, nor irregularity. This is especially true when certain domain knowledge is available in describing possibly some frequent and not so irregular behaviour patterns that are deemed to be salient, e.g. jumping a red-light at a traffic junction. There are also other situations in which a particular behaviour class of interest is not well-defined visually, easily confused by another class of similar behaviour not of particular interest.

To model behavioural saliency of known interest under those conditions, supervised model learning is required where samples of the particular behaviour class of interest are labelled in the training data set. A fully supervised learning requires exhaustive labelling of all the data, not just those of the behaviours of interest. This is laborious, often rendered impractical given the vast amount of video data to be processed. It could also be inconsistent and error prone, as discussed in Chap. 8. As a compromise, semi-supervised learning can be exploited with minimal data labelling (Zhu 2007; Zhu et al. 2003).

To perform semi-supervised learning with a Markov clustering topic model, a few examples of the behaviours of interest need be labelled in a training data set. For model learning, a few modifications to Algorithm 9.1 are required as follows:

1. Prior to model learning, the labelled examples l are initialised to their known behaviour class z_l rather than randomly.
2. During model learning, the labelled examples are considered being observed, and not updated by the sampler given by (9.3). All the action layer variables defined by (9.2) are still updated in the same way as before so to find a good intermediate action representation under the additional constraint of the specified behaviour labels.
3. The model thus samples from $p(z_{\setminus l}, \mathbf{y}_{1:T} | z_l, \mathbf{x}_{1:T})$ rather than $p(z_{1:T}, \mathbf{y}_{1:T} | \mathbf{x}_{1:T})$.

It is possible to label some examples of a specific behaviour of interest, and allow the model to determine the best clustering of other behaviour patterns under this constraint, or to define all classes of behaviour and to label a few examples for each class.

To compare semi-supervised learning with unsupervised learning for modelling behavioural saliency of choice, let us consider an example from the street intersection scene. In this scenario, there are four legal traffic behaviour patterns: left, right, vertical, and turning vertical. There is also a 'common-but-illegal' behaviour, consisting of vertical traffic turning through the oncoming traffic flow. When a model is trained by unsupervised learning with $N_z = 5$ behaviour clusters, the model cannot learn the common-but-illegal behaviour class. Because there are other statistically probable ways to partition the training data set into five different behaviour clusters. Figures 9.7(a)–(e) show examples of the five behaviour classes learned from unsupervised clustering.

Fig. 9.7 Comparing
behaviour clusters obtained
by unsupervised learning
(**a**)–(**e**) and semi-supervised
learning (**f**)–(**j**)

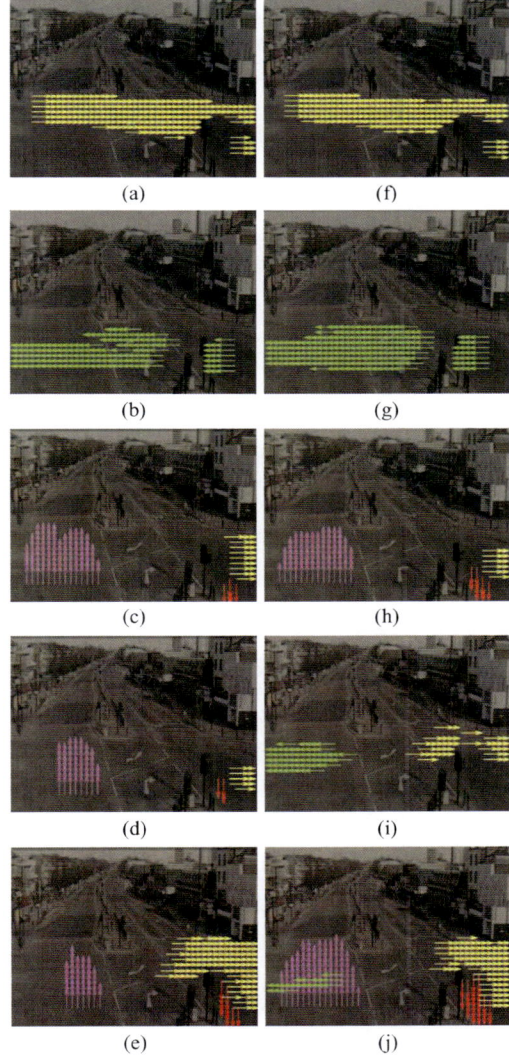

A model can be trained by semi-supervised learning to learn this common-but-illegal behaviour from data. This requires manually labelling a few example video clips of the vertical traffic turning through the oncoming traffic flow. The model learning is biased by domain knowledge so that a known interesting behaviour class can be modelled explicitly. Figures 9.7(f)–(j) show examples of the five behaviour classes learned from semi-supervised learning, including the common-but-illegal behaviour class shown in Fig. 9.7(j). Of course, a few examples of all clusters can also be specified to define an exhaustive set of specific behaviours if the statistics of the whole set of behaviours are of interest. This also alleviates the label switching problem in Markov chain Monte Carlo (Gilks et al. 1995; Griffiths and Steyvers 2004), so that the common-but-illegal turn-through behaviour is represented by the same specified behaviour cluster.

The potential for semi-supervised learning of a Markov clustering topic model is important. This makes the model very flexible. It can be learned either from fully automated model selection and profiling without any prior domain knowledge, or by semi-supervised labelling of some or all of the behaviours of interest when prior domain knowledge is available. This enables a model to detect behavioural saliency by both 'known but interesting' (9.7) and 'unknown and unusual' (9.9).

9.6 Discussion

We considered a dynamic topic model, known as the Markov clustering topic model, for hierarchical clustering of local motion events into actions and behaviours. The model addresses two basic functions in visual interpretation of behaviour: modelling object behavioural characteristics at different spatial and temporal levels, and on-line behaviour screening for saliency detection. The Markov clustering topic model addresses three fundamental challenges in modelling complex behaviours in public spaces: robustness, sensitivity and computational tractability. The model draws on machine learning theories on probabilistic topic models and dynamic Bayesian networks to achieve a robust hierarchical model of behaviours and their dynamics. It has a three-layer hierarchical architecture for extracting visual information and describing behavioural dynamics.

In essence, a dynamic topic model is a hybrid of a probabilistic topic model (PTM) and a dynamic Bayesian network (DBN), designed to strike the right balance between sensitivity and robustness in salient behaviour discovery. Specifically, the model aims to retain the strength of a PTM on robustness against noise due to its bag-of-words nature, and that of a DBN on sensitivity due to its explicit modelling of temporal order. Care must be taken on designing an efficient learning and inference algorithm to make the model tractable and scalable, as a hierarchical model can have a complex structure described by a large number of model parameters for group activity modelling. There are other ways of introducing temporal order modelling capability into a topic model. One of such alternative ways will be considered later in Chap. 15.

A dynamic topic model such as a Markov clustering topic model can be learned using different learning strategies including unsupervised learning, supervised learning, and semi-supervised learning, depending on the availability of human supervision. This flexibility is important for behavioural anomaly detection, especially when the definition of abnormality is beyond statistical rarity or irregularity, and cannot be specified using visual information alone. Due to the typically large quantity of video data for visual analysis of behaviour, a fully supervised learning strategy is often infeasible. Learning a model with partial supervision becomes critical. Semi-supervised learning is one way of learning with partial supervision. Later in this book, we shall consider learning strategies suitable for situations where human supervision is both partial and 'weak', that is, incomplete, or situations where human supervision is in the form of active feedback.

In a complex dynamic scene, semantic interpretation of behaviour is context dependent. That is, to understand behaviour, a model needs to incorporate information about the context in which objects behave. Behavioural context includes information on when and where an activity takes place, and how the activity correlates with objects in the scene. The behaviour model described in this chapter has the ability to model implicitly context by clustering co-occurring events and actions. However, it is also possible to model context explicitly in order to decompose and analyse complex behaviour more effectively. In the next chapter, behavioural context is defined comprehensively and explicit context models are considered.

References

Belhumeur, P.N., Hespanha, J.P., Kriegman, D.J.: Eigenfaces vs. fisherfaces: recognition using class specific linear projection. IEEE Trans. Pattern Anal. Mach. Intell. **19**(7), 711–720 (1997)

Bishop, C.M.: Pattern Recognition and Machine Learning. Springer, Berlin (2006)

Blei, D.M., Ng, A.Y., Jordan, M.I.: Latent Dirichlet allocation. J. Mach. Learn. Res. **3**, 993–1022 (2003)

Fisher, R.A.: The statistical utilisation of multiple measurements. Ann. Eugen. **8**, 376–386 (1938)

Gilks, W.R., Richardson, S., Spiegelhalter, D. (eds.): Markov Chain Monte Carlo in Practice: Interdisciplinary Statistics. Chapman & Hall/CRC Press, London/Boca Raton (1995)

Griffiths, T.L., Steyvers, M.: Finding scientific topics. Proc. Natl. Acad. Sci. USA **101**, 5228–5235 (2004)

Griffiths, T.L., Steyvers, M., Blei, D.M., Tenenbaum, J.: Integrating topics and syntax. In: Advances in Neural Information Processing Systems, pp. 537–544 (2005)

Gruber, A., Rosen-Zvi, M., Weiss, Y.: Hidden topic Markov models. In: Artificial Intelligence and Statistics, San Juan, Puerto Rico (2007)

Hospedales, T., Gong, S., Xiang, T.: A Markov clustering topic model for behaviour mining in video. In: IEEE International Conference on Computer Vision, Kyoto, Japan, October 2009, pp. 1165–1172 (2009)

Kuettel, D., Breitenstein, M.D., Van Gool, L., Ferrari, V.: What's going on? discovering spatio-temporal dependencies in dynamic scenes. In: IEEE Conference on Computer Vision and Pattern Recognition, San Francisco, USA, June 2010, pp. 1951–1958 (2010)

Li, J., Gong, S., Xiang, T.: Global behaviour inference using probabilistic latent semantic analysis. In: British Machine Vision Conference, Leeds, UK, September 2008, pp. 193–202 (2008)

Rosen-Zvi, M., Griffiths, T.L., Steyvers, M., Smyth, P.: The author-topic model for authors and documents. In: Uncertainty in Artificial Intelligence, Banff, Canada, pp. 487–494 (2008)

Wallach, H., Murray, I., Salakhutdinov, R., Mimno, D.: Evaluation methods for topic models. In: International Conference on Machine Learning, pp. 1105–1112 (2009)

Wang, X., Ma, X., Grimson, W.E.L.: Unsupervised activity perception by hierarchical Bayesian models. IEEE Trans. Pattern Anal. Mach. Intell. **31**(3), 539–555 (2009)

Zhu, X.: Semi-supervised learning literature survey. Technical Report 1530, University of Wisconsin-Madison Department of Computer Science (2007)

Zhu, X., Ghahramani, Z., Lafferty, J.: Semi-supervised learning using Gaussian fields and harmonic functions. In: International Conference on Machine Learning, pp. 912–919 (2003)

Chapter 10
Learning Behavioural Context

Visual context is the environment, background, and settings within which objects and associated behaviours are observed visually. A human observer employs visual context extensively for object recognition in images and behavioural interpretation in image sequences. Extensive cognitive, physiological and psychophysical studies have shown that visual context plays a critical role in the human visual system (Bar 2004; Bar and Aminof 2003; Bar and Ullman 1993; Biederman et al. 1982; Palmer 1975). Motivated by these studies, there is an increasing interest in exploiting contextual information for computer vision based object detection (Carbonetto et al. 2004; Galleguillos et al. 2008; Gupta and Davis 2008; Heitz and Koller 2008; Kumar and Hebert 2005; Murphy et al. 2003; Rabinovich et al. 2007; Wolf and Bileschi 2006; Zheng et al. 2009a), action recognition (Marszalek et al. 2009), tracking (Ali and Shah 2008; Gong and Buxton 1993; Loy et al. 2009; Sherrah and Gong 2000; Yang et al. 2008; Zheng et al. 2009b), and trajectory analysis (Buxton and Gong 1995; Gong and Buxton 1992).

For understanding object behaviour in a crowded space, a semantic interpretation of behaviour depends largely on knowledge of spatial and temporal context defining where and when it occurs, and correlational context specifying the expectation on correlated other objects in its vicinity. In the following, we examine the question of how to model computationally behavioural context for context-aware behavioural anomaly detection in a visual scene. Specifically, we consider models for learning three types of behavioural context.

Behaviour "spatial context" provides situational awareness about *where* a behaviour is likely to take place. A public space serving any public function such as a road junction or a train platform can often be segmented into a number of distinctive zones. Behaviours of certain characteristics are expected in one zone but differ from those observed in other zones. We call these behaviour sensitive zones "semantic regions". For instance, in a train station, behaviours of passengers on the train platform and those in front of a ticket machine can be very different. Another example can be seen in Fig. 10.1 where different traffic zones and lanes play an important role in defining how objects are expected to behave. We address the problem of modelling behaviour spatial context in Sect. 10.1.

S. Gong, T. Xiang, *Visual Analysis of Behaviour*,
DOI 10.1007/978-0-85729-670-2_10, © Springer-Verlag London Limited 2011

(a) The two traffic phases

(b) A behavioural anomaly

Fig. 10.1 A behaviour in a street intersection scene needs be understood by taking into account its spatial context, that is, where it occurs, temporal context, that is, when it takes place, and correlational context, that is, how other correlated objects behave in the same shared space. In (**b**) a fire engine moved horizontally and broke the vertical traffic flow. This is an anomaly because it is incoherent with both the temporal context as it happens during the vertical traffic phase, and with the correlational context because horizontal and vertical traffics are not expected to occur simultaneously

Behaviour "correlational context" specifies *how* the interpretation of a behaviour can be affected by behaviours of other objects either nearby in the same semantic region or further away in other regions. Object behaviours in a complex scene are often correlated and need be interpreted together rather than in isolation. For example, behaviours are contextually correlated when they occur concurrently in a scene. Figure 10.1 shows some examples of moving vertical traffic flow with standby horizontal flow typically co-occurring, with the exception of emergency vehicles jumping red-light (Fig. 10.1(b)). There are methods capable of modelling correlational context implicitly using either dynamic Bayesian networks (Chap. 7) or dynamic topic models (Chap. 9). For modelling explicitly context of a complex scene, a model needs to cope with multiple semantic regions at multiple scales. For instance, behaviour correlational context can be considered at two scales: local context within each region, and global context that represents behaviour correlations across regions. Multi-scale behavioural context can be modelled by a cascaded topic model. We shall describe this model in Sect. 10.2.

Behaviour "temporal context" provides information regarding *when* different behaviours are expected to happen both inside each semantic region and across regions. This is shown by examples in Fig. 10.1 where behaviour temporal context is determined by traffic light phases. More specifically, semantic interpretation of vehicle behaviour needs to take into account temporal phasing of the traffic light. For example, a vehicle is expected to be moving if the traffic light governing its lane is green, and stopped if red. Even if the traffic light is not directly visible in the scene, this contextual constraint translates visually to an observer's expectation and interpretation on how other vehicles move in the scene. Similar to behaviour correlational context, we consider temporal context at two scales corresponding to within-region and cross-region context. The Markov clustering topic model, described in the preceding chapter, can be used for learning implicitly behaviour temporal context at the cross-region scale. There is an alternative approach for explicitly learning behaviour temporal context at multiple scales, which we shall discuss in Sect. 10.2.

Behavioural contextual information learned from visual observations is useful for both increasing the sensitivity of a model in detecting abnormal behaviours and improving the specificity of a model on differentiating visually subtle anomalies. We consider a context-aware anomaly detection model in Sect. 10.3.

10.1 Spatial Context

We consider that behaviour spatial context is given as distributed semantic regions where behaviour patterns observed within each region are similar to each other whilst being dissimilar to those occurring in other regions. The aim for a model is to discover such semantic regions automatically and unsupervised from visual observational data directly without top-down manual insertion. This problem can be addressed using a two-step process as follows:

1. Each pixel location in a scene is labelled by a behaviour-footprint measuring how different behaviour patterns occur over time. This pixel-wise location behaviour-footprint is necessarily a distribution measurement, for example, a histogram of contextual event classes occurring at each location over time (Fig. 10.2).
2. Given behaviour-footprints computed at all pixel locations of a scene, a spectral clustering algorithm is employed to segment all pixel locations by their behaviour-footprints into different non-overlapping regions with the optimal number of regions determined automatically.

10.1.1 Behaviour-Footprint

A behaviour-footprint is computed for each pixel location in a scene. Using the method described in Sect. 7.1, behaviours are represented by object-independent

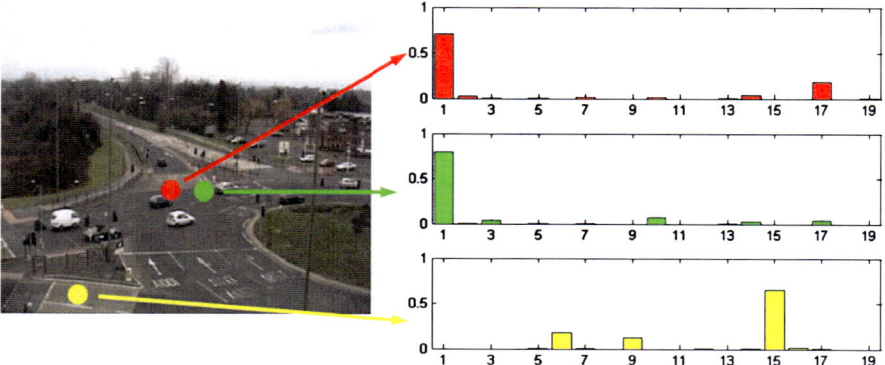

Fig. 10.2 Examples of behaviour-footprint. Three random pixel locations in a traffic roundabout scene and their corresponding behaviour-footprints are shown *colour-coded*. It is evident that the *red* and *green locations* have similar behaviour-footprints that differ from the *yellow location*

contextual events categorised in space and over time. Suppose there are V classes of contextual events, to measure how different behaviours have taken place at a pixel location in a video, its behaviour-footprint is computed as a histogram of V bins:

$$\mathbf{p} = [p_1, \ldots, p_v, \ldots, p_V] \qquad (10.1)$$

where p_v counts for the number of occurrence of the vth event class at this pixel location. Figure 10.2 shows examples of computed behaviour-footprints at different locations of a scene.

10.1.2 Semantic Scene Decomposition

Given pixel-wise behaviour-footprints, the problem of learning behaviour spatial context is treated as a segmentation problem where the image space is segmented by pixel-wise feature vectors given by the V-dimensional behaviour-footprints (10.1). Various image segmentation methods can be applied. We consider a spectral clustering based segmentation model (Zelnik-Manor and Perona 2004). Given a scene with N pixel locations, an $N \times N$ affinity matrix $\mathbf{A}$ is constructed. The similarity between the behaviour-footprints at the ith and jth locations is computed as

$$\mathbf{A}(i, j) = \begin{cases} \exp(-\frac{(d(\mathbf{p}_i, \mathbf{p}_j))^2}{\sigma_i \sigma_j}) \exp(-\frac{(d(\mathbf{x}_i, \mathbf{x}_j))^2}{\sigma_x^2}), & \text{if } \|\mathbf{x}_i - \mathbf{x}_j\| \leq r \\ 0, & \text{otherwise} \end{cases} \qquad (10.2)$$

where $\mathbf{p}_i$ and $\mathbf{p}_j$ are the behaviour-footprints at the ith and the jth pixel loci, d represents Euclidean distance, σ_i and σ_j correspond to the scaling factors for the feature vectors at the ith and the jth positions, $\mathbf{x}_i$ and $\mathbf{x}_j$ are their image coordinates and σ_x

is a spatial scaling factor; r is the radius indicating a circle, within which similarity is computed. Using this model, two pixel locations will have strong similarity and thus be grouped into one semantic region if they have similar behaviour-footprints and are also close to each other spatially, for example, the red and green dots in Fig. 10.2.

For a spectral clustering based segmentation method, choosing a correct scaling factor is critical for obtaining a meaningful segmentation. The Zelnik–Perona model computes σ_i using a pre-defined constant distance between the behaviour-footprint at the ith pixel location and its neighbours, which is rather arbitrary. Li et al. (2008b) show experimental results to suggest that this may result in model under-fitting, yielding fewer segments than optimum for scene decomposition. To address this problem, the model can be modified as follows:

1. σ_i is computed as the standard deviation of behaviour-footprint distances between the ith pixel location and all locations within a given radius r. Similarly, the spatial scaling factor σ_x is computed as the mean of the spatial distances between all locations within radius r and the centre.
2. The affinity matrix is then normalised according to

$$\bar{\mathbf{A}} = \mathbf{L}^{-\frac{1}{2}} \mathbf{A} \mathbf{L}^{-\frac{1}{2}} \tag{10.3}$$

where $\mathbf{L}$ is a diagonal matrix with

$$\mathbf{L}(i, i) = \sum_{j=1}^{N} \big(\mathbf{A}(i, j)\big) \tag{10.4}$$

3. $\bar{\mathbf{A}}$ is then used as the input to the Zelnik–Perona's algorithm which automatically determines the number of clusters and performs segmentation.

This procedure groups pixel locations into Q optimal regions by their behaviour-footprints for a given scene. Since scene segmentation is performed by behaviours, those locations without, or with few, object movement detected by background subtraction are absent from the segmentation process. That is, the N pixel locations to be clustered using the above algorithm do not include those where no or few foreground objects have been detected.

Figure 10.3 shows two examples of scene segmentation for learning spatial context. Both examples are from crowded traffic scenes including a street intersection and a roundabout. These two scenes feature large numbers of objects including pedestrians, variable sized vehicles, and cyclists. The behaviours of these objects in the two scenes are controlled by traffic lights and constrained by other objects co-existing in the scenes. The spectral clustering algorithm decomposes automatically the intersection scene into six regions (Fig. 10.3(c)) and the roundabout scene into nine regions (Fig. 10.3(f)). The scene decomposition process shows that semantically meaningful regions are obtained. In particular, the decomposed regions correspond to traffic lanes and waiting zones where object behaviours tend to follow certain similar patterns.

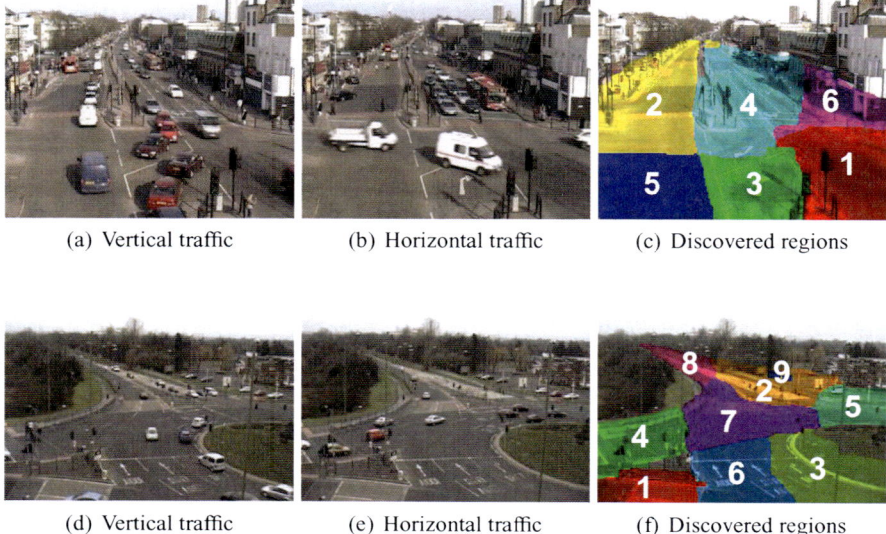

(a) Vertical traffic   (b) Horizontal traffic   (c) Discovered regions

(d) Vertical traffic   (e) Horizontal traffic   (f) Discovered regions

Fig. 10.3 Examples of semantic scene segmentation for a street intersection scene (*top row*) and a roundabout scene (*bottom row*)

10.2 Correlational and Temporal Context

We also consider how a model can be constructed to learn behaviour correlational and temporal context from data directly. An automatic system needs to exploit the contextual constraints on how the behaviour of an object is correlated to and affected by those of other objects both nearby in the same semantic region and further away in other regions of a scene.

Given learned spatial behavioural context, correlational and temporal context are considered at two scales: regional context and global context. In order to discover and quantify context at both scales, a two-stage cascaded topic model is exploited. Figure 10.4 shows the structure of this model. The inputs to the first stage of the model are regional behaviours represented as labels of regional events (Sect. 10.2.1). We consider that the inferred topics from the first-stage modelling correspond to regional behaviour correlational context. They are utilised for computing regional temporal context that informs temporal phases of behaviours in each semantic region. The learned regional temporal context is further used to initialise a second-stage topic modelling, where global behaviour correlational context is discovered by inferring topics from the model. Finally, global temporal context is computed.

In each stage of the cascaded topic model, latent Dirichlet allocation (LDA) (Blei et al. 2003) is performed (Sect. 9.2). In a LDA model, a document $\mathbf{w}$ is a collection of N_w words, where $\mathbf{w} = \{w_1, \ldots, w_n, \ldots, w_{N_w}\}$, and can be modelled as a mixture of K topics $\mathbf{z} = \{z_1, \ldots, z_K\}$. Each topic is modelled as a multinomial distribution over a vocabulary consisting of V words, from which all words in $\mathbf{w}$ are sampled. Given the vocabulary, a document is represented as a V-dimensional feature vector,

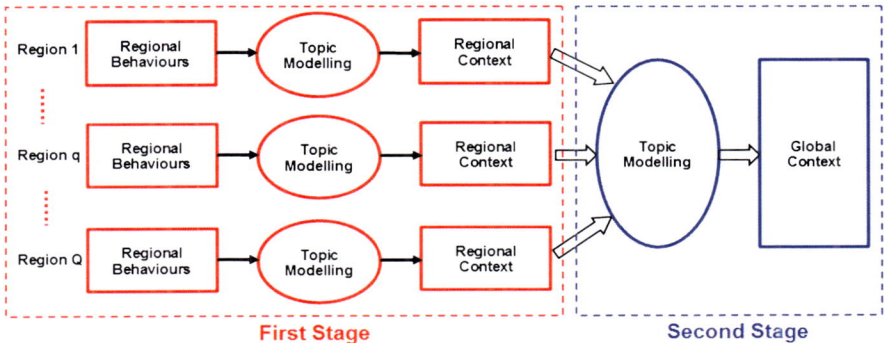

Fig. 10.4 The structure of a two-stage cascaded topic model

each element of which corresponds to the count of the number of times a specific word occurs in the document. With LDA models deployed for learning behavioural context both locally within each semantic region and globally across different regions, the cascaded topic model becomes a cascaded LDA model with two stages.

10.2.1 Learning Regional Context

Let us first consider how regional behaviours as inputs to a model are represented. Assume a complex scene has been decomposed into Q regions as a result of spatial context learning (Sect. 10.1). For each region, contextual events are detected and categorised (Sect. 7.1). Assume that V^q regional event classes have been obtained in each region q, where $1 \leq q \leq Q$. Note that each region may have different numbers of event classes and those numbers are determined by automatic model order selection. The regional event class labels are then used as inputs to the cascaded topic model. For modelling regional behaviour correlational and temporal context, regional events are considered to be video words. All events detected from a single region in a clip form a video document. Regional correlational context is defined by topics discovered from documents. The inferred topic profile for each document is used to categorise each document into temporal phases, and considered as regional temporal context.

More specifically, a training video, or a set of training videos, of a scene is considered as a video corpus. It is segmented temporally into equal-length and non-overlapping short video clips.[1] These short video clips are considered as video documents for training a cascaded topic model. Given Q semantic regions of a spatially segmented scene, each region is modelled using a LDA. Regional events detected in the qth region are video words that form a document denoted as d_t^q. The documents

---

[1] We alternate the use of terms 'video document' and 'video clip' in this book.

corresponding to all T clips in the qth region form the regional corpus $\mathcal{D}^q = \{d_t^q\}$, where $t = 1, \ldots, T$ is the clip index.

Assuming that there are V^q classes of regional events in the qth region, the size of a codebook as vocabulary is thus V^q. Each document is modelled as a mixture of K^q topics, which correspond to K^q different types of regional behaviour correlations. This represents the regional correlational context. In a standard LDA model, a document is represented as the counts of different video words. In contrast, a document of events in this model is represented as a binary V^q-dimensional feature vector, and each element of the vector is a binary value indicating whether a certain regional event class is present in that clip. This is due to two reasons: (1) conceptually, this model is more interested in what behaviours co-occur in each document as a video clip, rather than how often they occur. (2) This binary vector representation is more robust against noise and error from event detection (Li et al. 2008b).

A LDA model for the qth region has the following parameters:

1. α^q: A K^q-dimensional vector governing the Dirichlet distributions of topics in a training video corpus, including all video clips for the qth region.
2. β^q: A $K^q \times V^q$-dimensional matrix representing the multinomial distributions of words in a codebook as the vocabulary for all learned topics, where $\beta_{k,n}^q = P(w_n^q | z_k^q)$ and $\sum_{n=1}^{V^q} \beta_{k,n}^q = 1$.

Given these model parameters, video words in a local document can be repeatedly sampled as follows:

1. Sample a K^q-dimensional vector θ^q from the Dirichlet distribution governed by parameter α^q: $\theta^q \sim \mathrm{Dir}(\alpha^q)$. Vector θ^q contains the information about how the K^q topics are to be mixed in the document.
2. Sample words from topics:
   a. Choose a topic for w_n^q: $z_n^q \sim \mathrm{multinomial}(\theta^q)$.
   b. Choose a word w_n^q from the vocabulary of V^q words according to $P(w_n^q | z_n^q, \beta^q)$.

Following the conditional dependency of the components in the generative process, the log-likelihood of a document d_t^q given model parameters for the clip is computed as

$$\log p(d_t^q | \alpha^q, \beta^q) = \log \int p(\theta^q | \alpha^q) \left(\prod_{n=1}^{N_w^q} \sum_{z_n^q} p(z_n^q | \theta^q) p(w_n^q | z_n^q, \beta^q) \right) d\theta^q$$

(10.5)

and the log-likelihood of an entire video corpus $\mathcal{D}^q = \{d_t^q\}$ of T clips for the qth region is

$$\log p(\mathcal{D}^q) = \sum_{t=1}^{T} \log p(d_t^q | \alpha^q, \beta^q)$$

(10.6)

where t is the clip index and T is the total number of documents in the corpus, representing all the clips in a training data set.

The model parameters α^q and β^q are estimated by maximising the log-likelihood function $\log p(\mathcal{D}^q)$ in (10.6). However, there is no closed-form analytical solution to the problem. A variational expectation-maximisation (EM) algorithm can be employed (Blei et al. 2003). Specifically:

1. In the E-step of variational inference, the posterior distribution of the hidden variables $p(\theta^q, \{z_n^q\}|d_t^q, \alpha^q, \beta^q)$ in a specific document d_t^q is approximated by a variational distribution $p(\theta^q, \{z_n^q\}|\gamma^q, \phi^q)$, where γ^q and ϕ^q are document d_t^q specific variational parameters. They differ for different documents. Maximising the log-likelihood $\log p(d_t^q|\alpha^q, \beta^q)$ corresponds to minimising the Kullback–Leibler (KL) divergence between $q(\theta^q, \{z_n^q\}|\gamma^q, \phi^q)$ and $p(\theta^q, \{z_n^q\}|d_t^q, \alpha^q, \beta^q)$, resulting in $\log p(d_t^q|\alpha^q, \beta^q)$ being approximated by its maximised lower bound $L(\gamma^q, \phi^q; \alpha^q, \beta^q)$. By setting α^q and β^q as constants, the variational parameters for d_t^q are estimated according to the following pair of updating equations:

$$\phi_{n,k}^q \propto \beta_{k,v}^q \exp\left(\Psi\left(\gamma_k^q\right) - \Psi\left(\sum_{k=1}^{K^q} \gamma_k^q\right)\right) \tag{10.7}$$

$$\gamma_k^q = \alpha_k^q + \sum_{n=1}^{N_w^q} \phi_{n,k}^q \tag{10.8}$$

where $n = 1, \ldots, N_w^q$ indicates the nth word in d_t^q, $k = 1, \ldots, K^q$ indicates the kth regional topic, $v = 1, \ldots, V^q$ indicates the vth word in the regional vocabulary, and Ψ is the first order derivative of a log Γ function.

2. In the M-step, the learned variational parameters $\{\gamma^q\}$ and $\{\phi^q\}$ are set as constant and the model parameters α^q and β^q are learned by maximising the lower bound of the log-likelihood of the whole corpus:

$$L\left(\{\gamma_t^q\}, \{\phi_t^q\}; \alpha^q, \beta^q\right) = \sum_{t=1}^{T} L\left(\gamma_t^q, \phi_t^q; \alpha^q, \beta^q\right) \tag{10.9}$$

The learned model parameters β^q is given in a matrix specifying the conditional probabilities of words as regional events given all the topics. Therefore, this matrix captures how different types of events are correlated with topics in each semantic region and represents the regional behaviour correlational context.

To compute regional temporal context reflecting any intrinsic temporal phases of behaviours occurring in each region, document clustering is performed on the video clips from the training data set. Specifically, the Dirichlet parameter γ_t^q represents a document in the topic simplex and can be viewed as the topic profile in the qth region and tth clip, that is, it represents how likely different topics are combined. The set of topic profiles $\{\gamma_t^q\}$ for a regional corpus $\mathcal{D}^q$ consist of T documents as video clips in the qth region. These documents are grouped into C^q categories $\mathbf{h}^q = \{h_c^q\}$, where $c = 1, \ldots, C^q$. Such a clustering process can be readily performed

Fig. 10.5 Different topics are discovered in six different semantic regions of a street intersection scene. Each topic is illustrated by the *top* two most likely local event classes in each region. An event class is depicted using an *ellipse*, whose size and orientation correspond to the spatial distribution of the detected events belonging to that class modelled as a Gaussian (see Sect. 7.1). The mean motion magnitude and direction for each event class is also illustrated using a *red arrow* originating from the *centre* of *each ellipse*

by k-means. As a result, regional behaviours occurring within each document as a video clip are uniquely assigned a temporal phase using the class label of that document.

An example of the learned regional behaviour context is illustrated in Fig. 10.5 using the street intersection scene. Six regions are automatically obtained by a model

(Fig. 10.3(c)). The number of topics is set to four for all six regions in the scene.[2] Each learned topic is regarded as the regional behaviour correlational context that captures one specific type of commonly observed co-occurring object behaviours in a particular traffic phase. For instance, in Region-1 shown in Fig. 10.5, both Topic-1 and Topic-4 correspond to the vertical traffic flow in that region. Whilst Topic-2 represents leftward horizontal flow of pedestrians, Topic-3 captures the rightward horizontal flow of pedestrians waiting for road crossing.

10.2.2 Learning Global Context

A second-stage LDA, referred to as global context LDA, is constructed to learn inter-regional behaviour correlational and temporal context. For this LDA, a document is a collection of words representing temporal phases of different semantic regions in the scene. A temporal phase of a semantic region represents a common regional pattern of change at a time instance. Temporal phases of a region can be regional topics discovered by the first-stage LDA, where each topic is a group of co-occurring events at a time instance in the region. Given a total number of $C = \sum_{q=1}^{Q} C^q$ regional temporal phases discovered from the first-stage LDA in all Q regions of a scene, a codebook of regional topics as temporal phases is a list of index given as

$$\mathcal{H} = \left[h_1^1, \ldots, h_{C^1}^1, \ldots, h_1^q, \ldots, h_{C^q}^q, \ldots, h_1^Q, \ldots, h_{C^Q}^Q \right] \qquad (10.10)$$

Each element of $\mathcal{H}$ corresponds to the index of a regional temporal phase.

A document as a video clip $d = \{w_n\}$, where $n = 1, \ldots, N_w$, can then be represented by a bag-of-words discovered from different semantic regions, with each word being a regional topic of co-occurring events, referred to as a regional temporal phase. Any word w_n in a document is sampled from codebook $\mathcal{H}$ and each word in this codebook is only allowed to be observed once in the document. We consider an element of the global correlational and temporal context is an inter-regional topic of co-occurring local regional topics in different regions, to be discovered by the second-stage LDA. This is essentially a process of discovering hierarchical topics.

The learning and inference processes of this global context LDA are identical to those for regional context LDA. Suppose K types of global correlational context are discovered from a corpus $\mathcal{D} = \{d_t\}$, where $t = 1, \ldots, T$, the learned parameters β is a $K \times C$ matrix representing the probabilities of each of the C categories of regional temporal phases occurring in K topics of global correlations. This matrix defines the overall global behavioural context of a scene. In the same process, the second-stage LDA also infers global topic profiles $\{\gamma_t\}$. That is, the process determines a mixture distribution of global topics for each document d_t. These topic profiles can be used

---

[2]It can also be learned automatically as discussed in Sect. 9.4.

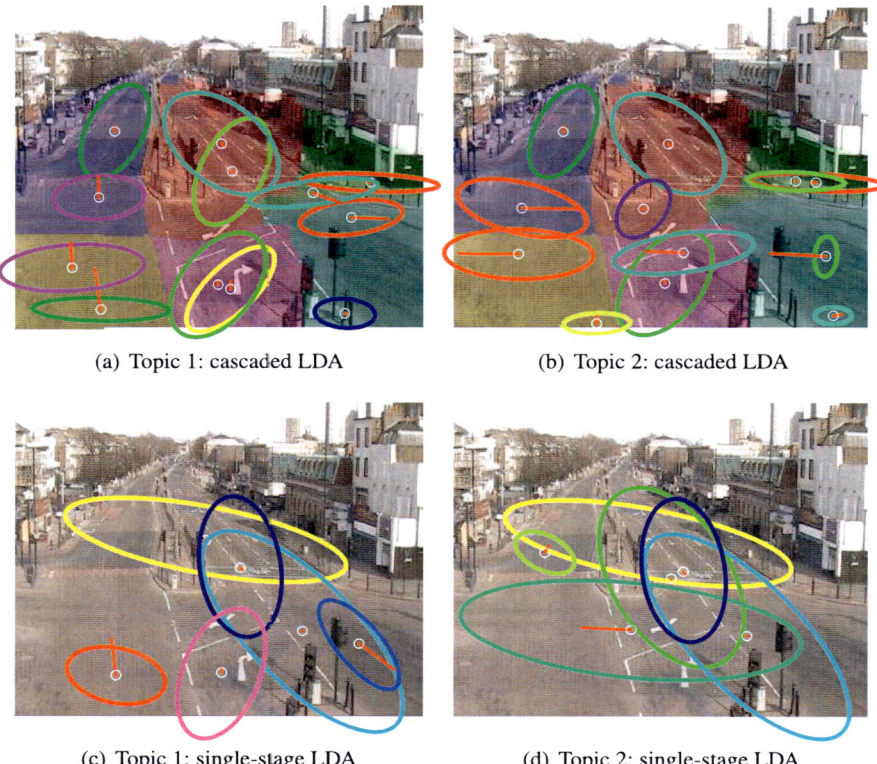

(a) Topic 1: cascaded LDA   (b) Topic 2: cascaded LDA

(c) Topic 1: single-stage LDA   (d) Topic 2: single-stage LDA

Fig. 10.6 Learning global behaviour context. (**a**)–(**b**): Topics learned using the second-stage LDA of the model, which correspond to inter-regional topic correlations in different traffic phases. Each topic is illustrated by the *top* two most likely regional event classes in that topic. (**c**)–(**d**): Topics learned using a single-stage conventional LDA without the cascade structure. The *top* six most likely event classes, visualised in the same way as in Fig. 10.5, are shown for two global topics

to classify documents as video clips into different temporal phases corresponding to the global temporal context.

After learning a cascaded topic model, the model can be applied to interpret behaviours captured in unseen videos. In particular, the model can be employed to infer a topic profile using either the regional LDA or the global LDA. The former reveals what types of regional behaviour correlation exist. The latter informs existence of any global behaviour correlations. Moreover, the inferred regional and global temporal phases are assigned as labels to clips. They can be used to temporally segment unseen videos by topics. In contrast to LDA, there are other alternative topic models that can be exploited for learning behavioural context. One of which is probabilistic latent semantic analysis (Hofmann 1999a,b; Li et al. 2008a).

Figure 10.6 compares the global behavioural context learned for the street intersection scene using the cascaded topic model to that by a standard single-stage LDA. Topic-1 (Fig. 10.6(a)) represents co-occurring object behaviours in different regions during a vertical traffic flow phase, whilst Topic-2 (Fig. 10.6(b)) corresponds to those during a horizontal traffic flow phase. Topics learned using a standard single-stage LDA capture less meaningful correlations among object behaviours compared to those from a cascaded LDA model. This highlights that the limitations of learning behavioural context from local contextual events directly, as these local events do not capture effectively correlational and temporal relationships between behaviours in different regions of a scene.

10.3 Context-Aware Anomaly Detection

Given a behavioural context model based on the cascaded LDA, behavioural anomalies can be detected as *contextually incoherent* behaviours that cannot be explained away or predicted by behavioural context learned from data. Based on this principle, a method can be formulated to not only detect when an anomaly occurs in a video document at a specific time, but also locate which semantic region(s) cause the anomaly with what local events. This is attractive for a topic based behaviour model, as it provides a powerful feature of being location sensitive in detecting anomalies, especially given its inherent bag-of-words representation without explicit object segmentation and tracking. More precisely, a context-aware anomaly detection process consists of two steps:

1. Detection: A probe video is processed to generate a sequence of non-overlapping clips as video documents. A trained global context LDA model is employed to examine whether each video document contains abnormal behaviours. This is computed by evaluating the normalised log-likelihood of a clip given the trained LDA model. The log-likelihood of observing an unseen clip d^* is approximated by its maximised lower bound:

$$\log p(d^*) \approx L(\gamma^*, \phi^*; \alpha, \beta) \tag{10.11}$$

where γ^* are Dirichlet parameters representing a video clip d^* in a topic simplex, ϕ^* are the posterior probabilities of topics occurring in d^*, α and β are the conditional probabilities of words given topics as model parameters, computed by (10.9). Given a model, $L(\gamma^*, \phi^*; \alpha, \beta)$ is evaluated by setting α and β as constants and updating γ^* and ϕ^* until $L(\gamma^*, \phi^*; \alpha, \beta)$ is maximised, using (10.7) and (10.8). A normality score for the probe clip is defined as

$$S(d^*) = \frac{L(\gamma^*, \phi^*; \alpha, \beta)}{N_w^*} \tag{10.12}$$

where N_w^* is the number of words in d^*. In this case, it corresponds to the number of all the regional temporal phases identified in clip d^*. Any clip with a normality score lower than a threshold TH_c is considered abnormal. The value of TH_c is set to balance a model's detection rate against its false alarm rate.

2. Localisation: Once an abnormal video clip $d^* = \{w_n^*\}$ is detected, where $n = 1, \ldots, N_w^*$, the offending anomalous regions and events within the clip are identified by evaluating the log-likelihood of the regional temporal phases against the global behaviour correlational context, $p(w_n^*|d^*, \alpha, \beta)$. This likelihood value can be approximated by $\log p(w_n^*|d_{-n}^*, \alpha, \beta)$, where d_{-n}^* represents a document in which word w_n^* is removed from d^*. Therefore, $\log p(w_n^*|d_{-n}^*, \alpha, \beta)$ represents the log-likelihood of w_n^* co-occurring with all other distinct words in d^*, and is defined as

$$\log p(w_n^*|d_{-n}^*, \alpha, \beta) = \log p(w_n^*, d_{-n}^*|\alpha, \beta) - \log p(d_{-n}^*|\alpha, \beta)$$
$$= \log p(d^*|\alpha, \beta) - \log p(d_{-n}^*|\alpha, \beta) \qquad (10.13)$$

To compute (10.13), we take the following considerations. As d_{-n}^* is derived from d^* by removing a single word, it is reasonable to assume that they have the same topic profile γ^*. This suggests that the log-likelihood of d_{-n}^* can be estimated by γ^* learned from d^*. To that end, $\log p(d^*|\alpha, \beta)$ and $\log p(d_{-n}^*|\alpha, \beta)$ are estimated by

a. Step 1: Given α and β, compute $L(\gamma^*, \phi^*; \alpha, \beta)$ to maximise the lower bound of $\log p(d^*|\alpha, \beta)$ through iteratively updating (10.7) and (10.8), and storing the value of γ^*.
b. Step 2: Given document d_{-n}^*, set γ^* in (10.8) as constant and update ϕ in (10.7), so that the maximised lower bound of $\log p(d_{-n}^*|\alpha, \beta)$ is computed as $L(\gamma^*, \phi_{-n}^*; \alpha, \beta)$, where each element in ϕ_{-n}^* represents how likely it is to assign each global topic to a word in d_{-n}^*.

Finally, $\log p(w_n^*|d_{-n}^*, \alpha, \beta)$ is estimated as

$$\log p(w_n^*|d_{-n}^*, \alpha, \beta) \approx L(\gamma^*, \phi^*; \alpha, \beta) - L(\gamma^*, \phi_{-n}^*; \alpha, \beta) \qquad (10.14)$$

The semantic regions with the lowest values of $\log p(w_n^*|d_{-n}^*, \alpha, \beta)$ are considered to contain abnormal behaviours. In those regions, the above procedure can be further deployed to locate abnormal events by computing:

$$\log p\left(w_n^{q*}|d_{-n}^{q*}, \alpha^q, \beta^q\right) \approx L\left(\gamma^{q*}, \phi^{q*}; \alpha^q, \beta^q\right) - L\left(\gamma^{q*}, \phi_{-n}^{q*}; \alpha^q, \beta^q\right)$$
$$(10.15)$$

for all regional events w_n^{q*} using the regional LDAs. The abnormal regional events are those having the lowest log-likelihood values for co-occurring with all other events in the same region of the same clip.

Li et al. (2008b) show that learning behavioural context is a powerful mechanism for interpreting behaviours in complex dynamic scenes. By learning directly from data a probabilistic behavioural context model at multiple levels both spatially and temporally, the model semantically decomposes a dynamic scene by spatial, correlational and temporal distributions of object behaviours. This multi-layered behaviour context modelling provides a rich and structured source of cumulative information that facilitates not only local event and behaviour discovery, but also global behaviour correlation, anomaly detection and localisation, all in a single framework. This facilitates more effective interpretation of multi-object behaviours, giving higher specificity in detecting subtle anomalies and improved sensitivity therefore robustness to noise at the same time (see a discussion on model specificity and sensitivity in Sect. 8.2.2).

10.4 Discussion

Explicitly learning behavioural context can assist and improve the detection of more subtle and non-exaggerated abnormal behaviours. A behavioural anomaly captured in video from a crowded public space is only likely to be meaningful when interpreted in context. We consider that such context is intrinsically at different levels and cannot be easily specified by hard-wired top-down rules, either exhaustively or partially. In particular, two visually identical behaviours in a public space can be deemed one being normal and the other abnormal depending on when, where, and how they occur in relations to other objects. For instance, a person running on a platform with train approaching and all other people also running is normal, whilst the same person running on an empty platform with no train in sight is likely to be abnormal. In other words, a behavioural anomaly can be detected more reliably and accurately by measured how *contextually incoherent* it is. That is, a behaviour is more likely to be considered as being abnormal if it cannot be predicted nor explained away by context. A key question is how to quantify and represent behavioural contextual information extracted directly from visual observations.

We give a comprehensive treatment on modelling behavioural context and described a framework for learning three different types of behavioural context, including spatial context, correlational context, and temporal context. For learning spatial context, a semantic scene segmentation method is studied. A learned spatial context model is exploited to compute a cascaded topic model for further learning of correlational and temporal context at multiple scales. These behaviour context models are particularly useful for recognising non-exaggerated multi-object co-occurring behaviours. To that end, a method is considered for detecting subtle behavioural anomalies against learned context.

The methods and techniques for group behaviour modelling and anomaly detection described in the preceding chapters of Part III are predominantly based on unsupervised learning. They assume that an unlabelled training data set is available containing only or largely normal behaviour patterns. After a model for normal behaviour is learned, the anomaly detection problem is treated as an outlier detection problem. We started to introduce the notion of semi-supervised learning by human intervention in assisting a model to learn more effectively for anomaly detection given sparse training examples (Sect. 9.5). Usually, abnormal behaviours that are difficult to be detected are also statistically rare and visually subtle, that is, difficult to be distinguished from a normal behaviour. These abnormal behaviours will cause serious problems for an outlier detection based model, learned without supervision. In this situation, utilising some labelled examples, however few, for model learning becomes more critical. A model for describing rare behaviours may have to be learned from very limited, sparse training examples with only coarse labelling information available. By coarse labelling, it implies that human knowledge is limited to only capable of indicating when an anomaly may have occurred but without any specific knowledge on where and what objects have caused it. To address this problem, a weakly supervised approach to model learning is introduced in the next chapter. Such a learning strategy aims to yield a plausible model from poorly sampled training data for describing a class of rare behaviours that are visually similar to normal behaviours, with the extreme case of having only one training example for model learning.

References

Ali, S., Shah, M.: Floor fields for tracking in high density crowd scenes. In: European Conference on Computer Vision, Marseille, France, October 2008, pp. 1–14 (2008)

Bar, M.: Visual objects in context. Nat. Rev., Neurosci. **5**, 617–629 (2004)

Bar, M., Aminof, E.: Cortical analysis of visual context. Neuron **38**, 347–358 (2003)

Bar, M., Ullman, S.: Spatial context in recognition. Perception **25**, 343–352 (1993)

Biederman, I., Mezzanotte, R.J., Rabinowitz, J.C.: Scene perception: detecting and judging objects undergoing relational violations. Cogn. Psychol. **14**, 143–177 (1982)

Blei, D.M., Ng, A.Y., Jordan, M.I.: Latent Dirichlet allocation. J. Mach. Learn. Res. **3**, 993–1022 (2003)

Buxton, H., Gong, S.: Visual surveillance in a dynamic and uncertain world. Artif. Intell. **78**(1–2), 431–459 (1995)

Carbonetto, P., de Freitas, N., Barnard, K.: A statistical model for general contextual object recognition. In: European Conference on Computer Vision, Prague, Czech Republic, May 2004, pp. 350–362 (2004)

Galleguillos, C., Rabinovich, A., Belongie, S.: Object categorization using co-occurrence, location and appearance. In: IEEE Conference on Computer Vision and Pattern Recognition, Alaska, USA, June 2008, pp. 1–8 (2008)

Gong, S., Buxton, H.: On the visual expectations of moving objects: a probabilistic approach with augmented hidden Markov models. In: European Conference on Artificial Intelligence, Vienna, August 1992, pp. 781–786 (1992)

Gong, S., Buxton, H.: Bayesian nets for mapping contextual knowledge to computational constraints in motion segmentation and tracking. In: British Machine Vision Conference, Guildford, UK, September 1993, pp. 229–238 (1993)

Gupta, A., Davis, L.S.: Beyond nouns: exploiting prepositions and comparative adjectives for learning visual classifier. In: European Conference on Computer Vision, Marseille, France, October 2008, pp. 16–29 (2008)

Heitz, G., Koller, D.: Learning spatial context: using stuff to find things. In: European Conference on Computer Vision, Marseille, France, October 2008, pp. 30–43 (2008)

Hofmann, T.: Probabilistic latent semantic analysis. In: Uncertainty in Artificial Intelligence, pp. 43–52 (1999a)

Hofmann, T.: Probabilistic latent semantic indexing. In: The Annual International SIGIR Conference on Research and Development in Information Retrieval, Berkley, USA, pp. 50–57 (1999b)

Kumar, S., Hebert, M.: A hierarchical field framework for unified context-based classification. In: IEEE International Conference on Computer Vision, Beijing, China, October 2005, pp. 1284–1291 (2005)

Li, J., Gong, S., Xiang, T.: Global behaviour inference using probabilistic latent semantic analysis. In: British Machine Vision Conference, Leeds, UK, September 2008, pp. 193–202 (2008a)

Li, J., Gong, S., Xiang, T.: Scene segmentation for behaviour correlation. In: European Conference on Computer Vision, Marseille, France, October 2008, pp. 383–395 (2008b)

Loy, C.C., Xiang, T., Gong, S.: Multi-camera activity correlation analysis. In: IEEE Conference on Computer Vision and Pattern Recognition, Miami, USA, June 2009, pp. 1988–1995 (2009)

Marszalek, M., Laptev, I., Schmid, C.: Actions in context. In: IEEE Conference on Computer Vision and Pattern Recognition, Miami, USA, June 2009, pp. 2929–2937 (2009)

Murphy, K.P., Torralba, A., Freeman, W.T.: Using the forest to see the tree: a graphical model relating features, objects and the scenes. In: Advances in Neural Information Processing Systems, Vancouver, Canada (2003)

Palmer, S.: The effects of contextual scenes on the identification of objects. Mem. Cogn. **3**, 519–526 (1975)

Rabinovich, A., Vedaldi, A., Galleguillos, C., Wiewiora, E., Blelongie, S.: Objects in context. In: IEEE International Conference on Computer Vision, Rio de Janeiro, Brazil, October 2007, pp. 1–8 (2007)

Sherrah, J., Gong, S.: Tracking discontinuous motion using Bayesian inference. In: European Conference on Computer Vision, Dublin, Ireland, June 2000, pp. 150–166 (2000)

Wolf, L., Bileschi, S.: A critical view of context. Int. J. Comput. Vis. **69**(2), 251–261 (2006)

Yang, M., Wu, Y., Hua, G.: Context-aware visual tracking. IEEE Trans. Pattern Anal. Mach. Intell. **31**(7), 1195–1209 (2008)

Zelnik-Manor, L., Perona, P.: Self-tuning spectral clustering. In: Advances in Neural Information Processing Systems, pp. 1601–1608 (2004)

Zheng, W., Gong, S., Xiang, T.: Quantifying contextual information for object detection. In: International Conference on Computer Vision, Kyoto, Japan, September 2009, pp. 932–939 (2009a)

Zheng, W., Gong, S., Xiang, T.: Associating groups of people. In: British Machine Vision Conference, London, UK, September 2009b

Chapter 11
Modelling Rare and Subtle Behaviours

One of the most desired capabilities for automated visual analysis of behaviour is the identification of rarely occurring and subtle behaviours of genuine interest. This is of practical value because the behaviours of greatest interest for detection are normally rare, for example civil disobedience, shoplifting, driving offenses, and may be intentionally disguised to be visually subtle compared to more obvious ongoing behaviours in a public space. These are also the reasons why this problem is challenging. Rare behaviours by definition have few examples for a model to learn from. The most interesting rare behaviours which are also subtle do not exhibit abundance of strong visual features in the data that describe them. Consider, for example, the scene illustrated by Fig. 11.1. The traffic violations shown in the figure are simple and rare. However, they are also hard to be singled out amongst the ongoing typical behaviours. This also highlights the need for effective classification. Different rare behaviours may indicate situations of different severity, for instance, a traffic-light turning violation and a collision require different response.

In Chap. 7, a supervised method is studied for group activity recognition by which a generative model is learned for each activity class using labelled training examples. Given an unseen behaviour pattern, the model which can best explain the data will determine the class label. However, a model learned from few examples using a supervised method will suffer from model over-fitting, resulting in the model incapable of generalising well to unseen data. A number of unsupervised behaviour modelling methods are also considered in Chaps. 8 and 9, in order to examine their suitability for modelling and detecting rare and irregular behaviours. These methods learn generative models of normal behaviour and can thereby potentially detect rare behaviours as outliers. However, unsupervised methods have a number of limitations in detecting rare and subtle behaviours:

1. Their performance could be sub-optimal due to not exploiting the few positive examples that may be available for model training.
2. As outlier detectors, they are not able to *categorise* different types of rare behaviours.
3. They are ineffective in cases where a rare behaviour is non-separable from normal behaviours in a feature space. That is, if visual observation or feature extraction

S. Gong, T. Xiang, *Visual Analysis of Behaviour*,
DOI 10.1007/978-0-85729-670-2_11, © Springer-Verlag London Limited 2011

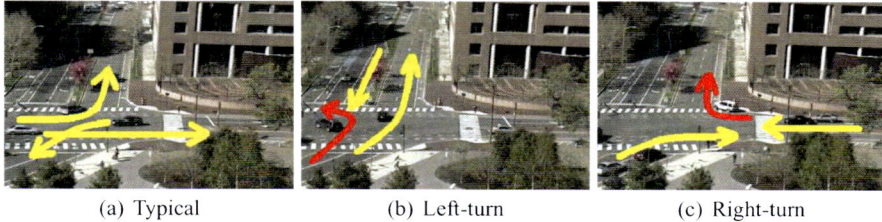

    (a) Typical   (b) Left-turn   (c) Right-turn

Fig. 11.1 A public scene usually contains: (**a**) numerous examples of typical behaviours, (**b**) and (**c**) sparse examples of rare behaviours. Rare behaviours, shown in *red*, also usually co-occur with other typical behaviours, shown in *yellow*, making them visually subtle and difficult to be singled out

is limited such that a rare behaviour is indistinguishable in its representational feature space from some typical behaviours, then outlier detectors trained from unsupervised learning will not be able to detect the rare behaviour of interest reliably. This situation is especially true when the rare behaviour is visually subtle.

In the following, we consider the problem of learning behaviour models from rare and subtle examples. By rare, we mean as few as one example. A similar problem in image and object categorisation is also known as 'one-shot learning' (Fei-Fei et al. 2006). By subtle, we mean weak visual evidence. There may only be a few pixels associated with a behaviour of interest captured in video data, and a few more pixels differentiating a rare behaviour from a typical one. To eliminate the prohibitive manual labelling cost, both in time and inconsistency, required by traditional supervised methods, we consider a weakly supervised framework, in which a user needs not, or cannot, explicitly locate the target behaviours of interest in the training video data. In particular, we examine a weakly supervised joint topic model introduced by Li et al. (2010), with a learning algorithm that learns jointly a model for all the classes using a partially shared common basis. The intuition behind this model is that well-learned common behaviours highlight implicitly the few features relevant to characterising the rare and subtle behaviours of interest, thereby permitting them to be learned without explicit supervision. Moreover, the shared common basis helps to alleviate the statistical insufficiency problem, therefore enabling model learning from few examples. This statistical insufficiency problem is particularly acute in modelling rare behaviours by supervised learning.

11.1 Weakly Supervised Joint Topic Model

11.1.1 Model Structure

Similar to the behaviour representation adopted by the Markov clustering topic model in Chap. 9, a low-level local motion event-based representation is adopted here. A video is temporally segmented into N_d non-overlapping short clips of a constant duration, with each clip considered as a video document. In each document, spatial and directional quantisation of optical flow vector from each image cell of

Fig. 11.2 (**a**) LDA (Blei
et al. 2003) and (**b**) the
weakly supervised joint topic
model graphical model
structures, where only one
rare class is shown for
illustration. *Shaded nodes* are
observed

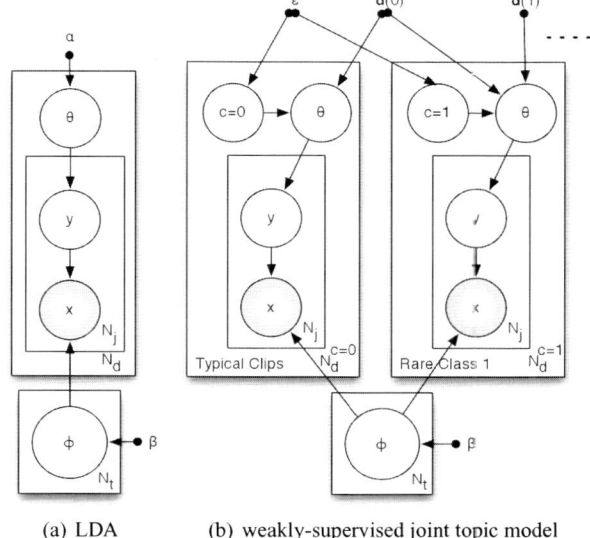

(a) LDA   (b) weakly-supervised joint topic model

a fixed size are computed as local motion events, considered as video words. The
video words computed from each clip represent that video document. We further
denote a video corpus of N_d documents as

$$\mathbf{X} = \{\mathbf{x}_j\}_{j=1}^{N_d} \tag{11.1}$$

in which each document $\mathbf{x}_j$ is represented by a bag of N_j words $\mathbf{x}_j = \{x_{j,i}\}_{i=1}^{N_j}$. This
video corpus represents the entire training set.

A weakly supervised joint topic model builds on the latent Dirichlet allocation
(LDA) model (Blei et al. 2003), and is related to the Markov clustering topic model
introduced in Chap. 9 and the cascaded topic model introduced in Chap. 10. Applied
to unsupervised behaviour modelling (Hospedales et al. 2009; Li et al. 2009; Wang
et al. 2009), LDA learns a generative model of video clips $\mathbf{x}_j$ (Fig. 11.2(a)). Each
event $x_{j,i}$ in a clip is distributed according to a discrete distribution $p(x_{j,i}|\boldsymbol{\phi}_{y_{j,i}}, y_{j,i})$
with parameter $\boldsymbol{\Phi}$ indexed by its unknown parent behaviour $y_{j,i}$. Object behaviours
captured in the video are distributed as $p(\mathbf{y}|\boldsymbol{\theta}_j)$ according to a per-clip Dirichlet
distribution $\boldsymbol{\theta}_j$. In essence, learning in this model clusters co-occurring local motion
events in $\mathbf{X}$ and thereby discovers 'regular' behaviours $\mathbf{y}$ in a data set given by a
video or some videos.

In contrast to a standard LDA, a weakly supervised joint topic model
(Fig. 11.2(b)) has two objectives: (1) learning robust and accurate representations
for both typical behaviours which are statistically sufficiently represented, and a
number of rare behaviour classes for which few non-localised examples are available; (2) classifying behaviours in some unseen test data using the learned model. A
key to both model learning and deployment is jointly modelling the shared aspects
of typical and rare behaviours.

Given a database of N_d clips $\mathbf{X} = \{\mathbf{x}_j\}_{j=1}^{N_d}$, assume that $\mathbf{X}$ can be divided into $N_c + 1$ classes: $\mathbf{X} = \{\mathbf{X}^c\}_{c=0}^{N_c}$ with N_d^c clips per class. In order to permit rare behaviour modelling in the presence of other overwhelming normal behavioural distractors, a crucial assumption made by a weakly supervised joint topic model is that clips $\mathbf{X}^0$ of class 0 contain only typical behaviours, whilst class $c > 0$ clips $\mathbf{X}^c$ may contain *both* typical and class c rare behaviours. This modelling assumption is enforced by partially switching the generative model of clip according to its class (Fig. 11.2(b)).

Specifically, let T_0 be the N_t^0 element list of typical behaviours and T_c be the N_t^c element list of rare behaviours unique to each rare class c. Then, typical clips $\mathbf{x} \in \mathbf{X}^0$ are composed by a mixture of behaviours from T_0 (Fig. 11.2(b), left). Clips $x \in \mathbf{X}^c$ in each rare class c are composed from a mixture of behaviours $T_{0,c} \triangleq T_0 \cup T_c$ (Fig. 11.2(b), right). Whilst there are $N_t = \sum_{c=0}^{N_c} N_t^c$ behaviours in total, each clip may be explained by a class-specific subset of behaviours. The proportions and locations of that subset are unknown, which need be determined by the model.

In contrast to the standard LDA, which uses a fixed, and usually uniform, behaviour hyper-parameter α, the dimension of the per-clip behaviour proportions θ and prior $\boldsymbol{\alpha}$ in the weakly supervised joint topic model are class dependent and non-uniform. So if $\boldsymbol{\alpha}(0)$ are the priors of typical behaviours, $\boldsymbol{\alpha}(c)$ are the priors unique to class c, and $\boldsymbol{\alpha} \triangleq [\boldsymbol{\alpha}(0), \boldsymbol{\alpha}(1), \ldots, \boldsymbol{\alpha}(N_c)]$ is the list of all the behaviour priors; then typical clips $c = 0$ are generated with parameters $\boldsymbol{\alpha}^{c=0} \triangleq \boldsymbol{\alpha}(0)$ and rare clips $c > 0$ with $\boldsymbol{\alpha}^c \triangleq [\boldsymbol{\alpha}(0), \boldsymbol{\alpha}(c)]$. In this way, a shared space is established explicitly between common and rare behaviours. This shared space enables the model to differentiate subtle differences between typical and rare behaviours, therefore overcoming the inherent problem of modelling sparse rare behaviours in the presence of overwhelming distractors, and improving behaviour classification accuracy.

The generative process of a weakly supervised joint topic model is as follows:

1. For each behaviour k, $k = 1, \ldots, N_t$;
   a. Draw a Dirichlet event-behaviour distribution $\boldsymbol{\phi}_k \sim \text{Dir}(\beta)$;
2. For each clip j, $j = 1, \ldots, N_d$;
   a. Draw a class label $c_j \sim \text{Multi}(\varepsilon)$.
   b. Choose the shared prior $\boldsymbol{\alpha}^{c=0} \triangleq \boldsymbol{\alpha}(0)$ or $\boldsymbol{\alpha}^{c>0} \triangleq [\boldsymbol{\alpha}(0), \boldsymbol{\alpha}(c)]$.
   c. Draw a Dirichlet class-constrained behaviour distribution $\boldsymbol{\theta}_j \sim \text{Dir}(\boldsymbol{\alpha}^c)$.
   d. For observed events $i = 1, \ldots, N_w^j$ in clip j:
      i. Draw an behaviour $y_{j,i} \sim \text{Multi}(\boldsymbol{\theta}_j)$;
      ii. Sample a word $x_{j,i} \sim \text{Multi}(\boldsymbol{\phi}_{y_{j,i}})$.

The probability of variables $\{\mathbf{x}_j, \mathbf{y}_j, c_j, \boldsymbol{\theta}_j\}$ and parameters $\boldsymbol{\Phi}$ given hyper-parameters $\{\boldsymbol{\alpha}, \beta, \epsilon\}$ in a clip j is

$$p(\mathbf{x}_j, \mathbf{y}_j, \boldsymbol{\theta}_j, \boldsymbol{\Phi}, c_j | \boldsymbol{\alpha}, \beta, \varepsilon)$$

$$= \prod_{t=1}^{N_t} p(\boldsymbol{\phi}_t | \beta) \prod_{i=1}^{N_w^j} p(x_{j,i} | y_{j,i}, \boldsymbol{\phi}_{y_{j,i}}) p(y_{j,i} | \boldsymbol{\theta}_j) p(\boldsymbol{\theta}_j | \boldsymbol{\alpha}^{c_j}) p(c_j | \varepsilon) \quad (11.2)$$

The probability $p(\mathbf{X}, \mathbf{Y}, \mathbf{c} | \boldsymbol{\alpha}, \beta, \epsilon)$ of a video data set $\mathbf{X} = \{\mathbf{x}_j\}_{j=1}^{N_d}$, $\mathbf{Y} = \{\mathbf{y}_j\}_{j=1}^{N_d}$, $\mathbf{c} = \{c_j\}_{j=1}^{N_d}$ can be factored as

$$p(\mathbf{X}, \mathbf{Y}, \mathbf{c} | \boldsymbol{\alpha}, \beta, \epsilon) = p(\mathbf{X}|\mathbf{Y}, \beta) p(\mathbf{Y}|\mathbf{c}, \boldsymbol{\alpha}) p(\mathbf{c}|\varepsilon) \qquad (11.3)$$

The first two terms above are products of Polya distributions over actions k and clips j, respectively:

$$p(\mathbf{X}|\mathbf{Y}, \beta) = \int p(\mathbf{X}|\mathbf{Y}, \boldsymbol{\Phi}) p(\boldsymbol{\Phi}|\beta) d\boldsymbol{\Phi}$$

$$= \prod_{k=1}^{N_t} \frac{\Gamma(N_v \beta)}{\prod_v \Gamma(\beta)} \frac{\prod_v \Gamma(n_{k,v} + \beta)}{\Gamma(\sum_v n_{k,v} + \beta)} \qquad (11.4)$$

$$p(\mathbf{Y}|\mathbf{c}, \boldsymbol{\alpha}) = \prod_{c=0}^{N_c} \prod_{j=1}^{N_d^c} \int p(\mathbf{y}_j|\boldsymbol{\theta}_j) p(\boldsymbol{\theta}_j|\boldsymbol{\alpha}, c_j) d\boldsymbol{\theta}_j$$

$$= \prod_{c=0}^{N_c} \prod_{j=1}^{N_d^c} \frac{\Gamma(\sum_k \alpha_k)}{\prod_k \Gamma(\alpha_k)} \frac{\prod_k \Gamma(n_{j,k} + \alpha_k)}{\Gamma(\sum_k n_{j,k} + \alpha_k)} \qquad (11.5)$$

where $n_{k,v}$ and $n_{j,k}$ indicate the counts of action-word and clip-action associations and k ranges over behaviours $T_{0,c}$ permitted by the current document class c_j.

We next show how to learn a weakly supervised joint topic model by training, and use the learned model to classify new data by testing. For training, we assume weak supervision in the form of labels c_j, and the goal is to learn the model parameters $\{\boldsymbol{\Phi}, \boldsymbol{\alpha}, \beta\}$. For testing, parameters $\{\boldsymbol{\Phi}, \boldsymbol{\alpha}, \beta\}$ are fixed and the model infers the unknown class of test clips $\mathbf{x}^*$.

11.1.2 Model Parameters

The first problem to address is to learn the parameters of a weakly supervised joint topic model from a training data set $\{\mathbf{X}, \mathbf{c}\}$. This is non-trivial because of the correlated unknown behaviour latents $\{\mathbf{Y}, \boldsymbol{\theta}\}$ and model parameters $\{\boldsymbol{\Phi}, \boldsymbol{\alpha}, \beta\}$. The correlation between these unknowns is intuitive as they broadly represent 'which behaviours are present where' and 'what each behaviour looks like'. A standard expectation-maximisation (EM) approach to model learning with latent variables is to alternate between inference, computing $p(\mathbf{Y}|\mathbf{X}, \mathbf{c}, \boldsymbol{\alpha}, \beta)$, and hyper-parameter estimation, optimising $\{\boldsymbol{\alpha}, \beta\} \leftarrow \text{argmax}_{\boldsymbol{\alpha}, \beta} \sum_{\mathbf{Y}} \log p(\mathbf{X}, \mathbf{Y}|\mathbf{c}, \boldsymbol{\alpha}, \beta) p(\mathbf{Y}|\mathbf{X}, \mathbf{c}, \boldsymbol{\alpha}, \beta)$. Neither of these sub-problems have analytical solutions in this case. However, approximate solutions can be developed.

Inference

Similarly to LDA, exact inference in weakly supervised joint topic model is intractable. Alternatively, it is possible to derive a collapsed Gibbs sampler (Gilks et al. 1995) to approximate $p(\mathbf{Y}|\mathbf{X}, \mathbf{c}, \boldsymbol{\alpha}, \beta, \epsilon)$. The Gibbs sampling update for action $y_{j,i}$ is derived by integrating out the parameters $\boldsymbol{\Phi}$ and $\boldsymbol{\theta}$ in its conditional probability given the other variables:

$$p(y_{j,i}|\mathbf{Y}_{-j,i}, \mathbf{X}, \mathbf{c}, \boldsymbol{\alpha}, \beta) \propto \frac{n_{y,x}^{-j,i} + \beta}{\sum_v (n_{y,v}^{-j,i} + \beta)} \frac{n_{j,y}^{-j,i} + \alpha_y}{\sum_k (n_{j,k}^{-j,i} + \alpha_k)} \tag{11.6}$$

where $\mathbf{Y}_{-j,i}$ denotes all behaviours excluding $y_{j,i}$, $n_{y,x}$ denotes the counts of feature $x_{j,i}$ being associated to action $y_{j,i}$, $n_{j,y}$ denotes the counts of behaviour $y_{j,i}$ in clip j, and superscript $-j, i$ denotes counts excluding item (j, i).

An iteration of (11.6) draws samples from the posterior $p(\mathbf{Y}|\mathbf{X}, \mathbf{c}, \boldsymbol{\alpha}, \beta)$. In contrast to the standard LDA, the topics which may be allocated by this joint topic model are constrained by clip class c to be in $T_0 \cup T_c$. Behaviours $T_{c=0}$ will be well constrained by the abundant typical data. Clips of some rare class $c > 0$ may use extra behaviours T_c in their representation. These will therefore come to represent the unique characteristics of interesting class c.

Each sample of behaviours $\mathbf{Y}$ entails Dirichlet distributions over the behaviour-word parameter $\boldsymbol{\Phi}$ and per-clip behaviour parameter $(\boldsymbol{\theta}_j - p(\boldsymbol{\Phi}|\mathbf{X}, \mathbf{Y}, \beta))$ and $p(\boldsymbol{\theta}|\mathbf{Y}, \mathbf{c}, \boldsymbol{\alpha})$. These can then be point-estimated by the mean of their Dirichlet posteriors:

$$\hat{\phi}_{k,v} = \frac{n_{k,v} + \beta}{\sum_v (n_{k,v} + \beta)} \tag{11.7}$$

$$\hat{\theta}_{j,k} = \frac{n_{j,k} + \alpha_k}{\sum_k (n_{j,k} + \alpha_k)} \tag{11.8}$$

Hyper-parameter Estimation

The Dirichlet prior hyper-parameters $\boldsymbol{\alpha}$ and β play an important role in governing the behaviour-event $p(\mathbf{X}|\mathbf{Y}, \beta)$ and clip-behaviour $p(\mathbf{Y}|\mathbf{c}, \boldsymbol{\alpha})$ distributions; β describes the prior expectation about the 'size' of the behaviours. This size is the number of local motion events expected in each behaviour. More crucially, elements of $\boldsymbol{\alpha}$ describe the relative dominance of each behaviour within a clip of a particular class. That is, in a class c clip, how frequently are observations related to rare behaviours T_c compared to ongoing normal behaviours T_0.

Direct optimisation $\{\boldsymbol{\alpha}, \beta\} \leftarrow \text{argmax}_{\boldsymbol{\alpha}, \beta} \sum_{\mathbf{Y}} \log p(\mathbf{X}, \mathbf{Y}|\mathbf{c}, \boldsymbol{\alpha}, \beta) p(\mathbf{Y}|\mathbf{X}, \mathbf{c}, \boldsymbol{\alpha}, \beta)$ using the EM framework is computationally intractable, because the sum over $\mathbf{Y}$ has exponentially many terms. However, we can use N_s Gibbs samples $\mathbf{Y}_s \sim$

$p(\mathbf{Y}|\mathbf{X}, \mathbf{c}, \boldsymbol{\alpha}, \beta)$ drawn during inference (11.6) to define a Gibbs-EM algorithm (Andrieu et al. 2003; Wallach 2006) which approximates the required optimisation as

$$\{\boldsymbol{\alpha}, \beta\} \leftarrow \underset{\boldsymbol{\alpha}, \beta}{\mathrm{argmax}} \frac{1}{N_s} \sum_s^{N_s} \log\big(p(\mathbf{X}|\mathbf{Y}_s, \beta)p(\mathbf{Y}_s|\mathbf{c}, \boldsymbol{\alpha})\big) \qquad (11.9)$$

This enables β to be learned by substituting (11.4) into (11.9) and maximising for β. The gradient g with respect to β is

$$g = \frac{d}{d\beta} \frac{1}{N_s} \sum_s \log p(\mathbf{X}|\mathbf{Y}_s, \beta)$$

$$= \frac{1}{N_s} \sum_s \sum_k \Big(N_v \Psi(N_v\beta) - N_v\Psi(\beta) + \sum_v \Psi(n_{k,v} + \beta)$$

$$- N_v \Psi(n_{k,\cdot} + N_v\beta) \Big) \qquad (11.10)$$

where $n_{k,v}$ is the matrix of topic-word counts for each E-step sample, $n_{k,\cdot} \triangleq \sum_v n_{k,v}$, and Ψ is the digamma function. This can be iteratively updated (Minka 2000; Wallach 2006) by the following:

$$\beta^{\mathrm{new}} = \beta \frac{\sum_s \sum_k \sum_v \Psi(n_{k,v} + \beta) - N_t N_v \Psi(\beta)}{N_v \sum_s (\sum_k \Psi(n_{k,\cdot} + N_v\beta) - N_t \Psi(N_v\beta))} \qquad (11.11)$$

Compared to β, learning the hyper-parameters $\boldsymbol{\alpha}$ is harder because they are class dependent. A simple approach is to define a completely separate $\boldsymbol{\alpha}^c$ for each class and maximise $p(\mathbf{Y}^c|\boldsymbol{\alpha}^c)$ independently for each c. However, this leads to poor estimates for the frequency of typical behaviours from the point of view of each rare class. This is because the rare class $\boldsymbol{\alpha}^{c>0}$ parameter updates would take into account only a few, possibly one, clips to constrain the typical elements of $\boldsymbol{\alpha}^c$, despite much more data about typical behaviour being actually available.

To alleviate the problem of statistical insufficiency in learning $\boldsymbol{\alpha}$, the shared-structure of a weakly supervised joint topic model (Sect. 11.1.1 and Fig. 11.2(b)) is exploited to develop a suitable learning algorithm. Specifically, by defining $\boldsymbol{\alpha}^c \triangleq [\boldsymbol{\alpha}(0), \boldsymbol{\alpha}(c)]$, a shared space of typical behaviours $\boldsymbol{\alpha}(0)$ is established for all classes. This will alleviate the sparsity problem by allowing data from all clips to help constrain these parameters.

In the following, K^0 is used to represent the indices into $\boldsymbol{\alpha}$ of the N_t^0 typical behaviours, K^c is used to represent the N_t^c indices of the rare behaviours in class c, and $K^{c,0} = K^0 \cup K^c$ are the indices for both typical and class c behaviours. Hyper-parameters $\boldsymbol{\alpha}$ are then learned by fixed point iterations of the following form:

$$\alpha_k^{\mathrm{new}} = \alpha_k \frac{a}{b} \qquad (11.12)$$

For typical behaviours $k \in K^0$, the terms are computed by

$$a = \sum_{s=1}^{N_s} \sum_{j=1}^{N_d} \left(\Psi(n_{j,k} + \alpha_k) - \Psi(\alpha_k) \right) \tag{11.13}$$

$$b = \sum_{s=1}^{N_s} \sum_{c=1}^{N_c} \sum_{j=1}^{N_d^c} \left(\Psi\left(n_{j,\cdot}^{0,c} + \alpha_{\cdot}^{0,c}\right) - \Psi\left(\alpha_{\cdot}^{0,c}\right) \right) + \sum_{s=1}^{N_s} \sum_{j=1}^{N_d^0} \left(\Psi\left(n_{j,\cdot}^0 + \alpha_{\cdot}^0\right) - \Psi\left(\alpha_{\cdot}^0\right) \right) \tag{11.14}$$

where

$$\alpha_{\cdot}^0 \triangleq \sum_{k \in K^0} \alpha_k$$

$$n_{j,\cdot}^0 \triangleq \sum_{k \in K^0} n_{j,k}$$

$$\alpha_{\cdot}^{0,c} \triangleq \sum_{k \in K^{0,c}} \alpha_k$$

$$n_{j,\cdot}^{0,c} \triangleq \sum_{k \in K^{0,c}} n_{j,k}$$

For class c rare behaviours $k \in K^c$, the terms are computed by

$$a = \sum_{s=1}^{N_s} \sum_{j=1}^{N_d^c} \left(\Psi(n_{j,k} + \alpha_k) - \Psi(\alpha_k) \right) \tag{11.15}$$

$$b = \sum_{s=1}^{N_s} \sum_{j=1}^{N_d^c} \left(\Psi\left(n_{j,\cdot}^{0,c} + \alpha_{\cdot}^{0,c}\right) - \Psi\left(\alpha_{\cdot}^{0,c}\right) \right) \tag{11.16}$$

Iteration of (11.11) and (11.12)–(11.16) estimates the hyper-parameters $\{\alpha, \beta\}$. This is carried out periodically during the sampling (11.6) in order to complete the Gibbs-EM algorithm.

We illustrate the ability of weakly supervised joint topic model to model rare behaviour given extremely sparse examples using a traffic junction scene from the MIT traffic data set (Wang et al. 2009). The scene featured numerous objects exhibiting complex behaviours concurrently. Figure 11.3(a) illustrates the behaviours that typically occur in the scene. There are two types of rare behaviours identified in the scene (Figs. 11.3(b) and (c)), corresponding to illegal left-turn and right-turn at different locations of the scene, shown by the red arrows. The video data are segmented into non-overlapping video clips of 300 frames long at 25 FPS, therefore each video document lasting 12 seconds in time. Given a labelled training set where most clips contains typical behaviours only and very few contains one of the two

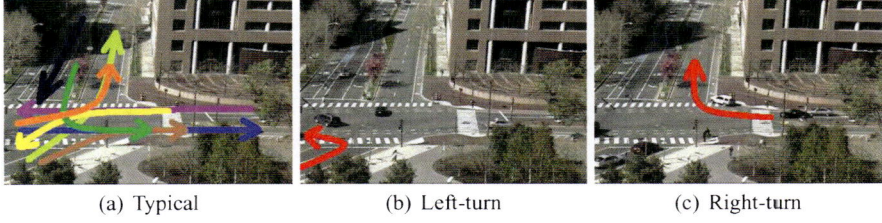

(a) Typical   (b) Left-turn   (c) Right-turn

Fig. 11.3 Examples of typical and rare behaviours in a traffic junction scene. Traffic flows of different directions are shown in *different colours*. The *red arrows* correspond to rare flow patterns

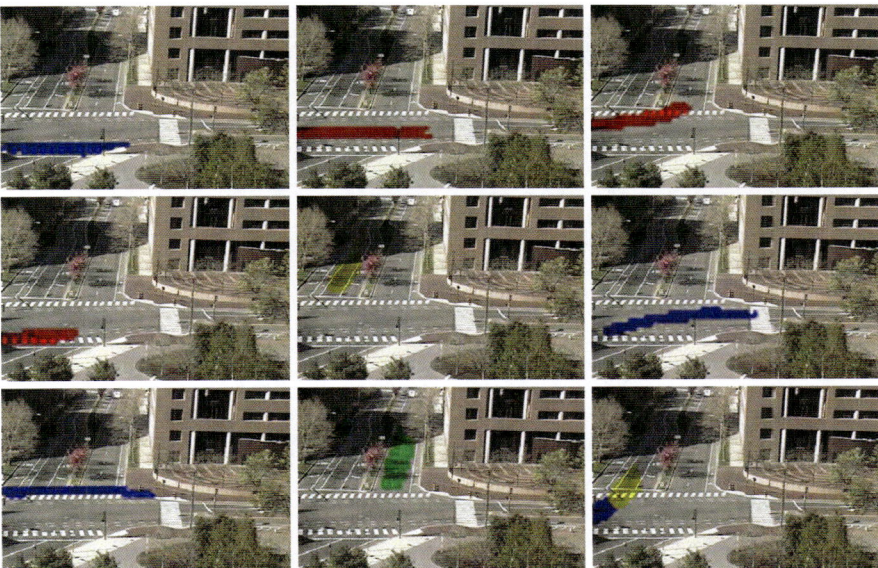

Fig. 11.4 Typical behaviours learned from the traffic junction scene. *Red*: *Right*, *Blue*: *Left*, *Green*: *Up*, *Yellow*: *Down*

rare behaviours, a weakly supervised joint topic model is learned to model both the typical and rare behaviour jointly.

The dominant events, given by large values of $p(x|\boldsymbol{\phi}_y)$, are used to illustrate some of the learned typical and rare behaviours $\boldsymbol{\phi}_y$ from a training set of 200 clips of typical behaviours, and one clip for each of the two rare behaviour classes. That is, the model is trained by 'one-shot learning' for describing each rare behaviour class. Figure 11.4 shows typical behaviours learned in this scene, including pedestrians crossing (left column), straight traffic (centre column) and turning traffic (right column). Figure 11.5 shows that the learned rare behaviours match the examples in Figs. 11.3(b) and (c). These behaviours are generally disambiguated by the model from ongoing typical behaviours (Fig. 11.3(a)). This is despite that

Fig. 11.5 'One-shot learning' of rare behaviours in the traffic junction traffic scene

(a) Rare 1: Left Turn   (b) Rare 2: Right Turn

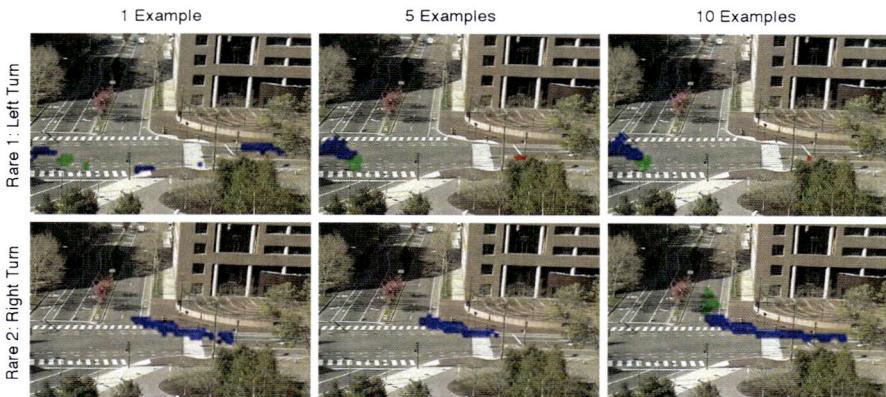

Fig. 11.6 Improvement on learning rare behaviour models with increasing number of training examples

1. each rare behaviour class is learned by 'one-shot learning' using a single example;
2. the rare behaviours are overwhelmed by abundant typical behaviours co-occurring in the same space or vicinity (Fig. 11.3(a));
3. only small visual differences exist between the rare and typical behaviours, seen by arrow segments in Figs. 11.3(b) and (c) overlapping those in Fig. 11.3(a).

These examples are based on 'one-shot learning'. The model can also explore additional rare class examples to improve the learned representation for each rare behaviour class. Figure 11.6 shows the learned rare behaviour models given an increasing number of examples. It indicates that when more training examples become available, a more meaningful and accurate model can be learned for each rare behaviour class.

To understand more about the effect of weakly supervised learning in a joint topic learning framework, we examine a standard LDA model learned in a supervised manner. That is, different LDA models are learned for different behaviour classes. It is observed that with one example for each of the two rare behaviour classes, it is simply insufficient to learn a behaviour from such sparse data. However, given 10 examples per rare class, a standard supervised LDA is able to learn a fair model of each class (Fig. 11.7). The learned rare behaviours include the correct cases, as shown by Behaviour-7 in Fig. 11.7(b) and Behaviour-5 in Fig. 11.7(c). However,

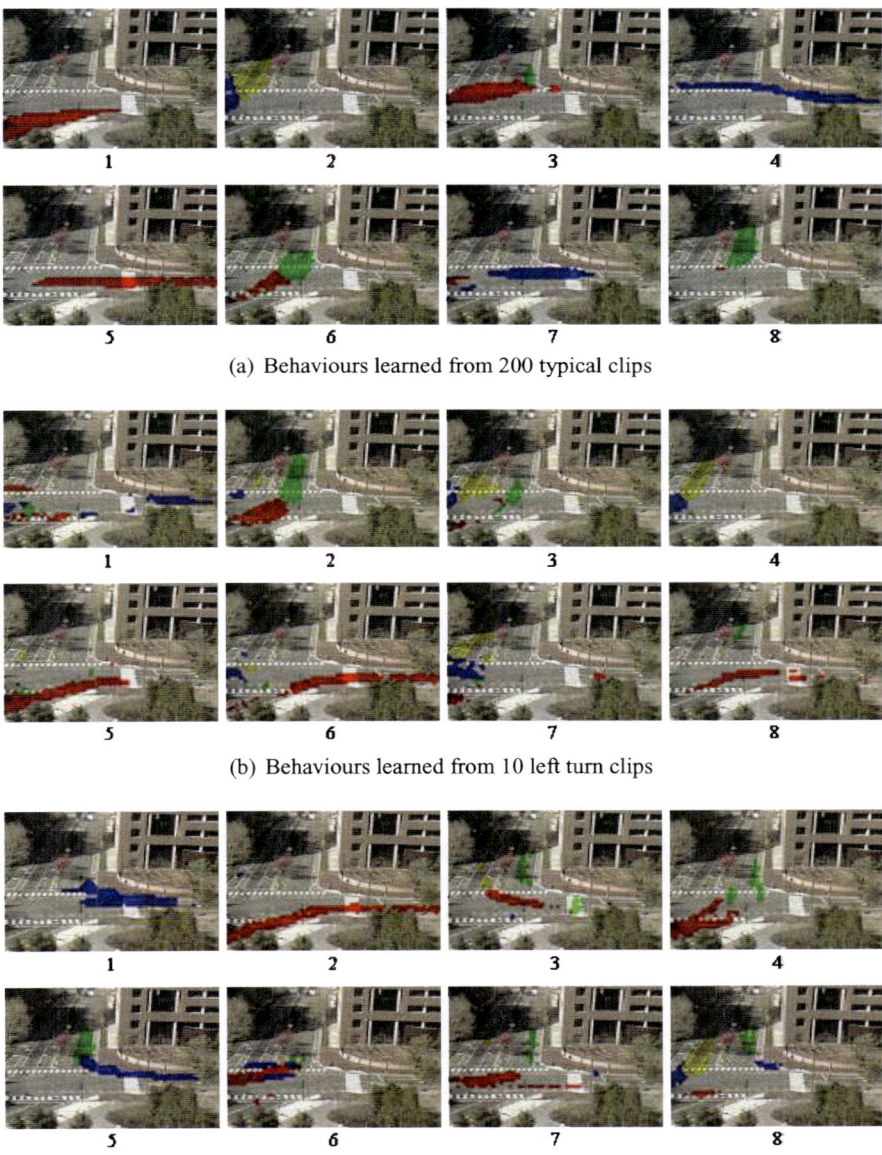

(a) Behaviours learned from 200 typical clips

(b) Behaviours learned from 10 left turn clips

(c) Behaviours learned from 10 right turn clips

Fig. 11.7 Examples of behaviours learned using a standard LDA model for the traffic junction scene

a key factor limiting the modelling power of LDA is that each LDA model has to learn independently the ongoing typical behaviours. For example, Behaviour-1 in Fig. 11.7(a), Behaviour-5 in Fig. 11.7(b) and Behaviour-2 in Fig. 11.7(c) all represent 'right-turn from below'. By learning separate models for the same typical behaviours, LDA introduces a significant source of noise. In contrast, by learning a single model for ongoing typical behaviours which is shared between all classes (Fig. 11.4), a weakly supervised joint topic model avoids this source of noise. Moreover, by permitting the rare class models to leverage a well-learned typical behaviour model, they can learn a more precise description of the rare behaviour classes. For example, the Behaviour-7 learned by a LDA in Fig. 11.7(b) is noisier than the left-turn behaviour in Fig. 11 6(a), learned by a weakly supervised joint topic model.

11.2 On-line Behaviour Classification

Given a learned joint topic model $\{\boldsymbol{\alpha}, \boldsymbol{\Phi}\}$, we consider the problem of model inference on unseen videos. Specifically, the model is deployed to classify each test clip $\mathbf{x}^*$ by determining whether it is better explained as a clip containing only typical behaviours ($c = 0$), or containing typical behaviours with some rare behaviours c ($c > 0$). In contrast to the E-step inference problem (11.6), where the posterior of *behaviours* $\mathbf{y}$ was computed by Gibbs sampling, we are now computing the posterior *class* c, which requires the harder task of integrating out behaviours $\mathbf{y}$. More specifically, the desired class posterior is given by

$$p(c|\mathbf{x}^*, \boldsymbol{\alpha}, \varepsilon, \boldsymbol{\Phi}) \propto p(\mathbf{x}^*|c, \boldsymbol{\alpha}, \boldsymbol{\Phi}) p(c|\varepsilon) \tag{11.17}$$

$$p(\mathbf{x}^*|c, \boldsymbol{\alpha}, \boldsymbol{\Phi}) = \int \sum_{\mathbf{y}} p(\mathbf{x}^*, \mathbf{y}|\theta, \boldsymbol{\Phi}) p(\theta|\boldsymbol{\alpha}) \, d\theta \tag{11.18}$$

The challenge is to compute accurately and efficiently the class-conditional marginal likelihood in (11.18). Due to the intractable sum over correlated $\mathbf{y}$ in (11.18), an importance sampling approximation can be defined to the marginal likelihood (Buntine 2009; Wallach et al. 2009):

$$p(\mathbf{x}^*|c) \approx \frac{1}{S} \sum_{s} \frac{p(\mathbf{x}^*, \mathbf{y}^s|c)}{q(\mathbf{y}^s|c)}, \quad \mathbf{y}^s \sim q(\mathbf{y}|c) \tag{11.19}$$

where we drop conditioning on the parameters for clarity. Different choices of proposal $q(\mathbf{y}|c)$ induce different estimation algorithms. The optimal proposal $q_o(\mathbf{y}|c)$, which is unknown, is $p(\mathbf{x}^*, \mathbf{y}|c)$. A mean field approximation $q_{mf}(\mathbf{y}|c) = \prod_i q_i(y_i|c)$ can be developed with minimal Kullback–Leibler (KL) divergence to the optimal proposal by iterating

$$q_i(y_i|c) \propto \left(\alpha_y^c + \sum_{l \neq i} q_l(y_l|c) \right) \hat{\phi}_{y_i, x_i} \tag{11.20}$$

This importance sampling proposal given by (11.20) results in faster and more accurate estimation of the marginal likelihood in (11.18) than the standard harmonic mean approximation approach using posterior Gibbs sampling (Griffiths and

Steyvers 2004; Wallach et al. 2009). The harmonic mean approximation of the marginal likelihood suffers from (1) requiring a slow Gibbs sampler at test time, therefore being prohibitive for any on-line computation, and (2) having a high variance of the harmonic mean estimator, therefore making classification inaccurate. For on-line model inference, a faster inference method is crucial as the relevance of behaviour classification is now determined by the speed and accuracy of computing marginal likelihood of (11.18).

In essence, for classifying a new unseen clip, an on-line model utilises the importance sampler of (11.19) and (11.20) to compute the marginal likelihood of (11.18) for each behaviour class c, including both a typical class ($c = 0$) and a number of rare classes ($c = 1, 2, \ldots$). This will provide the class posterior (11.17) for classifying the probe clip. Interestingly, behaviour classification using this framework is essentially a process of generative model selection (Mackay 1991). That is, the classification process must determine, in the presence of numerous latent variables $\mathbf{y}$, which one of the two generative model structures provides the better explanation of the data: a simple model explanation involving only typical behaviours (left in Fig. 11.2(b)); or a model explanation involving both typical and rare behaviours (right in Fig. 11.2(b)). The simpler typical-only model explanation is automatically preferred by Bayesian Occam's razor (Mackay 1991). However, if there is any evidence of a rare behaviour, the more complex model explanation is uniquely able to allocate the associated rare topic and thereby yield a higher likelihood value.

Li et al. (2010) show that even if each rare behaviour class is learned using one example, a weakly supervised joint topic model can yield good classification accuracy on unseen video clips. The model is able to reliably detect a rare behaviour in a clip and classify it into different types of rare behaviours. In contrast, alternative topic models such as LDA and supervised LDA (Blei and McAuliffe 2007) are unable to detect rare behaviours if only a few clips per rare class are available for model training.

11.3 Localisation of Rare Behaviour

Once a clip has been detected and classified as containing a particular rare behaviour, the behaviour in question may need be further located in space and time by its triggering local motion events. In contrast to the unsupervised outlier detection approach (Chaps. 8 and 9), a weakly supervised joint topic model provides a more principled means to achieve this. Specifically, for a test clip j of type c, the behaviour profile for a clip, $p(\mathbf{y}_j | \mathbf{x}_j, c, \boldsymbol{\alpha}, \hat{\boldsymbol{\Phi}})$, can be determined by iterative Gibbs sampling:

$$p(y_{j,i} | \mathbf{y}_{-j,i}, \mathbf{x}_j, c, \boldsymbol{\alpha}, \hat{\boldsymbol{\Phi}}) \propto \hat{\phi}_{y,x} \frac{n_{j,y}^{-j,i} + \alpha_y}{\sum_k (n_{j,k}^{-j,i} + \alpha_k)} \tag{11.21}$$

Fig. 11.8 Examples of localisation of rare behaviours by weakly supervised joint topic model

The local motion events i are listed for the sampled behaviour $y_{j,i}$ which belongs to a class c rare behaviour, i.e.

$$\mathcal{I} = \{i\} \quad \text{s.t.} \quad y_{j,i} \in T_c \tag{11.22}$$

The indices of these local motion events $\mathcal{I}$ within the clip provide an approximate spatio-temporal segmentation of the detected rare behaviour of interest. Because parameters α and Φ are already estimated, this Gibbs localisation procedure needs much fewer iterations than the initial model learning and is hence much faster.

Figure 11.8 shows examples of localising specific behaviours by their corresponding local motion events within classified clips for the traffic junction scene. The brighter areas in each image are specific local motion events which are labelled as triggering a rare behaviour. The segmentation is approximate, due to noisy optical flow features for local motion event detection and the Markov chain Monte Carlo (MCMC)-based labelling. The bounding boxes of rare behaviour triggering events, shown in red rectangles, nevertheless provide a useful behaviour localisation in space and time.

11.4 Discussion

Rare and subtle behaviour learning has significant practical usage for behaviour-based video analysis, as behaviours of greater interest for being identified are also usually both rare and visually subtle. Moreover, the ability for a model to accommodate weak supervision is important given the need for automated screening of large quantity of video data with a lack of human knowledge on where and what to look for. A combination of factors conspire to make modelling rare and subtle behaviours challenging, including:

1. crowdedness with cluttered background and severe occlusions;

2. statistical insufficiency caused by relative imbalance between fewer examples of rare behaviours and abundant examples of typical behaviours;
3. absolute sparsity of rare behaviour examples, with the most extreme case of having a single example;
4. rare behaviours of visual subtlety disguised amidst commonly seen behaviours.

A weakly supervised joint topic model leverages its model of typical behaviours to help locate and classify individual rare and subtle behaviours. The specific model learning and inference mechanism studied makes effective learning possible despite a challenging combination of weak supervision and sparse data sampling. Classification using a weakly supervised joint topic model is enabled by Bayesian model selection. Classification accuracy is enhanced by jointly modelling typical behaviour classes, and by hyper-parameter learning to overcome the lack of examples for modelling rare classes. Such a framework enables one-shot learning of rare and subtle behaviours from a single unlocalised example, and facilitates on-line detection, classification and localisation of rare behaviours.

For weakly supervised learning, manual efforts remain to be required to visually examine data and detect what are considered to be rare behaviours of interest. However, rare behaviours are unpredictable and by definition sparse. To exhaustively search rare behaviours of interest by manual means is both unreliable and costly. It is highly desirable to have a model capable of automatically selecting some subsets of data that are deemed plausible candidates worthy further examination by a human observer. The challenge of exhaustively searching for rare behaviours is reduced to a problem of how to best utilise limited human feedback to maximise model learning which in turn optimises the selection of plausible candidates for further human feedback. What is needed, in essence, is a 'man-in-the-loop' iterative process of human-guided computational model learning coupled with machine-assisted human search. To that end, we shall consider in the next chapter an active learning approach to man-in-the-loop guided data mining and machine-assisted behaviour search.

References

Andrieu, C., De-Freitas, N., Doucet, A., Jordan, M.I.: An introduction to MCMC for machine learning. Mach. Learn. **50**, 5–43 (2003)

Blei, D.M., McAuliffe, J.: Supervised topic models. In: Advances in Neural Information Processing Systems, Vancouver, Canada, December 2007

Blei, D.M., Ng, A.Y., Jordan, M.I.: Latent Dirichlet allocation. J. Mach. Learn. Res. **3**, 993–1022 (2003)

Buntine, W.: Estimating likelihoods for topic models. In: Asian Conference on Machine Learning, Nanjing, China, November 2009, pp. 51–64 (2009)

Fei-Fei, L., Fergus, R., Perona, P.: One-shot learning of object categories. IEEE Trans. Pattern Anal. Mach. Intell. **28**(4), 594–611 (2006)

Gilks, W.R., Richardson, S., Spiegelhalter, D. (eds.): Markov Chain Monte Carlo in Practice. Chapman & Hall/CRC Press, London/Boca Raton (1995)

Griffiths, T.L., Steyvers, M.: Finding scientific topics. Proc. Natl. Acad. Sci. USA **101**, 5228–5235 (2004)

Hospedales, T., Gong, S., Xiang, T.: A Markov clustering topic model for behaviour mining in video. In: IEEE International Conference on Computer Vision, Kyoto, Japan, October 2009, pp. 1165–1172 (2009)

Li, J., Gong, S., Xiang, T.: Discovering multi-camera behaviour correlations for on-the-fly global activity prediction and anomaly detection. In: IEEE International Workshop on Visual Surveillance, Kyoto, Japan, October 2009

Li, J., Hospedales, T., Gong, S., Xiang, T.: Learning rare behaviours. In: Asian Conference on Computer Vision, Queenstown, New Zealand, November 2010

Mackay, D.: Bayesian interpolation. Neural Comput. 4(3), 415–447 (1991)

Minka, T.: Estimating a Dirchlet distribution. Technical report, Microsoft (2000)

Wallach, H.: Topic modeling: beyond bag-of-words. In: International Conference on Machine Learning, Pittsburgh, USA, pp. 977–984 (2006)

Wallach, H., Murray, I., Salakhutdinov, R., Mimno, D.: Evaluation methods for topic models. In: International Conference on Machine Learning, pp. 1105–1112 (2009)

Wang, X., Ma, X., Grimson, W.E.L.: Unsupervised activity perception by hierarchical Bayesian models. IEEE Trans. Pattern Anal. Mach. Intell. 31(3), 539–555 (2009)

Chapter 12
Man in the Loop

Video data captured from a public space are typically characterised by highly imbalanced behaviour class distribution. Most of the captured data examples reflect normal behaviours. Unusual behaviours, either because of being rare or abnormal, only constituent a small portion in the observed data. In addition, normal behaviours are often known a priori, whilst unusual behaviours are usually unknown and unpredictable. An unsupervised learning based model can be constructed to detect unusual behaviours through a process of outlier detection, by which a model is learned from normal behaviour examples using unsupervised one-class learning to describe a class of all normal behaviours. Any unseen behaviours in new test video data that deviate statistically from the learned normal behaviour model are deemed unusual through binary classification (Chaps. 8 and 9). This strategy offers a practical way of bypassing the difficulties in statistical modelling of imbalanced class distributions with inadequate unusual behaviour training examples.

However, an outlier detection based model from unsupervised learning is fundamentally limited in a number of ways.

1. The model has difficulties in detecting subtle unusual behaviours. Subtle unusual behaviours often refer to behaviours that undergo small and local changes (Boiman and Irani 2007; Li et al. 2010) from the norm, or give rise to unusual temporal correlations among multiple objects (Xiang and Gong 2008). The visual subtlety of unusual behaviours in a crowded public space is often reflected by their visual feature distributions. In particular, unusual behaviour patterns in a feature space may overlap with normal behaviour patterns, making it inseparable from a large number of normal behaviour patterns co-existing in the same feature space (Fig. 12.1(b)). This makes it difficult to learn an outlier detection model for binary classification.
2. The model does not exploit information from detected unusual behaviours. Image noise and errors in feature extraction could be mistaken as genuine unusual behaviours of interest, giving false alarms in detection. If there is no subsequent exploitation of detected unusual behaviours, it is difficult for a model to learn from mistakes so that a model can refine its decision boundaries to better separate

S. Gong, T. Xiang, *Visual Analysis of Behaviour*,
DOI 10.1007/978-0-85729-670-2_12, © Springer-Verlag London Limited 2011

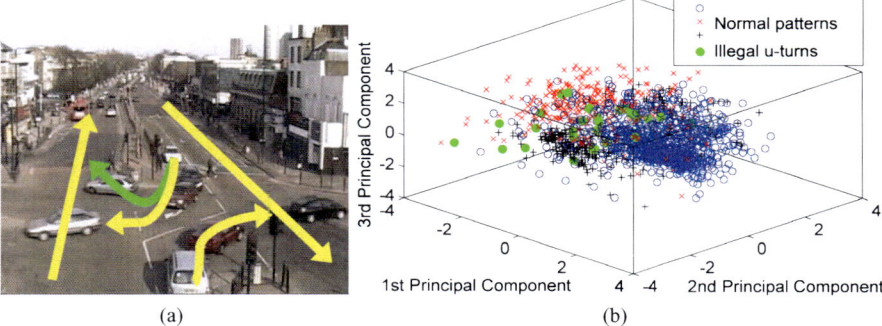

Fig. 12.1 (**a**) An example of a subtle illegal u-turn behaviour, shown by the *green arrow*, in a street intersection scene. It is subtle due to its visual similarity to a large number of co-occurring normal behaviour patterns in the same scene, shown by the *yellow arrows*. (**b**) By representing each behaviour pattern as a data point in a feature space (Sect. 12.2), behaviour patterns from a data set form distributions of normal and unusual behaviour patterns, shown in the first three principal components of the feature space. The u-turn patterns, plotted by *green dots*, overlap partially with normal behaviour patterns

genuine unusual behaviours from normal behaviours despite noisy observations and distribution overlapping in the feature space.

3. A large amount of rare but benign behaviours give rise to false alarms in unusual behaviour detection. Normal behaviour patterns in a public scene are diverse. There are normal behaviour patterns which are rarely seen but also not of any significance. They are rarely seen benign behaviours. It is unlikely that an exhaustive set of well-defined normal behaviour data is available for off-line model training. Therefore, unsupervised learning of a one-class behaviour model to encompass *all* normal behaviours is inherently difficult if not impossible. This results in some outlying regions of the learned normal class distribution being falsely and consistently detected as being unusual.

In the preceding chapter, we considered how to overcome some of these limitations by weakly supervised learning of a joint model describing normal and rare behaviours. However, weakly labelling all rare behaviours is in itself a challenging task (Sect. 11.4). Furthermore, since rare and unusual behaviours are mostly unpredictable, it cannot be assumed that examples of all rare behaviours exist in a training data set.

Human knowledge can be exploited in a different way to address this problem. Instead of giving supervision on the *input* to model learning by labelling the training data, human knowledge can be more effectively utilised by giving selective feedback to model *output*. This process reduces significantly the burden on human effort in assisting model training. Moreover, it selects deliberately more relevant human knowledge required to improve the model in discovering, by mining, the underlying structures of behaviour data so to yield optimal class decision boundaries. In particular, a human observer can provide relevance feedback in the form of giving approval to a model detected unusual behaviour by judging whether it is benign.

This becomes especially useful when model confidence is low due to difficulties in recognising an equivocal behaviour by visual features alone.

One way for a model to capture human relevance feedback is active learning (Settles 2010). This model learning strategy offers an alternative to conventional supervised, unsupervised, semi-supervised and weakly supervised learning strategies. An active learning method is designed especially for seeking human feedback on certain model selected instances of detection output, whilst the selection criteria is internal to the model rather than decided by the human observer. This is to ensure that not only the most relevant human feedback is sought, but also that the model is not subject to human bias without data support.

It should be noted that active learning differs from a semi-supervised learning strategy in a subtle but fundamental way. A semi-supervised learning strategy utilises labelling of *randomly* selected data. The labelling process thus treats all examples *equally*. However, not all data examples are equal to a model for learning the correct decision boundaries between overlapping class distributions. In essence, active learning enables a model to select automatically which particular data examples are more informative and necessary for seeking independent feedback from a human observer. This feedback is in turn utilised by the model to learn better class decision boundaries when data examples are incomplete.

12.1 Active Behaviour Learning Strategy

There are two different settings in active learning, pool-based and stream-based. A pool-based active learning method requires access to a pre-determined fixed-size pool of unlabelled data for searching the most informative example instances for querying human feedback. Unusual behaviours are often unpredictable. Beyond a controlled laboratory environment, preparing a pool of unlabelled data that encompasses implicitly unusual behaviours of interest is neither practical nor guaranteed. In contrast, a stream-based setting does not require a pre-determined data pool. Instead, it requests labelling on-line based on cumulative sequential observations. For analysing live video on-line, a stream-based active learning method has clear advantages over a pool-based scheme. However, stream-based learning is also intrinsically more difficult to compute. This is because the model has to make decisions on-the-fly without complete knowledge on the underlying data distribution (Ho and Wechsler 2008; Kapoor and Horvitz 2007).

For active learning, a model has to choose a suitable query criterion for selecting query examples. A number of criteria can be considered (Settles 2010), including uncertainty criterion (Ho and Wechsler 2008), likelihood criterion (Pelleg and Moore 2004), expected model change (Settles et al. 2007), and expected error reduction (Roy and McCallum 2001). The uncertainty criterion selects those example instances that a model is uncertain about their class labels. They correspond to ambiguous data points near the class decision boundaries in a feature space. The likelihood criterion selects examples of data points either furthest away from those data points with clear class labels, or having the lowest likelihood from a model. Different criteria put an emphasis on different aspects of learning a model. For instance, the likelihood criterion tends to lead to quick discovery of new classes, whilst the uncertainty criterion focuses more on refining decision boundaries between existing classes.

For stream-based active learning of behaviour models, it may be necessary to employ different query criteria simultaneously. The reason is that some behaviour classes, especially unusual behaviour classes, have to be discovered as new classes. This is known as model exploration. They may not be captured in the observed data during the early stage of model learning. Simultaneously, a model can also be improved incrementally by refining the decision boundaries of known classes, as more data become available. This is known as model exploitation. The exploration and exploitation aspects of model learning can be better satisfied by using different criteria in a single framework (Hospedales et al. 2011).

In the following, we consider a stream-based adaptive active learning method for unusual behaviour detection (Loy et al. 2010). Formally, given an unlabelled example behaviour $\mathbf{x}_t$ observed at time step t from an input data stream $\mathcal{X} = (\mathbf{x}_1, \ldots, \mathbf{x}_t, \ldots)$, a classifier C_t is required to determine on-the-fly whether to request human assistance in deciding for a label y_t for this example, or to discard this example $\mathbf{x}_t$ as it is deemed uninteresting. The goal is therefore to select critical examples from $\mathcal{X}$ for human annotation in order to achieve two computational tasks simultaneously:

1. to discover unusual behaviour classes, or unknown regions of existing classes in the feature space,
2. to refine the classification boundaries[1] of existing classes, with higher priorities given to feature space regions surrounding the unusual classes, so to improve detection accuracy of unusual behaviours.

In this framework, the data presented to a human observer are chosen based on an adaptive selection process involving two criteria.

1. The first query criterion is a likelihood criterion, which favours behaviour examples that have a low likelihood with respect to the current model. As a result, unknown classes or unexplored regions of existing classes in the feature space can be discovered. This process does not assume the availability of predefined classes. Once a new class is discovered, the model will expand itself automatically to accommodate the new class.
2. The second criterion is an uncertainty criterion based on a modified query-by-committee algorithm (Argamon-Engelson and Dagan 1999; McCallum and Nigam 1998; Seung et al. 1992). This criterion is used to refine decision boundaries by selecting controversial examples in uncertain feature space regions that causes the most disagreement among existing classes, with more emphasis given to the regions surrounding the unusual behaviour classes in order to address the problem of imbalance class distribution.

---

[1]In this book, we alternate the use of the terms 'classification boundaries' and 'decision boundaries'.

The two query criteria are dynamically and adaptively re-weighted based on measuring relative entropy or KL divergence (Cover and Thomas 2006) on the model before and after it is updated by a queried example. The rationale behind this adaptive weighting scheme is to favour the criterion that is more likely to return an informative example for query that brings most impact to the current model. Intuitively, the likelihood criterion is preferred at the beginning of a learning process since unknown behaviours will cause a greater impact to changes in model parameters. When sufficient examples from different classes are discovered, the uncertainty criterion will dominate. This is because the feature space is well explored, and any further exploration by the likelihood criterion is less likely to change significantly the model parameters.

12.2 Local Block-Based Behaviour

A schema of a stream-based active learning approach is depicted in Fig. 12.2. In this framework, behaviour patterns are represented by low-level local motion events (Fig. 12.2(a)–(d)). A Bayesian classifier is employed to model behaviour patterns (Fig. 12.2(e)) and it is subsequently used for unusual behaviour detection (Fig. 12.2(g)). The behaviour model would query for human verification on any behaviour pattern that fulfil the query constraint defined by the likelihood and uncertainty criteria (Fig. 12.2(f)).

Behaviour patterns are represented using location-specific motion information over a temporal window without relying on object segmentation and tracking. This is achieved through the following steps.

1. Local Block Behaviour Patterns: Given an input video, the spatial context learning model (Sect. 10.1) is employed to decompose automatically a dynamic scene into n regions, where $\mathcal{R} = \{\mathcal{R}_i | i = 1, \ldots, n\}$, according to the spatial and temporal distribution of motion patterns observed. In particular, the image space is first divided into equal-sized blocks with 8×8 pixels each (Fig. 12.2(a)). Optical flow in each pair of consecutive frames is extracted using the Lucas–Kanade method (Lucas and Kanade 1981). Flow vectors are averaged within each block to obtain a single local motion vector for that block at that time instance. By computing this average flow vector at each local block over time, we construct a pair of time series for each block from its horizontal and vertical average motion components, denoted as $\mathbf{u_b}$ and $\mathbf{v_b}$, where $\mathbf{b}$ specifies the two-dimensional coordinates of a block in the image space (Fig. 12.2(b)). Jointly, $\mathbf{u_b}$ and $\mathbf{v_b}$ give a bivariate time-series function. We refer to this as a local block behaviour pattern.

2. Scene Decomposition: Correlation distances are computed among local block behaviour patterns to construct an affinity matrix for the entire scene. Spectral clustering (Zelnik-Manor and Perona 2004) is carried out on this affinity matrix for scene decomposition (Fig. 12.2(c)). This is similar to the scene decomposition method described in Sect. 10.1.2.

3. Dominant Motion Directions: Motion direction of each moving pixel computed from optical flow in each segmented region is quantised into four directions, and

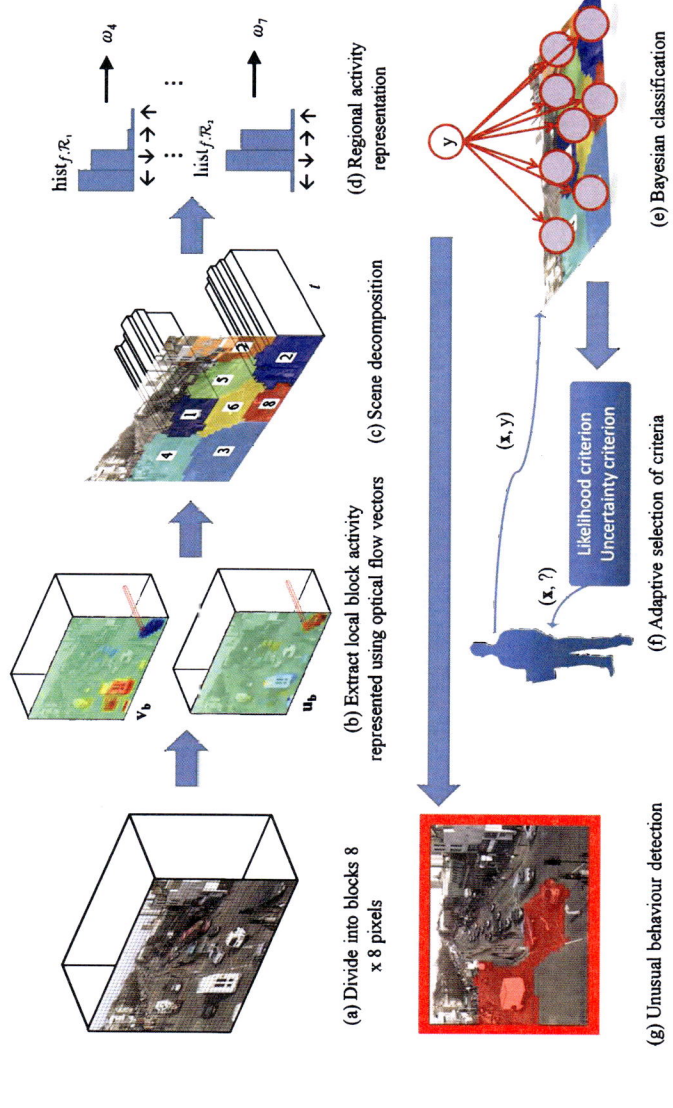

hist$_{f,\mathcal{R}_1}$ ω_4

hist$_{f,\mathcal{R}_2}$ ω_7

(d) Regional activity representation

(c) Scene decomposition

(b) Extract local block activity represented using optical flow vectors

$\mathbf{v}_b$

$\mathbf{u}_b$

(a) Divide into blocks 8 × 8 pixels

(e) Bayesian classification

y

$(\mathbf{x}, y)$

Likelihood criterion
Uncertainty criterion

$(\mathbf{x}, ?)$

(f) Adaptive selection of criteria

(g) Unusual behaviour detection

Fig. 12.2 A diagram illustrating a stream-based active unusual behaviour detection framework

put into four bins. A histogram $\text{hist}_{f,\mathcal{R}_i}$ with a size of four bins is constructed for each region $\mathcal{R}_i$ at each time frame f. The input video sequence is uniformly divided into non-overlapping short clips, with each clip having 50 frames in length. Individual bins of a regional histogram within each clip t are accumulated as

$$\text{hist}_{t,\mathcal{R}_i} = \sum_{f \in \text{clip } t} \text{hist}_{f,\mathcal{R}_i} \qquad (12.1)$$

This process is illustrated in Fig. 12.2(d).

The four directional bins are ranked in a descending order based on their values. The dominant motion directions are identified from the top ranked bins that account for a chosen percentage $P \in [0, 1]$ of the total bin values, that is, the area under the histogram. For instance, P can be set to 0.8. Non-dominant motion directions are then removed by setting the values of the corresponding bins to zero.

4. Motion Direction Profile: Finally, the motion directional histograms of all the regions in the scene $\{\mathcal{R}_i\}$ are discretised to construct a codebook with $r = 16$ words:

$$\{\omega_j | j = 1, \ldots, 16\} \qquad (12.2)$$

representing the combination (4×4) of possible dominant motion directions in a region. Each region is described by a discrete word value. For example, word ω_1 represents a motionless region, word ω_2 means only direction bin 1 is observed, and word ω_4 indicates the occurrence of both direction bins 1 and 2, as depicted in Fig. 12.2(d). In general, the ith region of the tth clip is represented by a variable $x_{i,t}$ of 16 possible discrete values according to its dominant motion direction profile. A clip is denoted as $\mathbf{x}_t = (x_{1,t}, \ldots, x_{n,t})$.

12.3 Bayesian Classification

Given a n-dimensional observed vector $\mathbf{x} = (x_1, \ldots, x_n)$ representing a video clip, one wishes to assign the vector into one of the K classes, where a class variable is represented by $y = k \in \{1, \ldots, K\}$. The size of K is unknown. This problem can be addressed by Bayesian classification. For stream-based active learning, a model has to be learned incrementally given streamed input. To facilitate efficient incremental learning, a naïve Bayesian classifier (Fig. 12.3) can be used. Conditional independence is assumed among the distributions of input attributes $x_1, \ldots, x_n$ given the class label. In the case of representing behaviour patterns in a scene, this assumes that the distribution of motion direction profile in each region of the scene is independent from those of other regions. This assumption of conditional independence may not be always valid. Nevertheless, even if the assumption is not precisely satisfied, a model is still capable of providing a sufficiently satisfactory description of the data (Hand and Yu 2001; Rish 2001).

A naïve Bayesian classifier is quantified by a parameter set specifying the conditional probability distributions. A separate multinomial distribution $p(x_i|y)$ on each

Fig. 12.3 A graphical representation of a naïve Bayesian classifier. Each variable in the observed vector $\mathbf{x} = (x_1, \ldots, x_n)$ are conditioned on the class variable y

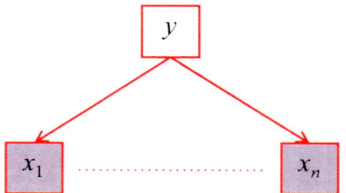

x_i for each class label is assumed. Consequently, $\boldsymbol{\theta}_{x_i|y} = \{\theta_{x_i=j|y}\}$ is used to represent parameters for the multinomial distribution $p(x_i|y)$, where $0 \leq \theta_{x_i=j|y} \leq 1$ and $\sum_{j=1}^{r} \theta_{x_i=j|y} = 1$. Since a video clip is represented by a vector of regional motion direction profiles, each of which given by a variable x_i with 16 possible discrete values (12.2), we have $r = 16$.

The conditional probability $p(\mathbf{x}|y = k)$ for an observed vector given class $y = k$ is computed by

$$p(\mathbf{x}|y = k) = \prod_{i=1}^{n} p(x_i|y = k) \tag{12.3}$$

The posterior conditional distribution $p(y|\mathbf{x})$ is computed by Bayes' rule. A class y^* that best explains $\mathbf{x}$ can then be obtained as follows:

$$y^* = \underset{k \in \{1, \ldots, K\}}{\arg\max} \; p(y = k|\mathbf{x}) = \underset{k \in \{1, \ldots, K\}}{\arg\max} \; p(y = k)p(\mathbf{x}|y = k) \tag{12.4}$$

For stream-based active learning, to solve (12.4) requires a tractable incremental learning scheme. Since observation at each time step is complete, that is, each $(\mathbf{x}_t, y_t)$ assigns values to all the variables of interest, conjugate prior can be employed to facilitate efficient Bayesian learning. The conjugate prior of a multinomial distribution with parameters $\boldsymbol{\theta}_{x_i|y}$ follows the Dirichlet distribution (Bishop 2006), given by

$$\mathrm{Dir}(\boldsymbol{\theta}_{x_i|y} \mid \boldsymbol{\alpha}_{x_i|y}) \propto \prod_{j=1}^{r} [\theta_{x_i=j|y}]^{\alpha_{x_i=j|y}-1} \tag{12.5}$$

where $\boldsymbol{\alpha}_{x_i|y}$ denotes $(\alpha_{x_i=1|y}, \ldots, \alpha_{x_i=r|y})$ and $\alpha_{x_i=j|y} \in \mathbb{R}^+$ are parameters of the Dirichlet distribution. It should be noted that with this formulation, a naïve Bayesian classifier is capable of expanding itself automatically whenever a new class is discovered during the active learning process. This is achieved by expanding the class variable y to $K + 1$ states and augmenting the conditional probability distribution of each x_i with $\boldsymbol{\theta}_{x_i|y=K+1}$.

12.4 Query Criteria

In a stream-based setting, a query decision is typically determined by a query score p^{query} derived from a query criterion $\mathcal{Q}$. The query score is then compared against a threshold Th. Specifically, if $p^{\mathrm{query}} \geq \mathrm{Th}$, a query is made; otherwise $\mathbf{x}_t$ is discarded. The likelihood criterion and the uncertainty criterion are complementary. They can be exploited jointly for simultaneous unknown behaviour discovery and classification boundary refinement.

12.4.1 Likelihood Criterion

To use this criterion, a data point is selected by comparing its likelihood against current distribution modelled by a classifier. In particular, given a data example $\mathbf{x}$, the class y^* that best explains the example according to (12.4) is first identified. Then, for each feature component x_i of $\mathbf{x}$, a normalised probability score of x_i given y^* is computed as

$$\hat{p}(x_i|y^*) = \frac{p(x_i|y^*) - \mathrm{E}[p(x_i = j|y^*)]}{\sqrt{\mathrm{E}[p(x_i = j|y^*) - \mathrm{E}[p(x_i = j|y^*)]]}}$$

This normalised probability score is bounded to ensure $-0.5 \leq \hat{p}(x_i|y^*) \leq 0.5$. Finally, the likelihood score at time step t for requesting human feedback on example $\mathbf{x}$ is calculated as

$$p_t^l = 1 - \left(\frac{1}{2} + \frac{1}{n} \sum_{i=1}^{n} \hat{p}(x_i|y^*) \right) \tag{12.6}$$

This likelihood score lies in $[0, 1]$. If the p_t^l value of an example is closer to 1, it is then more likely to be queried.

12.4.2 Uncertainty Criterion

For choosing the uncertainty criterion, we start with the query-by-committee algorithm (Argamon-Engelson and Dagan 1999; McCallum and Nigam 1998). In addition, we also consider to address the problem of conflicting classes in order to yield a balanced example selection. There are two main steps for computing an uncertainty score using the query-by-committee algorithm.

1. Generating a committee: Given a classifier C_t and some training data S_t, M committee members corresponding to hypotheses $\mathbf{h} = \{h_i\}$ of the hypotheses space $\mathcal{H}_t$ are generated, where each hypothesis is consistent with the training data seen so far (Seung et al. 1992), namely

$$h_i \in \mathcal{H}_t \,|\, \forall (\mathbf{x}, y) \in S_t, \quad h_i(\mathbf{x}) = y \tag{12.7}$$

For a naïve Bayes classifier with multinomial conditional probability distributions (12.4), committee members could be generated by sampling new parameters from the posterior Dirichlet distribution of the classifier (Argamon-Engelson and Dagan 1999; McCallum and Nigam 1998). On the other hand, Devroye

(1986) shows that a random vector $\boldsymbol{\theta} = (\theta_1, \ldots, \theta_r)$ from the Dirichlet distribution with parameter $\boldsymbol{\alpha} = (\alpha_1, \ldots, \alpha_r)$ (12.5) can be generated directly from the Gamma distributions, which is conceivably a simpler solution.

Assume r random examples $\varepsilon_{x_i=1|y}, \ldots, \varepsilon_{x_i=r|y}$ are generated from the Gamma distributions, each with density

$$\text{Gam}(\varepsilon_{x_i=j|y} \mid \alpha_{x_i=j|y}) = \frac{1}{\Gamma(\alpha_{x_i=j|y})} e^{-\varepsilon_{x_i=j|y}} (\varepsilon_{x_i=j|y})^{\alpha_{x_i=j|y}-1} \qquad (12.8)$$

where $\Gamma(\cdot)$ denotes the Gamma function and $\alpha_{x_i=j|y}$ plays the role of a shape parameter in the Gamma distribution. The parameter of a committee member is then estimated as

$$\hat{\theta}_{x_i=j|y} = \frac{\varepsilon_{x_i=j|y} + \lambda}{\sum_j (\varepsilon_{x_i=j|y} + \lambda)} \qquad (12.9)$$

where λ is a weight added to compensate for data sparseness. That is designed to prevent zero probabilities for infrequently occurring values $x_i = j$.

2. Measure disagreement: For measuring the level of disagreement among committee members, two measures can be considered.

   a. The first measure is the 'vote entropy' (Argamon-Engelson and Dagan 1999):

   $$p_t^{\text{VE}} = -\sum_{k=1}^{K} \frac{V(y=k)}{M} \log \frac{V(y=k)}{M} \qquad (12.10)$$

   where $V(y=k)$ represents the number of votes that a label receives from the predictions of the committee members. The higher the entropy p_t^{VE}, the more uncertain a data point is.

   b. Another measure is the 'average Kullback–Leibler (KL) divergence', also known as the 'KL divergence to the mean' (McCallum and Nigam 1998), given by

   $$p_t^{\text{KL}} = \frac{1}{M} \sum_{i=1}^{M} \sum_{k=1}^{K} p_{\Theta_i}(y=k\|\mathbf{x}_t) \log \frac{p_{\Theta_i}(y=k\|\mathbf{x}_t)}{\overline{p}(y=k\|\mathbf{x}_t)} \qquad (12.11)$$

   where Θ_i represents a set of parameters of a particular model in the committee; $\overline{p}(y=k\|\mathbf{x}_t)$ denotes the mean of all distributions, given by

   $$\overline{p}(y=k\|\mathbf{x}_t) = \frac{1}{M} \sum_{i=1}^{M} p_{\Theta_i}(y=k\|\mathbf{x}_t)$$

   This is considered as the 'consensus' probability that k is the correct label to be nominated (Settles 2010).

Both measures given by (12.10) and (12.11) are not designed to identify the corresponding classes that cause the most disagreement. To overcome this limitation, a third alternative uncertainty score is considered as follows:

a. A class disagreement score is computed over all possible class labels:

$$s_{y=k,t} = \left\{ \max_{h_i \in \mathcal{H}_t, h_j \in \mathcal{H}_t} \left[p_i(y=k|\mathbf{x}_t) - p_j(y=k|\mathbf{x}_t) \right] \right\} \qquad (12.12)$$

where $i \neq j$.

b. The top two classes that return the highest $s_{y=k,t}$ are identified as c_1 and c_2. The final uncertainty score is computed as

$$p_t^u = \frac{1}{2} \cdot \gamma_u \cdot [s_{y=c_1,t} + s_{y=c_2,t}] \qquad (12.13)$$

where γ_u is the prior introduced to favour the learning of classification boundary for unusual classes. Specifically, γ_u is set to a low value if c_1 and c_2 are both normal behaviour classes, and a high value if any one of c_1 and c_2 is a unusual behaviour class. If the p_t^u value of an example is closer to 1, it is more likely to be queried. Loy et al. (2010) show experimental results to suggest that this uncertainty score is more suitable for active behaviour learning.

12.5 Adaptive Query Selection

Adaptive selection from multiple query criteria is necessary for a joint process of unknown behaviour discovery and classification boundary refinement (Sect. 12.1). More precisely, given multiple query criteria $\mathcal{Q} \in \{\mathcal{Q}_1, \ldots, \mathcal{Q}_a, \ldots, \mathcal{Q}_A\}$, a weight $w_{a,t}$ is assigned to each query criterion $\mathcal{Q}_a$ at time instance t. A criterion is chosen by sampling from a multinomial distribution given as

$$a \sim \text{Mult}(\mathbf{w}_t) \qquad (12.14)$$

where $\mathbf{w}_t \in \{w_{1,t}, \ldots, w_{a,t}, \ldots, w_{A,t}\}$. Intuitively, a criterion with higher weight is more likely to be chosen. The weight $w_{a,t}$ is updated in every time step t as follows:

$$w_{a,t} = \beta w_{a,t-1} + (1-\beta) \frac{\overline{\mathcal{KL}}_a(\Theta \parallel \tilde{\Theta})}{\sum_{a=1}^{\mathcal{A}} \overline{\mathcal{KL}}_a(\Theta \parallel \tilde{\Theta})} \qquad (12.15)$$

where β denotes an update coefficient that controls the updating rate of weights, whilst Θ and $\tilde{\Theta}$ represent, respectively, parameters of a naïve Bayes classifier $\mathcal{C}_t$ and an updated classifier $\mathcal{C}_{t+1}$ trained using $\mathcal{S}_t \cup \{(\mathbf{x}_t, y_t)\}$.

In (12.15), the change of weight is guided by a symmetric KL divergence,[2] which is yielded by a query criterion $\mathcal{Q}_a$ when it last triggered a query. In essence, $\overline{\mathcal{KL}}_a(\Theta \parallel \tilde{\Theta})$ measures the difference between two distributions $p_{\Theta}(\mathbf{x})$ and $p_{\tilde{\Theta}}(\mathbf{x})$ modelled by the naïve Bayes classifier before and after it is updated using a newly queried example. Therefore, it measures the impact of the queried example on the

---

[2] A symmetric Kullback–Leibler divergence $\overline{\mathcal{KL}}(\Theta \parallel \tilde{\Theta})$ is required because in general, KL divergence is non-symmetric, where $\mathcal{KL}(\Theta \parallel \tilde{\Theta}) \neq \mathcal{KL}(\tilde{\Theta} \parallel \Theta)$. A non-symmetric KL divergence is not a true distance metric.

Algorithm 12.1: Stream-based active behaviour learning.

Input: Data stream $\mathcal{X} = (\mathbf{x}_1, \dots, \mathbf{x}_t, \dots)$, an initial classifier $\mathcal{C}_0$ trained with a small set of labelled examples from known classes

Output: A set of labelled examples $\mathcal{S}$ and a classifier $\mathcal{C}$ trained with $\mathcal{S}$

1   Set $\mathcal{S}_0 =$ a small set of labelled examples from known classes;

2   **for** t *from* $1, 2, \dots$ *until the data stream runs out* **do**

3   Receive $\mathbf{x}_t$;

4   Compute p_t^l (12.6);

5   Compute p_t^u (12.13);

6   Select query criterion by sampling $a \sim \text{Mult}(\mathbf{w})$, assign p_t^{query} based on the selected criterion;

7   **if** $p_t^{\text{query}} \geq \text{Th}$ **then**

8   Request y_t and set $\mathcal{S}_t = \mathcal{S}_{t-1} \cup \{(\mathbf{x}_t, y_t)\}$;

9   Obtain classifier $\mathcal{C}_{t+1}$ by updating classifier $\mathcal{C}_t$ with $\{(\mathbf{x}_t, y_t)\}$;

10   Update query criteria weights $\mathbf{w}$ (12.15);

11   **else**

12         $\mathcal{S}_t = \mathcal{S}_{t-1}$;

13   **end**

14   **end**

15   Unusual behaviour is detected if $p(y = \text{unusual}|\mathbf{x})$ is higher than $\text{Th}_{\text{unusual}}$;

---

model. Intuitively, a criterion is preferred if it is assigned a higher weight, because the model is seeking examples that give greater impact to the existing distribution modelled by the classifier.

A symmetric KL divergence $\overline{\mathcal{KL}}(\Theta \parallel \tilde{\Theta})$ is computed as follows:

$$\overline{\mathcal{KL}}(\Theta \parallel \tilde{\Theta}) = \frac{1}{2}\big[\mathcal{KL}(\Theta \parallel \tilde{\Theta}) + \mathcal{KL}(\tilde{\Theta} \parallel \Theta)\big] \tag{12.16}$$

where $\mathcal{KL}(\Theta \parallel \tilde{\Theta})$ is defined as

$$\mathcal{KL}(\Theta \parallel \tilde{\Theta}) = \sum_{\mathbf{x}} p_{\Theta}(\mathbf{x}) \ln \frac{p_{\Theta}(\mathbf{x})}{p_{\tilde{\Theta}}(\mathbf{x})} \tag{12.17}$$

This can be further decomposed as follows (Tong and Koller 2000):

$$\mathcal{KL}(\Theta \parallel \tilde{\Theta}) = \sum_{i=1}^{n} \mathcal{KL}\big(p_{\Theta}(x_i|y) \parallel p_{\tilde{\Theta}}(x_i|y)\big)$$

$$= \sum_{i=1}^{n} \sum_{k=1}^{K} p(y = k)\mathcal{KL}\big(p_{\Theta}(x_i|y = k) \parallel p_{\tilde{\Theta}}(x_i|y = k)\big) \tag{12.18}$$

Algorithm 12.1 summaries the stream-based active learning scheme. Similar to other stream-based active learning methods (Dagan and Engelson 1995; Ho and

Wechsler 2008), the computational cost of this approach is low. It is suitable for real-time processing.[3]

Figure 12.4 shows some examples of queries being selected during a stream-based active behaviour learning process for a street intersection scene. It is evident that different new classes, both usual and unusual, are discovered in the learning process. In particular, unusual behaviours such as queries 1 and 17 were selected for labelling request based on likelihood and uncertainty criteria, respectively. The active learning process also refines the model by discovering new instances of known classes. For example, queries 21 belongs to a normal vertical traffic flow class, but it encompasses a unique behaviour pattern that was previously unseen by the model.

Figure 12.5 shows an example of how different query criteria are selected over time during an active learning process. It is evident that the likelihood criterion dominates at the beginning of the process, since it discovers more unknown behaviours at this stage of the model learning process that give rise to greater changes in the parameter values of the model classifier. The model switches incrementally from the likelihood criterion to the uncertainty criterion to place more emphasis on refining the classification boundaries when exploratory learning is no longer fruitful. In this particular example, approximately 90% of total behaviour classes were discovered after 80 query examples.

Loy et al. (2010) show quantitatively that the stream-based active learning method is capable of yielding more robust and accurate detection of subtle unusual behaviours when compared to conventional supervised and unsupervised learning strategies.

12.6 Discussion

The use of active learning to utilise human relevance feedback is exploited to incrementally construct a more accurate model for on-line unusual behaviour detection. The rationales for active behaviour model learning are two-fold. The first is to select behaviour examples corresponding to data points either located close to the decision boundaries between normal and unusual classes, or poorly represented by a current model. The second is to utilise human feedback on the selected examples for incrementally improving the current model. The focus is placed on stream-based active behaviour modelling because it offers the possibility for improving a model on-the-fly as more data become available. This on-line process for model incremental learning also enables a model to be more adaptive to the changes in both context and definition of normality and abnormality. Model learning under such conditions is challenging because the model needs to overcome imbalanced data distributions whilst learning to discover new unknown classes. To that end, balancing different query criteria becomes critical for simultaneous unknown behaviour class discovery and decision boundary refinement.

---

[3]Loy et al. (2010) show that completing lines 2–8 of Algorithm 12.1 takes an average time of 0.026 seconds to compute with a Matlab implementation on a single-core 2.8 GHz machine.

Fig. 12.4 Examples of queried behaviours in a stream-based active behaviour learning process for a street intersection scene. The *label below* each frame provides the index for a series of queries. The *label* on each frame's *top-left corner* shows the class tag returned by a human observer in response to each query. The first frame with a *dashed border* is the initial example clip given to the model to initialise the active learning process. A query with a *blue border* indicates a new normal class discovered, whilst a query with a *red border* denotes a new unusual class found during the learning process. An additional number inside a bracket located at the frame's *top-left corner* denotes the criterion used to pick the query (1 = likelihood criterion, 2 = uncertainty criterion)

Fig. 12.5 An example of different query criteria being selected over time during an active behaviour learning process

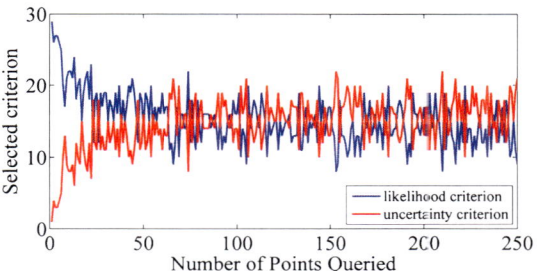

Although a naïve Bayesian classifier is considered in the active learning process, any alternative generative model with an incremental learning algorithm is equally suitable. In particular, when unusual behaviours are also visually subtle, overlapping with normal behaviours in the feature space, the weakly supervised joint topic model discussed in Chap. 11 can be considered. Its ability to cope with weak-supervision is an advantage, as the human feedback often does not specify when and where an unusual behaviour occurs. That is, it may only offer weak labelling.

In the course of Part III, we have considered different representations and learning strategies for modelling group behaviours, detecting rare and abnormal behaviours. As the number of objects engaged in an activity grows, the complexity and uncertainty of object behaviours also increase. In order to cope with more complex and uncertain group activities, learning behavioural context, behaviour based semantic scene decomposition, and hierarchical modelling can all contribute towards increasing the tractability and accuracy of model learning and inference. For exploiting human knowledge on supervised model discovery of new and unknown behaviours, the questions of whether and how much supervision is necessary are addressed. Algorithmic solutions are explored to quantify when and how supervised model learning can be performed. The overall aim is to develop suitable mechanisms and devices that can facilitate the construction of robust and tractable computational models for the analysis and interpretation of multiple object behaviours in an unconstrained environment.

Up to this point, visual analysis of behaviour is largely limited to a single observational viewpoint. This has somewhat restrained behavioural context with spatially localised behavioural scale and temporally limited activity range. The scale and range of a behaviour can and often go beyond a single observational viewpoint in unconstrained environments. For example, when people move through distributed public areas such as in an underground station or at an airport, visual observations of people from different CCTV camera views provide a richer source of information for a greater degree of global situational awareness. It is also more challenging to make sense of these distributed observations coherently. Increasingly, large scale distributed camera networks are deployed in public spaces. This brings about new challenges for visual analysis of behaviour. In the next part of this book, we shall examine in more detail these challenges and consider solutions to address them.

References

Argamon-Engelson, S., Dagan, I.: Committee-based sample selection for probabilistic classifiers. J. Artif. Intell. Res. **11**, 335–360 (1999)

Bishop, C.M.: Pattern Recognition and Machine Learning. Springer, Berlin (2006)

Boiman, O., Irani, M.: Detecting irregularities in images and in video. Int. J. Comput. Vis. **74**(1), 17–31 (2007)

Cover, T.M., Thomas, J.A.: Elements of Information Theory. Wiley, New York (2006)

Dagan, I., Engelson, S.: Committee-based sampling for training probabilistic classifiers. In: International Conference on Machine Learning, pp. 150–157 (1995)

Devroye, L.: Non-uniform Random Variate Generation. Springer, Berlin (1986)

Hand, D.J., Yu, K.: Idiot's Bayes: Not so stupid after all? Int. Stat. Rev. **69**(3), 385–398 (2001)

Ho, S., Wechsler, H.: Query by transduction. IEEE Trans. Pattern Anal. Mach. Intell. **30**(9), 1557–1571 (2008)

Hospedales, T., Gong, S., Xiang, T.: Finding rare classes: adapting generative and discriminative models in active learning. In: Pacific-Asia Conference on Knowledge Discovery and Data Mining, Shenzhen, China, May 2011

Kapoor, A., Horvitz, E.: On discarding, caching, and recalling samples in active learning. In: Uncertainty in Artificial Intelligence, pp. 209–216 (2007)

Li, J., Hospedales, T., Gong, S., Xiang, T.: Learning rare behaviours. In: Asian Conference on Computer Vision, Queenstown, New Zealand, November 2010

Loy, C.C., Xiang, T., Gong S.: Stream-based active unusual event detection. In: Asian Conference on Computer Vision, Queenstown, New Zealand, November 2010

Lucas, B.D., Kanade, T.: An iterative image registration technique with an application to stereo vision. In: DARPA Image Understanding Workshop, April 1981, pp. 121–130 (1981)

McCallum, A., Nigam, K.: Employing EM in pool-based active learning for text classification. In: International Conference on Machine Learning, pp. 350–358 (1998)

Pelleg, D., Moore, A.: Active learning for anomaly and rare-category detection. In: Advances in Neural Information Processing Systems, pp. 1073–1080 (2004)

Rish, I.: An empirical study of the naive Bayes classifier. In: IJCAI Workshop on Empirical Methods in Artificial Intelligence, pp. 41–46 (2001)

Roy, N., McCallum, A.: Toward optimal active learning through sampling estimation of error reduction. In: International Conference on Machine Learning, pp. 441–448 (2001)

Settles, B.: Active learning literature survey. Technical report, University of Wisconsin Madison (2010)

Settles, B., Craven, M., Ray, S.: Multiple instance active learning. In: Advances in Neural Information Processing Systems, pp. 1289–1296 (2007)

Seung, H.S., Opper, M., Sompolinsky, H.: Query by committee. In: Conference on Learning Theory, pp. 287–294 (1992)

Tong, S., Koller, D.: Active learning for parameter estimation in Bayesian networks. In: Advances in Neural Information Processing Systems, pp. 647–653 (2000)

Xiang, T., Gong, S.: Video behaviour profiling for anomaly detection. IEEE Trans. Pattern Anal. Mach. Intell. **30**(5), 893–908 (2008)

Zelnik-Manor, L., Perona, P.: Self-tuning spectral clustering. In: Advances in Neural Information Processing Systems, pp. 1601–1608 (2004)

Part IV
Distributed Behaviour

Chapter 13
Multi-camera Behaviour Correlation

In Part III, we focused on visual analysis of behaviour from a single observational viewpoint. From this chapter, we begin to address the problem of understanding and recognising object behaviours from distributed multiple observational viewpoints. Currently, distributed viewpoints of wide-area spaces are typically observed by a network of multiple visual sensors such as a closed-circuit television (CCTV) network in a public space. However, there are emerging technologies for enabling remote deployment of scattered non-networked heterogeneous sensors, which have the potential to provide richer sources of 'oversampled' observation of spaces at a greater scale and range.

For monitoring large public spaces such as airports, underground stations, shopping complexes and road junctions, it has become increasingly necessary to deploy a network of multiple cameras instead of a single camera, or a set of independent stand-alone single cameras. The aim is to provide the necessary observational data for gaining a more holistic sense of situational awareness. One primary goal for a multi-camera system is to provide a more complete record and survey a trail of object activities in wide-area spaces, both individually and collectively. This allows for a global interpretation of objects' latent behaviour patterns and intent. In a multi-camera surveillance system, disjoint cameras with non-overlapping field of views are more prevalent, due to the desire to maximise spatial coverage in a wide-area scene whilst minimising the deployment cost (see Fig. 2.11 for an illustration of different overlaps between views). However, global behaviour analysis across multiple disjoint cameras in a public scene is hampered by a number of obstacles:

- Inter-Camera Visual Appearance Variation: Objects moving across camera views often experience drastic variations in their visual appearances owing to different illumination conditions, camera view angles, distances from the camera, and changes in object posture (Fig. 13.1). All these factors increase the uncertainties in activity and behaviour understanding.
- Unknown and Arbitrary Inter-camera Gaps: Unknown and arbitrary inter-camera gap between disjoint cameras is another inextricable factor that leads to uncertainty in behavioural interpretation. In particular, the unknown and often large separation of cameras in space causes temporal discontinuity and incompleteness

S. Gong, T. Xiang, *Visual Analysis of Behaviour*,
DOI 10.1007/978-0-85729-670-2_13, © Springer-Verlag London Limited 2011

Fig. 13.1 Object visual appearance can vary significantly across disjoint camera views. The three camera views show that a group of people (*highlighted* in *green boxes*) got off a train (*left*) and subsequently took an upward escalator (*middle*), which led them to the escalator exit view (*right*). These people exhibited drastic appearance variations due to changes in illumination (both intra- and inter-camera), camera viewpoint, and distances to cameras

in visual observations. A global activity can only be observed partially in different views, whilst elements of the activity sequence are unobservable due to these inter-camera gaps. Furthermore, the observed spaces may not be a 'closed-world'. This can potentially cause uncertainty and ambiguity in behaviour interpretation. For example, two widely separated camera views may not cover an arbitrary number of entries into the space and exits leaving the space under observation.

Due to temporal discontinuities in visual observations, a potentially abnormal behavioural interpreted in a global context, of which incomplete observations can span across different views, may appear to be perfectly normal in each isolated camera view. For instance, consider two camera views that monitor road junctions A and B, respectively, which are one mile apart. Assume that a vehicle passing A will typically reappear at B in 2 minutes, and it is normal to observe either large or small volumes of traffic in either views. However, if a large volume of traffic is observed at junction A, but two minutes later only few vehicles can be seen in junction B, it is likely that a road accident has occurred somewhere between junctions A and B which is not observed. Given multiple views of distributed spaces, individual inspection of each view in isolation would fail to detect such an unusual behaviour as local activities in each view appear perfectly normal. This problem renders typical single-view behaviour modelling techniques ineffective. Global behavioural interpretation across disjoint camera views cannot solely based on reasoning about visual observations from a single closed-world space.

• Lack of Visual Details and Crowdedness: In a typical public scene, the sheer number of objects can cause severe and continuous inter-object occlusions. There is also a general lack of pixel-level details about objects (see Fig. 13.2). An object-centred representation that requires explicit object segmentation and tracking based on unique and consistent object visual appearance is less feasible. Tracking is also compounded by the typically low frame rate in surveillance videos,[1] where large

---

[1] Many surveillance systems record videos at lower than 5 frames per second (FPS) and compromise image resolution to optimise data bandwidth and storage space (Cohen et al. 2006; Kruegle 2006).

Fig. 13.2 Object tracking is difficult when lacking visual details and consistency in a crowded environment. The *top* and *bottom rows* show three consecutive video frames captured from two different cameras at 0.7 frames per second (FPS), respectively. It is evident that an object can pass through the two camera views in just three frames at such a low frame rate. Severe inter-object occlusion is also the norm

spatial displacement is observed in moving objects between consecutive frames (see Fig. 13.2).

• Visual Context Variation: In an unconstrained visual environment, changes in visual context are ineluctable and may occur either gradually or abruptly. In particular, gradual context change may involve gradual behaviour drift over time, for example, different volumes of crowd flows at different time periods. On the other hand, abrupt context change indicates more drastic changes such as camera view angle adjustments, removals or additions of cameras from and to a camera network. Gradual and abrupt changes in visual context can cause transition and modification to inter-camera object behaviour relationships over time.

To overcome these obstacles, a key to visual analysis of multi-camera behaviour lies on how well a model can correlate partial observations of object behaviours from different locations in order to carry out 'joined-up reasoning'. This requires a computational process to establish a joined-up representation of a *synchronised* global space, within which local activities from different observational viewpoints can be analysed and interpreted holistically.

Specifically, global activities of an object in a public space are inherently context-aware, exhibited through constraints imposed by its physical scene layout and correlated activities of other objects both in the same camera view and other views. Consequently, the partial observations of a global activity are inherently correlated in that they take place following a certain unknown temporal order with unknown temporal gaps. It can be considered that these partial observations often form a probabilistic graph of inter-correlated spatio-temporal behaviour patterns, spanning across different regions in a global space. The computational challenge is to develop suitable mechanisms and models capable of discovering and quantifying these unknown correlations in temporal ordering and temporal delays. Such correlational information is a key for providing a latent context in a single *synchronised* global space, where joined-up reasoning and interpretation of distributed activities observed from disjoint multi-camera views can take place.

13.1 Multi-view Activity Representation

A wide-area public scene captured in each camera view inherently consists of multiple local regions, each of which encapsulates a unique set of activity patterns correlated with each other either explicitly or implicitly. We considered in Sect. 10.1 a method for learning behaviour spatial context by activity-based scene decomposition for the interpretation of behaviour in a single camera view. The same principle can be applied to multiple camera views. More precisely, given a set of training videos captured using M cameras in a camera network, the goal is to decompose automatically the M camera views into N regions according to spatial-temporal distribution of activity patterns ($N \gg M$). The decomposed regions of all the views are

$$\mathcal{R} = \{\mathcal{R}_n | n = 1, \ldots, N\} \tag{13.1}$$

where the mth camera view contains N_m regions, with N_m being determined automatically (also see Sect. 10.1), and $N = \sum_{m=1}^{M} N_m$.

13.1.1 Local Bivariate Time-Series Events

Similar to computing the local block behaviour patterns as introduced in Sect. 12.2, the image space of a camera view is divided into equal-sized blocks, for example, with 10×10 pixels each. Foreground pixels are detected by background subtraction (Russell and Gong 2006; Stauffer and Grimson 2000). The detected foreground pixels are categorised as either static or moving by frame differencing, for instance, sitting-down people are detected as static foreground whilst passing-by people are detected as moving foreground. Given an example image sequence, local activity events at each block are represented by a bivariate time-series pattern

$$\mathbf{u_b} = (u_{\mathbf{b},1}, \ldots, u_{\mathbf{b},t}, \ldots, u_{\mathbf{b},T}), \qquad \mathbf{v_b} = (v_{\mathbf{b},1}, \ldots, v_{\mathbf{b},t}, \ldots, v_{\mathbf{b},T}) \tag{13.2}$$

where $\mathbf{b}$ represents the two-dimensional mean coordinates of a block in the image space and T is the total number of video frames in the sequence. Note that, different from computing optical flow for constructing a local block behaviour pattern, $u_{\mathbf{b},t}$ and $v_{\mathbf{b},t}$ are the percentage of static and moving foreground pixels within the block at frame t, respectively. The duration T of the example video needs to be sufficiently large to cover enough repetitions of activity patterns, depending on the complexity of a scene.

Using these local bivariate time-series events to represent activity is more compact and robust than either the contextual event-based representation (Sect. 7.1) or the local motion event-based representation (Sect. 9.1). This is because that multiple camera video footages typically have low spatial and temporal resolution which make those representations overly brickle to image noise and unreliable. For instance, at a frame rate of 0.7 FPS, the image sequences shown in Fig. 13.2 render computing optical flow highly unreliable due to the difficulties in finding correspondence between pixels over two consecutive frames. Despite its simplicity, this bivariate time-series representation using $\mathbf{u_b}$ and $\mathbf{v_b}$ is shown to be effective and compact in capturing the temporal characteristics of different local activity patterns both in each camera view and across all camera views (Loy et al. 2009b).

13.1.2 Activity-Based Scene Decomposition

After local block-based event extraction, the blocks are grouped into regions according to the similarity of their spatio-temporal characteristics given by $\mathbf{u_b}$ and $\mathbf{v_b}$. To that end, a method similar to that described in Sect. 10.1 is considered here. Two blocks are considered to be similar and placed into the same group if they are in spatial proximity and exhibit strong spatio-temporal correlations over time. To compute the correlation between activity patterns of equal lengths from two blocks, let their activity patterns be represented as $\{\mathbf{u}_i, \mathbf{v}_i\}$ and $\{\mathbf{u}_j, \mathbf{v}_j\}$ (13.2). Two correlation distances for the pair are defined by a dissimilarity metric given as

$$\bar{r}_{ij}^{\mathbf{u}} = 1 - \left|r_{ij}^{\mathbf{u}}\right|, \qquad \bar{r}_{ij}^{\mathbf{v}} = 1 - \left|r_{ij}^{\mathbf{v}}\right| \tag{13.3}$$

where $\bar{r}_{ij}^{\mathbf{u}} = \bar{r}_{ij}^{\mathbf{v}} = 0$ if two blocks have strongly correlated local activity patterns, or $\bar{r}_{ij}^{\mathbf{u}} = \bar{r}_{ij}^{\mathbf{v}} = 1$ otherwise; $r_{ij}^{\mathbf{u}}$ and $r_{ij}^{\mathbf{v}}$ are the Pearson's correlation coefficients for correlating the static and moving foreground aspects of the two blocks, respectively, computed by

$$r_{ij}^{\mathbf{u}} = \frac{\sum_{t=1}^{T}(u_{i,t} - m_i^{\mathbf{u}})(u_{j,t} - m_j^{\mathbf{u}})}{\sqrt{\sum_{t=1}^{T}(u_{i,t} - m_i^{\mathbf{u}})^2}\sqrt{\sum_{t=1}^{T}(u_{j,t} - m_j^{\mathbf{u}})^2}} \tag{13.4}$$

where $m_i^{\mathbf{u}}$ and $m_j^{\mathbf{u}}$ are the mean values of $\mathbf{u}_i$ and $\mathbf{u}_j$, respectively, $r_{i,s}^{\mathbf{u}} \in [-1, 1]$. Similarly, $r_{ij}^{\mathbf{v}}$ is computed as in the case of $r_{ij}^{\mathbf{u}}$. Subsequently, an affinity matrix $\mathbf{A} = \{A_{ij}\} \in \mathbb{R}^{B \times B}$ is constructed, where B is the total number of blocks in all camera views and A_{ij} is defined as

$$A_{ij} = \begin{cases} \exp\left(-\frac{(\bar{r}_{ij}^{\mathbf{u}})^2}{2\sigma_i^{\mathbf{u}}\sigma_j^{\mathbf{u}}}\right)\exp\left(-\frac{(\bar{r}_{ij}^{\mathbf{v}})^2}{2\sigma_i^{\mathbf{v}}\sigma_j^{\mathbf{v}}}\right)\exp\left(-\frac{\|\mathbf{b}_i - \mathbf{b}_j\|^2}{2\sigma_{\mathbf{b}}^2}\right), \\ \quad \text{if } \|\mathbf{b}_i - \mathbf{b}_j\| \le R \text{ and } i \ne j \\ 0, \quad \text{otherwise} \end{cases} \tag{13.5}$$

where the correlation distances of $\mathbf{u_b}$ and $\mathbf{v_b}$ between block i and block j are given by $\bar{r}_{ij}^{\mathbf{u}}$ and $\bar{r}_{ij}^{\mathbf{v}}$, respectively, whilst $[\sigma_i^{\mathbf{u}}, \sigma_j^{\mathbf{u}}]$ and $[\sigma_i^{\mathbf{v}}, \sigma_j^{\mathbf{v}}]$ are the respective correlation scaling factors for $\bar{r}_{i_j}^{\mathbf{u}}$ and $\bar{r}_{ij}^{\mathbf{v}}$. A correlation scaling factor is defined as the mean correlation distance between the current block and all blocks within a radius R. The coordinates of the two blocks are denoted as $\mathbf{b}_i$ and $\mathbf{b}_j$. Similar to the correlation scaling factors, a spatial scaling factor $\sigma_{\mathbf{b}}$ is defined as the mean spatial distance between the current block and all blocks within the radius R. The affinity matrix is normalised as

$$\overline{\mathbf{A}} = \mathbf{L}^{-\frac{1}{2}} \mathbf{A} \mathbf{L}^{-\frac{1}{2}} \tag{13.6}$$

where $\mathbf{L}$ is a diagonal matrix and $L_{ii} = \sum_{j=1}^{B} A_{ij}$. Upon obtaining the normalised affinity matrix $\overline{\mathbf{A}}$. The subsequent process for spectral clustering of the normalised affinity matrix is identical to that described in Sect. 10.1.2.

We illustrate the effect of this scene composition process using a visual environment captured by a camera network installed at an underground station. The camera views and layouts of the underground station are illustrated in Fig. 13.3.

In this visual environment, the station has a ticket hall and a concourse leading to two train platforms through escalators. Three cameras are placed in the ticket hall and two cameras are positioned to monitor the escalator areas. Both train platforms are covered by two cameras each. Typically, passengers enter from the main entrance, walk through the ticket hall or queue up for tickets (Cam-1), enter the concourse through the ticket barriers (Cam-2 and Cam-3), take the escalators (Cam-4 and Cam-5), and enter one of the platforms. The opposite route is taken if they are leaving the station. Apart from the two platforms in Cam-6 and Cam-7 and Cam-8 and Cam-9, passengers may also proceed from the concourse to other platforms, which are not visible in the camera views, without taking the escalators. In other words, this is not a closed-world visual environment, typical for a camera system installed in a public space. In addition, after getting off a train, a passenger may also go to a different platform without leaving the station. Figure 13.4 shows that nine distributed camera views were automatically decomposed into 96 regions. These decomposed regions correspond to semantically meaningful physical zones despite of severe inter-object occlusions and low frame rate (Loy et al. 2009a).

This visual environment highlights a number of activity characteristics from multi-camera views that make learning behaviour correlation a challenging problem.

1. Complexity and diversity of activities: Activities observed in different scenes take place at ticket hall, concourse, train platforms and escalators. They are very different in nature.
2. Low temporal and spatial resolution in video data: The video frame rate is at 0.7 FPS.
3. The lighting conditions are very different across camera views.
4. Severe inter-object occlusions due to enormous number of objects in the scene, especially during travelling peak hours.

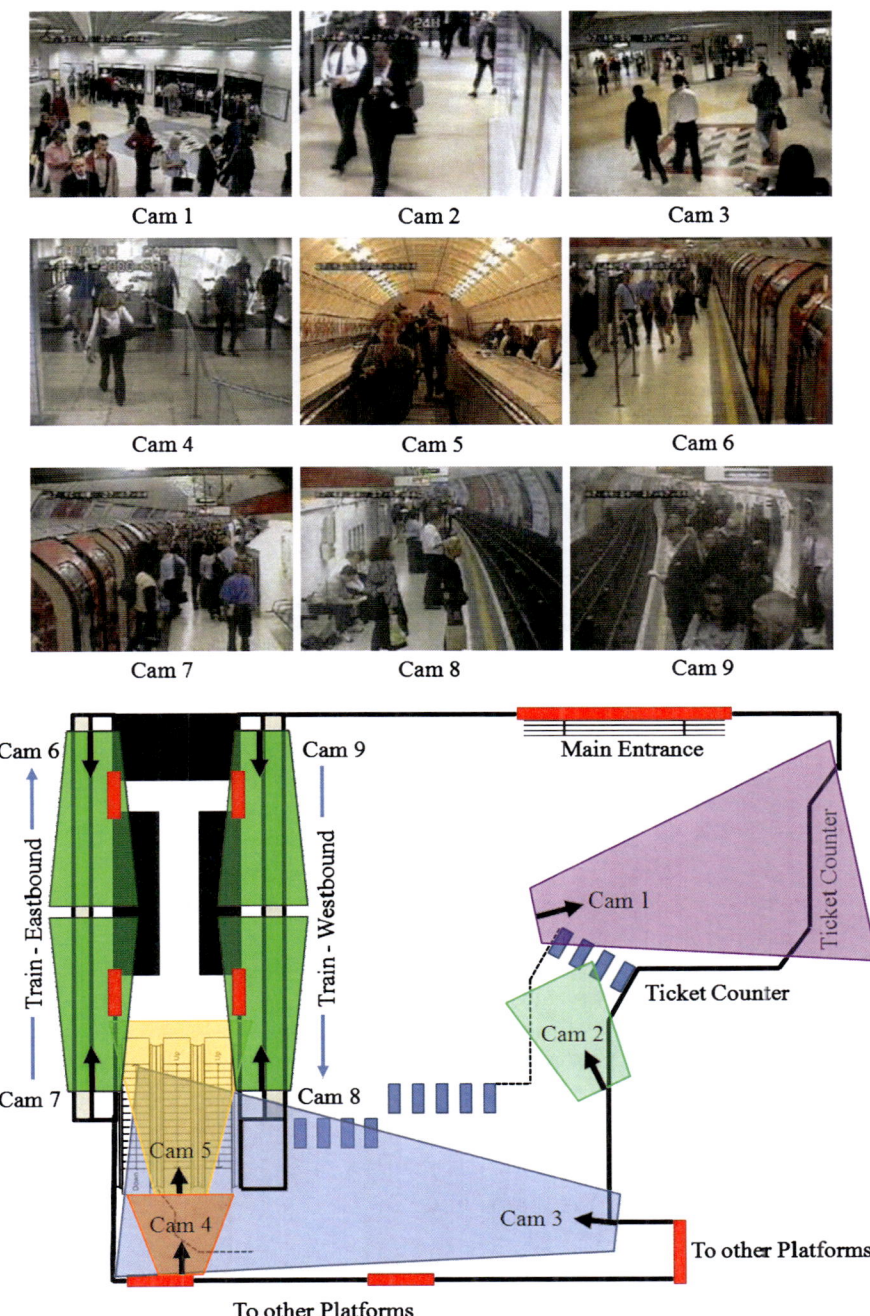

Fig. 13.3 An example of a crowded underground station, with illustrations on scene layout, camera topology and example camera views. The entry and exit points of the station are highlighted in *red bars*

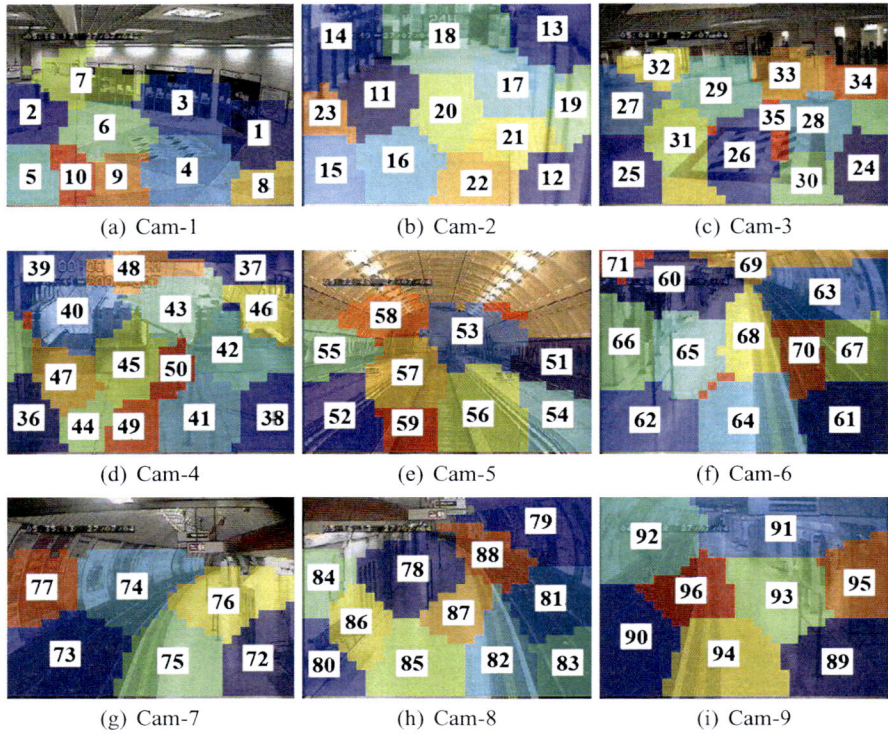

Fig. 13.4 An example of activity-based multi-camera scene decomposition. Nine camera views from an underground station are automatically decomposed into 96 regions corresponding to physical areas of different activity patterns

5. Complex crowd dynamics: Passengers may appear in a group or individually, remain stationary at any point of the scenes, or not get on an arrived train. People in public spaces exhibit many unusual but benign behaviours.
6. Only limited areas of this underground station are covered by cameras. There are multiple entry and exit points that are not visible in the camera views. This 'open-world' environment increases the uncertainty in the interpretation of the observed activities.

13.2 Learning Pair-Wise Correlation

Given multiple regions from scene decomposition, an average regional activity pattern is constructed to represent each region $\mathcal{R}_n$ as

$$\hat{\mathbf{u}}_n = \frac{1}{|\mathcal{R}_n|} \sum_{\mathbf{b} \in \mathcal{R}_n} \mathbf{u_b}, \qquad \hat{\mathbf{v}}_n = \frac{1}{|\mathcal{R}_n|} \sum_{\mathbf{b} \in \mathcal{R}_n} \mathbf{v_b} \qquad (13.7)$$

where $|\mathcal{R}_n|$ is the number of blocks belonging to region $\mathcal{R}_n$. To discard insignificant regions, one may wish to filter out 'low-activity regions' by removing any region with $\hat{\mathbf{u}}_n$ and $\hat{\mathbf{v}}_n$ of low variance over time, measured by the standard deviation.

Given a pair of regions, a model needs to establish the strength of activity correlation and any temporal relationships between the two regions. Two factors make solving this problem non-straightforward: (1) Different viewing angles of cameras can introduce pattern variations across camera views. That is, same activity may appear to be visually different due to a change of viewpoint. (2) The temporal gap between two related activities in two different regions is unknown and rather unpredictable.

In the following, we consider two methods, cross canonical correlation analysis (xCCA) (Loy et al. 2009a) and time-delayed mutual information analysis (Loy et al. 2009b), to model the correlation of two regional activities as a problem of fitting a function of an unknown time lag τ to a time series.

13.2.1 Cross Canonical Correlation Analysis

Cross canonical correlation analysis (xCCA) seeks a linear combination of time-series functions, that both maximises the correlation value and minimises the temporal dependency of arbitrary order between two time series. It is a generalisation of canonical correlation analysis (Hardoon et al. 2004). As activity patterns are represented by the percentage of static and moving foreground pixels (13.7), they reflect the crowd densities in the regions. If two regions are physically connected with correlated activity patterns, even though that connection may not be observed, it is expected that similar changes in crowd density across the two regions can be observed given a certain time delay. We assume that the correlations between activity patterns of any two regions are linear with unknown temporal delay, given the multi-camera activity representation chosen (Sect. 13.1).

Let $\mathbf{x}_i(t)$ and $\mathbf{x}_j(t)$ denote two regional activity time series observed in the ith and jth regions, respectively. Note that $\mathbf{x}_i(t)$ and $\mathbf{x}_j(t)$ are time series of N_f-dimensional variables, where $N_f = 2$ due to the two features $\hat{\mathbf{u}}$ and $\hat{\mathbf{v}}$ extracted from each region. We denote $\mathbf{y}(t) = \mathbf{x}_j(t + \tau)$ to represent shifting $\mathbf{x}_j(t)$ by τ time gap, we call τ a time delay index. For clarity and conciseness, we omit symbol t, so that $\mathbf{x}_i(t)$ becomes $\mathbf{x}_i$ and $\mathbf{y}(t)$ becomes $\mathbf{y}$.

At each time delay index τ, xCCA finds two sets of optimal basis vectors $\mathbf{w}_{x_i}$ and $\mathbf{w}_y$ for $\mathbf{x}_i$ and $\mathbf{y}$, such that correlation of the projections of $\mathbf{x}_i$ and $\mathbf{y}$ onto the basis vectors are mutually maximised. Let linear combinations of canonical variates be $x_i = \mathbf{w}_{x_i}^\mathsf{T}\mathbf{x}_i$ and $y = \mathbf{w}_y^\mathsf{T}\mathbf{y}$, the canonical correlation between the two time series is computed as

$$\rho_{\mathbf{x}_i, \mathbf{x}_j}(\tau) = \frac{\mathrm{E}[x_i\, y]}{\sqrt{\mathrm{E}[x_i^2]\mathrm{E}[y^2]}} = \frac{\mathrm{E}[\mathbf{w}_{x_i}^\mathsf{T}\mathbf{x}_i\mathbf{y}^\mathsf{T}\mathbf{w}_y]}{\sqrt{\mathrm{E}[\mathbf{w}_{x_i}^\mathsf{T}\mathbf{x}_i\mathbf{x}\mathsf{T}_i\mathbf{w}_{x_i}]}\sqrt{\mathrm{E}[\mathbf{w}_y^\mathsf{T}\mathbf{y}\mathbf{y}^\mathsf{T}\mathbf{w}_y]}}$$

$$= \frac{\mathbf{w}_{x_i}^{\mathsf{T}} \mathbf{C}_{x_i y} \mathbf{w}_y}{\sqrt{\mathbf{w}_{x_i}^{\mathsf{T}} \mathbf{C}_{x_i x_i} \mathbf{w}_{x_i}} \sqrt{\mathbf{w}_y^{\mathsf{T}} \mathbf{C}_{yy} \mathbf{w}_y}} \tag{13.8}$$

where $\mathbf{C}_{x_i x_i}$ and $\mathbf{C}_{yy}$ are the within-set covariance matrices of $\mathbf{x}_i$ and $\mathbf{y}$, respectively; $\mathbf{C}_{x_i y}$ is between-set covariance matrix (also see CCA as defined by (5.30)).

The maximisation of $\boldsymbol{\rho}_{\mathbf{x}_i, \mathbf{x}_j}(\tau)$ at each time delay index τ can be solved by setting the derivatives of (13.8) to zero, yielding the following eigenvalue equations:

$$\begin{cases} \mathbf{C}_{x_i x_i}^{-1} \mathbf{C}_{x_i y} \mathbf{C}_{yy}^{-1} \mathbf{C}_{yx_i} \mathbf{w}_{x_i} = \boldsymbol{\rho}_{\mathbf{x}_i, \mathbf{x}_j}^2(\tau) \mathbf{w}_{x_i} \\ \mathbf{C}_{yy}^{-1} \mathbf{C}_{yx_i} \mathbf{C}_{x_i x_i}^{-1} \mathbf{C}_{x_i y} \mathbf{w}_y = \boldsymbol{\rho}_{\mathbf{x}_i, \mathbf{x}_j}^2(\tau) \mathbf{w}_y \end{cases} \tag{13.9}$$

where the eigenvalues $\boldsymbol{\rho}_{\mathbf{x}_i, \mathbf{x}_j}^2(\tau)$ are the square canonical correlations, and the eigenvectors $\mathbf{w}_{x_i}$ and $\mathbf{w}_y$ are the basis vectors. One only needs to solve one of the eigenvalue equations, since the equations are related by

$$\begin{cases} \mathbf{C}_{x_i y} \mathbf{w}_y = \boldsymbol{\rho}_{\mathbf{x}_i, \mathbf{x}_j}(\tau) \lambda_{x_i} \mathbf{C}_{x_i x_i} \mathbf{w}_{x_i} \\ \mathbf{C}_{yx_i} \mathbf{w}_{x_i} = \boldsymbol{\rho}_{\mathbf{x}_i, \mathbf{x}_j}(\tau) \lambda_y \mathbf{C}_{yy} \mathbf{w}_y \end{cases} \tag{13.10}$$

where

$$\lambda_{x_i} = \lambda_y^{-1} = \sqrt{\frac{\mathbf{w}_y^{\mathsf{T}} \mathbf{C}_{yy} \mathbf{w}_y}{\mathbf{w}_{x_i}^{\mathsf{T}} \mathbf{C}_{x_i x_i} \mathbf{w}_{x_i}}} \tag{13.11}$$

The time delay index that maximises the canonical correlation between $\mathbf{x}_i$ and $\mathbf{x}_j$ is computed as

$$\hat{\tau}_{\mathbf{x}_i, \mathbf{x}_j} = \arg\max_\tau \frac{\sum^\Gamma \boldsymbol{\rho}_{\mathbf{x}_i, \mathbf{x}_j}(\tau)}{\Gamma} \tag{13.12}$$

where $\Gamma = \min(\text{rank}(\mathbf{x}_i), \text{rank}(\mathbf{x}_j))$. Note that the canonical correlation function is averaged with Γ to obtain a single correlation value at each time delay index. The associated maximum canonical correlation is then obtained by locating the peak value in the averaged canonical correlation function as

$$\hat{\rho}_{\mathbf{x}_i, \mathbf{x}_j} = \frac{\sum^\Gamma \boldsymbol{\rho}_{\mathbf{x}_i, \mathbf{x}_j}(\hat{\tau}_{\mathbf{x}_i, \mathbf{x}_j})}{\Gamma} \tag{13.13}$$

After computing the maximum canonical correlation and the associated time delay index for each pair of regional activity patterns, a regional activity affinity matrix is constructed as

$$\mathbf{P} = \{P_{ij}\} \in \mathbb{R}^{N \times N}, \quad P_{ij} = \hat{\rho}_{\mathbf{x}_i \mathbf{x}_j} \tag{13.14}$$

and a time delay matrix is computed as

$$\mathbf{D} = \{D_{ij}\} \in \mathbb{R}^{N \times N}, \quad D_{ij} = \hat{\tau}_{\mathbf{x}_i \mathbf{x}_j} \tag{13.15}$$

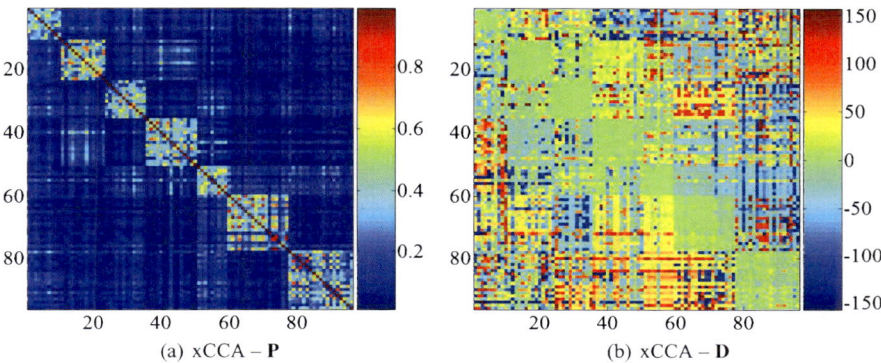

Fig. 13.5 An example of a regional activity affinity matrix **P**, normalised to [0, 1], and the associated time delay matrix **D** computed using xCCA for the underground station camera network views (Fig. 13.3)

The maximum correlation value is in the range of $0 \leq \hat{\rho}_{\mathbf{x}_i, \mathbf{x}_j} \leq 1$ with equality to 1 if, and only if, the two regional time series are identical. If $\tau = 0$, xCCA is equivalent to the standard canonical correlation analysis (CCA) (Sect. 5.4) on $\mathbf{x}_i(t)$ and $\mathbf{x}_j(t)$.

Figure 13.5 shows an example of a regional activity affinity matrix **P** (13.14) and the corresponding time delay matrix **D** (13.15) for the underground station camera network views. It is evident from the block structure along the diagonals of the **P** matrix that xCCA is able to discover strong correlations and relatively shorter time delays between regions in the same camera views. It also shows that the model is able to discover a number of other interesting cross-camera correlations. For instance, Fig. 13.5(a) shows a cluster of high correlation values with a time delay of 13 seconds (Fig. 13.5(b)) between Region-46 (Cam-4) and Region-51 (Cam-5). This corresponds to a frequently occurred inter-camera activity of passengers taking the upward escalator (Cam-5) and leaving from the escalator exit (Cam-4).

13.2.2 Time-Delayed Mutual Information Analysis

An alternative method for discovering and quantifying pair-wise regional activity correlation is to explore time-delayed mutual information. In contrast to xCCA, time-delayed mutual information has the potential to cope with non-linear correlations among activity patterns across different views, whilst xCCA only models linear correlations. Time-delayed mutual information is designed to measure the mutual information between a time series $\mathbf{x}(t)$ and a time shifted copy of itself, $\mathbf{x}(t + \tau)$ as a function of time delay τ (Fraser and Swinney 1986).

Let the regional activity patterns in the ith and jth regions be time series variables denoted as $\mathbf{x}_i(t)$ and $\mathbf{x}_j(t)$; the mutual information between them for a time delay τ

is computed as

$$I(\tau) = \sum_{\mathbf{x}_i(t)} \sum_{\mathbf{x}_j(t+\tau)} p\big(\mathbf{x}_i(t), \mathbf{x}_j(t+\tau)\big) \log_2 \frac{p(\mathbf{x}_i(t), \mathbf{x}_j(t+\tau))}{p(\mathbf{x}_i(t))p(\mathbf{x}_j(t+\tau))} \qquad (13.16)$$

where $p(\mathbf{x}_i(t))$ and $p(\mathbf{x}_j(t+\tau))$ are the marginal probability distribution functions of $\mathbf{x}_i(t)$ and $\mathbf{x}_j(t+\tau)$, respectively; $p(\mathbf{x}_i(t), \mathbf{x}_j(t+\tau))$ is the joint probability distribution function of $\mathbf{x}_i(t)$ and $\mathbf{x}_j(t+\tau)$. These probability distribution functions are approximated by constructing histograms with K_n equal-width bins.

The value range of this time-delayed mutual information is $I(\tau) \geq 0$, with equality if, and only if, $\mathbf{x}_i$ and $\mathbf{x}_j$ are independent. If $\tau = 0$, time-delayed mutual information is equivalent to mutual information between $\mathbf{x}_i$ and $\mathbf{x}_j$. The time delay between $\mathbf{x}_i$ and $\mathbf{x}_j$ is estimated as

$$\hat{\tau}_{\mathbf{x}_i\mathbf{x}_j} = \arg\max_{\tau} I(\tau) \qquad (13.17)$$

and the corresponding time-delayed mutual information is obtained as

$$\hat{I}_{\mathbf{x}_i\mathbf{x}_j} = I(\hat{\tau}_{\mathbf{x}_i\mathbf{x}_j}) \qquad (13.18)$$

Similar to xCCA, after computing the time delay and time-delayed mutual information for each pair of regional activity patterns, a time-delayed mutual information matrix $\mathbf{I} = [\hat{I}_{\mathbf{x}_i\mathbf{x}_j}]_{N \times N}$ and the associated time delay matrix $\mathbf{D} = [\hat{\tau}_{\mathbf{x}_i\mathbf{x}_j}]_{N \times N}$ are constructed. Loy et al. (2009b) show that compared to xCCA, pair-wise correlation learned using time-delayed mutual information is more powerful in capturing a global behaviour model due to its ability to model non-linear relationships between regional activities, whilst xCCA is more suitable for providing a contextual constraint for matching object across camera views (see Sect. 14.2).

13.3 Multi-camera Topology Inference

The regional activity correlation of arbitrary order, discovered and quantified between each pair of regions using either xCCA or time-delayed mutual information analysis, can be utilised to infer multi-camera topology. A straightforward method is to examine exhaustively each pair of camera views. Two cameras are deemed to be connected if strong correlations are discovered between regions across the two camera views. However, Loy et al. (2009a) show there is a bias towards finding strong correlations between camera pairs of constant crowd movement, even if they are not physically or logically connected. Moreover, these spurious correlation peaks are also usually associated with large time delays. Therefore, for more reliable multi-camera topology inference, both time delay and correlation strength need be considered. Specifically, two cameras are considered to be connected if they contain connected regions, defined as pairs of inter-camera regions with high correlation value and short time delay. Let us consider how to infer multi-camera topology by

Fig. 13.6 An example of estimated multi-camera topology compared to the actual topology of an underground station camera network. M = missing edges, S = spurious edges

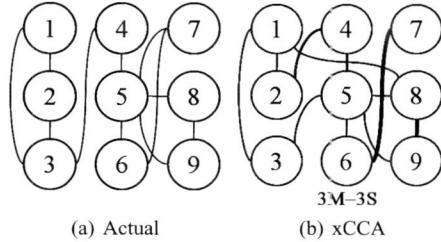

(a) Actual   (b) xCCA

xCCA correlation analysis. A similar procedure can be adopted for a time-delayed mutual information analysis-based approach.

A region connectivity matrix, defined as $\boldsymbol{\Psi} = \{\Psi_{ij}\} \in \mathbb{R}^{N \times N}$, can be constructed to represent how likely each pair of regions in a camera network are connected, by measuring the strength of their connectivity. More specifically, each element in a region connectivity matrix is computed as

$$\Psi_{ij} = \overline{\hat{\rho}_{\mathbf{x}_i,\mathbf{x}_j}} \left(1 - \overline{|\hat{\tau}_{\mathbf{x}_i,\mathbf{x}_j}|}\right) \tag{13.19}$$

where $\overline{\hat{\rho}_{\mathbf{x}_i,\mathbf{x}_j}}$ is computed from a normalised regional activity affinity matrix $\mathbf{P}$ (13.14), so that it has a value range of $[0, 1]$; $\overline{|\hat{\tau}_{\mathbf{x}_i,\mathbf{x}_j}|}$ is computed by normalising the absolute values of the elements of the time delay matrix $\mathbf{D}$ (13.15). These two normalisation steps ensure that $0 \le \Psi_{ij} \le 1$. A high value Ψ_{ij} indicates a strong connectivity between a pair of inter-camera regions.

Given an estimated region connectivity matrix, the multi-camera topology, defined by a camera connectivity matrix $\boldsymbol{\Phi} = \{\Phi_{ij}\} \in \mathbb{R}^{M \times M}$, can be estimated. Specifically, the strength of the connectivity between the ith and jth cameras is measured by averaging the regional activity connectivity strength (13.19) between each pair of regions across the two camera views. To limit the sensitivity to noise and spurious connectivities discovered in $\boldsymbol{\Psi}$, a model can start with searching for the strongest connectivity between a region in the ith camera view and all the regions in the jth camera view. This searching step is repeated for all regions in the ith camera view. The top N_e connectivities between the ith camera and the jth camera are averaged to estimate the value of Φ_{ij}, with N_e being set to half of the number of regions in the ith camera view. Finally $\boldsymbol{\Phi}$ is normalised to $[0,1]$. Two cameras are deemed to be connected if the corresponding Φ_{ij} value is greater than the average value of all the elements of $\boldsymbol{\Phi}$.

An example of estimated multi-camera topology for an underground station camera network (see Fig. 13.3) is shown in Fig. 13.6. The estimated topology is compared to the actual topology. The number of missing edges (M) and spurious edges (S) are also shown. It is evident that xCCA is able to estimate a fairly accurate multi-camera topology. Among the missing and spurious connections, the miss-detection of a connection between Cam-3 and Cam-4 is due to multiple entry and exit points (see Fig. 13.3). The spurious connection between camera pair Cam-3 and Cam-5 is caused by the proximity of Cam-3 to Cam-4 to Cam-5, resulting in activities in Cam-5 almost always correlating directly with activities in Cam-3 (see Fig. 13.3).

Therefore, one may consider that Cam-3 and Cam-5 are logically connected, even if they are not physically adjacent.

13.4 Discussion

We considered methods for discovering and modelling time-delayed correlations among multi-camera activities. Specifically, each camera view in a multi-camera system is decomposed automatically into many local regions in order to facilitate learning of inter-camera correlations based on distinctive regional activity patterns. Two methods are considered for modelling pair-wise behaviour correlations across camera views, namely cross canonical correlation analysis (xCCA), and time-delayed mutual information estimation. These methods do not rely on inter-camera or intra-camera tracking. They are better suited for visual analysis of behaviour in surveillance video data, which are typically featured with severe occlusions due to enormous number of objects in the scenes, and poor image quality caused by low video frame rate and image resolution.

Learning inter-camera activity correlation can assist in understanding multi-camera topology. This includes both the spatial topology and the temporal topology of a multi-camera network. In particular, to discover the temporal topology of a multi-camera system, a model is needed for estimating unknown time gaps between correlated partial (regional) observations taking place in different camera views. We considered how to utilise a learned correlation model to infer both spatial and temporal topology of a camera network.

Learning to establish multi-camera behaviour correlation is a first step towards multi-camera behavioural interpretation. Another important step towards global behavioural interpretation is to associate and re-identify objects of interest across different camera views. In particular, when an object of interest disappears from one camera view, it is necessary to re-establish a match of this object against other objects observed in a different camera view, after a certain time delay. We shall consider plausible solutions to this problem in the next chapter. In particular, we shall discuss how multi-camera behaviour correlation can facilitate object association and re-identification across camera views.

References

Cohen, N., Gatusso, J., MacLennan-Brown, K.: CCTV Operational Requirements Manual— Is Your CCTV System Fit for Purpose? Home Office Scientific Development Branch, version 4 (55/06) edition (2006)

Fraser, A.M., Swinney, H.L.: Independent coordinates for strange attractors from mutual information. Phys. Rev. **33**(2), 1134–1140 (1986)

Hardoon, D., Szedmak, S., Shawe-Taylor, J.: Canonical correlation analysis; an overview with application to learning methods. Neural Comput. **16**(12), 2639–2664 (2004)

Kruegle, H.: CCTV Surveillance: Video Practices and Technology. Butterworth/Heinemann, Stoneham/London (2006)

Loy, C.C., Xiang, T., Gong, S.: Multi-camera activity correlation analysis. In: IEEE Conference on Computer Vision and Pattern Recognition, Miami, USA, June 2009, pp. 1988–1995 (2009a)

Loy, C.C., Xiang, T., Gong, S.: Modelling activity global temporal dependencies using time delayed probabilistic graphical model. In: IEEE International Conference on Computer Vision, Kyoto, Japan, October 2009, pp. 120–127 (2009b)

Russell, D., Gong, S.: Minimum Cuts of a Time-varying Background. In: British Machine Vision Conference, Edinburgh, UK, September 2006, pp. 809–818 (2006)

Stauffer, C., Grimson, W.E.L.: Learning patterns of activity using real-time tracking. IEEE Trans. Pattern Anal. Mach. Intell. $22(8)$, 747–758 (2000)

Chapter 14
Person Re-identification

A fundamental task for a distributed multi-camera system is to associate people across camera views at different locations and times. This is known as the person re-identification problem. Specifically, to re-establish a match of the same person over different camera views located at different physical sites, the system aims to track individuals either retrospectively or on-the-fly when they move through different loci. In a crowded and uncontrolled environment observed by cameras from a distance, person re-identification relying upon biometrics such as face and gait is neither feasible nor reliable due to insufficient image details for extracting robust biometrics features. Alternatively, the visual appearance of people can be considered for extracting features for inter-camera matching. These features are less sensitive to, although not completely free from, variations in camera distance, lighting condition and view angle when compared to biometrics. However, visual appearance features, extracted mainly from clothing, are intrinsically weak for matching people. For instance, most people in public spaces wear dark clothes in winter. To further compound the problem, a person's appearance can change significantly between different camera views if large changes occur in view angle, lighting, background clutter and occlusion (see examples in Fig. 14.1). This will result in different people appearing more alike than that of the same person across different camera views.

To match automatically people across camera views by the similarity in their appearance, a model needs to take the following steps: (1) extracting some imagery feature sets, (2) constructing a template that describes an individual, and (3) matching the template, by measuring a distance metric, against other features extracted from some images of unknown people. The feature sets can include:

1. colours (Javed et al. 2005; Madden et al. 2007; Prosser et al. 2008a),
2. combinations of colour and texture (Farenzena et al. 2010; Gray and Tao 2008; Prosser et al. 2008b),
3. shape and spatio-temporal structure (Gheissari et al. 2006; Wang et al. 2007).

Typical distance measures include:

1. histogram based Bhattacharyya distance (Javed et al. 2005; Prosser et al. 2008b),
2. k-nearest neighbour classifiers (Hahnel et al. 2004),

Fig. 14.1 Typical examples of visual appearance of people changes caused by significant cross-view variations in view angle, lighting, background clutter and occlusion. *Each column* shows a pair of images of a person from two different camera views

3. L1-norm (Wang et al. 2006),
4. distance measures of relative proportions of colours (Madden et al. 2007),
5. ranking (Freund et al. 2003; Joachims 2002; Prosser et al. 2010),
6. probabilistic relative distance (Zheng et al. 2011).

Regardless of the choice of imagery features and distance measures, person re-identification by visual appearance information alone is challenging because there is often too much of an overlap between feature distributions of different objects. The problem is so acute that given a probe image,[1] an incorrect gallery image,[2] can appear to be more similar to the probe than a correct gallery image. To overcome this problem, we consider a method based on support vector ranking for learning the optimal matching distance criterion, regardless feature representation (Prosser et al. 2010). This approach to person re-identification shifts the burden of computation from finding some universally optimal imagery features to discovering a matching mechanism for selecting adaptively different features that are deemed *locally* optimal for each and every different pairs of matches. This problem is framed in general as a probabilistic relative distance comparison problem in model learning (Zheng et al. 2011).

Given that visual appearance information is inherently unreliable for person re-identification, contextual information can be exploited to reduce matching uncertainty (Gilbert and Bowden 2006; Loy et al. 2010; Zheng et al. 2009). We considered in the preceding chapter methods for discovering and quantifying correlations between activities observed in different camera views. The learned behaviour correlations hold useful spatio-temporal contextual information about expectations on

---

[1] A probe image is an image of unknown identity.

[2] A gallery image is an image of known identity.

where and when a person may re-appear in a networked visible spaces, given the location and time when this person had disappeared from. This information can be utilised together with appearance information for improving matching accuracy through a context-aware search.

14.1 Re-identification by Ranking

The person re-identification problem can be considered as a ranking problem when visual features are subject to significant degrees of uncertainty. More precisely, given a probe image of an unknown person, a model is required to find the most relevant matches from some observed candidates in a different camera view, with a focus on the highest ranked, most probable match. The main difference between a ranking based approach to matching and a conventional template matching is that the former is not concerned with comparing direct distance similarity scores between potential correct and incorrect matches. Instead, the sole interest of the model is on the relative ranking of these scores that reflects the relevance of each likely match to the probe image. By doing so, the person re-identification problem is converted from an absolute scoring problem to a relative ranking problem. The model thus avoids the need for seeking a maximum distance score and does not need to assume the existence of, or optimise for, large disparities therefore minimal overlaps between the distributions of different object classes in a feature space.

Ranking can be based on either boosting learning (Schapire and Singer 1999) or kernel learning, such as support vector machines. Freund et al. (2003) introduce a boosting learning based ranking method, known as RankBoost. A RankBoost model uses a set of weak rankers, rather than classifiers, for an iterative process of boosting them to form a strong ranker. However, the re-identification problem intrinsically suffers from a large degree of feature overlapping in a multi-dimensional feature space. Picking weak rankers by selecting different individual features is equivalent to selecting locally some individual sub-dimensions of the feature space. This is likely to lead to overly trivial weak rankers unsuitable for boosting. In contrast, a ranking method based on kernel learning seeks to learn a ranking function in a higher dimensional feature space by taking into consideration all the feature dimensions, therefore emphasises the selection of a good global combination of features. A global ranking method is potentially more effective for coping with highly overlapped feature distributions in person re-identification.

14.1.1 Support Vector Ranking

RankSVM is a kernel learning based ranking model (Joachims 2002). Person re-identification by ranking can be formulated as a RankSVM learning problem as follows. Assume that there exists a set of relevance ranks $\lambda = \{r_1, r_2, \ldots, r_\rho\}$, such that $r_\rho \succ r_{\rho-1} \succ \cdots \succ r$ where ρ is the number of ranks and $\succ$ indicates the order. In the person re-identification problem, there are only two relevance ranks, that of relevant and that of irrelevant observation feature vectors, corresponding to the correct and incorrect matches. Suppose a data set is denoted as $X = \{(x_i, y_i)\}_{i=1}^{m}$, where x_i is a multi-dimensional feature vector representing the visual appearance of a person captured in one view, y_i is its label and m is the number of training example images of people. Each vector $x_i (\in R^d)$ has an associated set of relevant observation feature vectors $\mathbf{d}_i^+ = \{x_{i,1}^+, x_{i,2}^+, \ldots, x_{i,m^+(x_i)}^+\}$ and related irrelevant observation feature vectors $\mathbf{d}_i^- = \{x_{i,1}^-, x_{i,2}^-, \ldots, x_{i,m^-(x_i)}^-\}$, corresponding to correct and incorrect matches from another camera view. Here $m^+(x_i)$ is the number of relevant observations, and $m^-(x_i)$ the number of the related irrelevant observations, for probe x_i. We have $m^-(x_i) = m - m^+(x_i) - 1$.

In general, there is $m^+(x_i) \ll m^-(x_i)$ because there are likely only a few instances of correct matches and many incorrect matches. The goal of ranking any paired image relevance is to learn a ranking function δ for all pairs of $(x_i, x_{i,j}^+)$ and $(x_i, x_{i,j'}^-)$ such that the relevance ranking score $\delta(x_i, x_{i,j}^+)$ is larger than $\delta(x_i, x_{i,j'}^-)$. Using support vector ranking, a model seeks to compute the ranking score δ in terms of a pair-wise example $(x_i, x_{i,j})$ by a linear function w as follows:

$$\delta(x_i, x_{i,j}) = w^\top |x_i - x_{i,j}| \tag{14.1}$$

where $|x_i - x_{i,j}| = (|x_i(1) - x_{i,j}(1)|, \ldots, |x_i(d) - x_{i,j}(d)|)^\top$ and d is the dimension of x_i. The term $|x_i - x_{i,j}|$ is known as the absolute difference vector.

For a probe feature vector x_i, one wishes to have the following rank relationship for a relevant feature vector $x_{i,j}^+$ and a related irrelevant feature vector $x_{i,j'}^-$:

$$w^\top \left(\left| x_i - x_{i,j}^+ \right| - \left| x_i - x_{i,j'}^- \right| \right) > 0 \tag{14.2}$$

Let $\hat{x}_s^+ = |x_i - x_{i,j}^+|$ and $\hat{x}_s^- = |x_i - x_{i,j'}^-|$. Then, by going through all examples x_i as well as the $x_{i,j}^+$ and $x_{i,j}^-$ in the data set X, a corresponding set of all pair-wise relevant difference vectors can be obtained in which $w^\top(\hat{x}_s^+ - \hat{x}_s^-) > 0$ is expected. This vector set is denoted by $P = \{(\hat{x}_s^+, \hat{x}_s^-)\}$. A RankSVM model is then defined as the minimisation of the following objective function:

$$\frac{1}{2}\|w\|^2 + C \sum_{s=1}^{|P|} \xi_s \tag{14.3}$$

$$\text{s.t.} \quad w^\top(\hat{x}_s^+ - \hat{x}_s^-) \geq 1 - \xi_s, \quad s = 1, \ldots, |P|, \xi_s \geq 0, s = 1, \ldots, |P|$$

where C is a positive parameter that trades margin size against training error.

14.1.2 Scalability and Complexity

One of the main problems with using a support vector machine to solve the ranking problem is the potentially large size of P in (14.3). In problems with a lot of queries,[3] or queries with a lot of associated observation feature vectors, a large size of P implies that forming the $\hat{x}_s^+ - \hat{x}_s^-$ vectors becomes computationally challenging. In the case of person re-identification, if a training data set consists of m person images in two camera views, the size of P is proportional to m^2. This number increases rapidly as m increases. Support vector machine (SVM) based methods also rely on parameter C in (14.3), which must be known before model training. In order to yield a reasonable model, one typically uses cross-validation to tune model parameters. This step requires the *rebuilding* of both the training and validation data sets at each iteration, and thus further increases the computational cost and memory usage. In general, the RankSVM model given by (14.3) is not computationally tractable for large scale problems due to both computational cost and memory usage.

As a potential solution to this problem, a primal-based RankSVM relaxes the constrained RankSVM (Chapelle and Keerthi 2010). It uses a non-constraint model as follows:

$$w = \arg\min_{w} \frac{1}{2}\|w\|^2 + C \sum_{s=1}^{|P|} \ell\big(0, 1 - w^\top\big(\hat{x}_s^+ - \hat{x}_s^-\big)\big)^2 \tag{14.4}$$

where C is a positive importance weight on the ranking performance and ℓ is a hinge loss function (Chapelle and Keerthi 2010). Moreover, a Newton optimisation method is introduced to reduce the training time of the SVM. It removes the need for an explicit computation of the $\hat{x}_s^+ - \hat{x}_s^-$ pairs through the use of a sparse matrix.

However, whilst the computational cost of RankSVM has been reduced significantly, the memory usage issue remains. Specifically, in the case of person re-identification, the spatial complexity, which contributes to excessive memory cost, of creating all the training samples is

$$O\left(\sum_{i=1}^{m} dm^+(x_i)m^-(x_i)\right) \tag{14.5}$$

where d is the feature space dimensionality and m is the number of training example images. Suppose there are L people in the training set and $\frac{m}{L}$ images for each person, we then have $m^+(x_i) = \frac{m}{L} - 1$ and the spatial complexity can be re-written as

$$O\left(d\left(\left(\frac{1}{L} - \frac{1}{L^2}\right)m^3 + \left(\frac{1}{L} - 1\right)m^2\right)\right) \tag{14.6}$$

This spatial complexity value becomes very high with a large number of training samples m and a feature space of high dimensions d. It cannot be reduced by a

---

[3] A 'probe' is sometimes also referred to as a 'query'.

primal-based RankSVM. In order to make RankSVM computationally tractable for coping with a large scale person re-identification problem, an ensemble RankSVM model can be considered to reduce the spatial complexity of RankSVM. An ensemble RankSVM exploits the use of a boosting principle on selecting weak primal-based RankSVMs in order to maintain sensible computational cost in a high-dimensional feature space (Prosser et al. 2010). An additional benefit of this model is that it integrates the parameter tuning step of a primal-based RankSVM into a boosting framework, removing the need to rebuild training and validation sets at every iteration in model learning.

14.1.3 Ensemble RankSVM

Rather than learning a RankSVM in a batch mode, an ensemble RankSVM aims to learn a set of weak RankSVMs, each computed from a small set of data, then combines them to build a stronger ranker using ensemble learning. More precisely, a strong ranker w_{opt} is constructed by a set of weak rankers w_i as follows:

$$w_{opt} = \sum_{i}^{N} \alpha_i w_i \tag{14.7}$$

Learning Weak Rankers

To learn the weak rankers, a data set is divided into groups and each weak ranker is learned based on one group of data. Suppose there are in total L people $C = \{C_1, \ldots, C_L\}$. They are equally divided into n groups $G_1, \ldots, G_n$ without overlapping, that is, $C = \bigcup_{i=1}^{n} G_i$ and $\forall i \neq j$, $G_i \cap G_j = \emptyset$. The training data set Z is then divided into n groups $Z_1, \ldots, Z_n$ as follows:

$$Z_i = \{(x_i, y_i) | y_i \in G_i\} \tag{14.8}$$

The simplest way to learn a weak ranker is to perform RankSVM on each subset Z_i. However, in order to avoid learning trivial weak rankers, one can learn each weak ranker from a combined data subset $\tilde{Z}_i = Z_i \cup O_i$, where O_i is a data subset of identical size to Z_i but randomly selected from $Z - Z_i$. In doing so, the weak rankers are learned on randomly overlapped but different subsets of data.

A weak ranker is learned from each $\tilde{Z}_i$ and for each candidate importance weight C. That is, if there are s candidate values for parameter C, $N = sn$ weak rankers are computed. By doing so, the selection of C in a primal-based RankSVM learning process is unified into the ensemble learning framework, without the need for an additional cross-validation step. For each $\tilde{Z}_i$, a weak ranker w_i is learned using a primal-based RankSVM, which is now tractable given the moderate size for each subset of data.

Algorithm 14.1: Ensemble RankSVM Boosting Learning

Input: Pair-wise relevant difference vector set P, Initial distribution
$$D_1 = \{D_1^s\}$$
Output: Output $w_{\text{opt}} = \sum_{t=1}^{T} \alpha_t w_{k_t}$

1 **begin**
2   **for** $t = 1, \ldots, T$ **do**
3   Select the best ranker w_{k_t} by (14.9);
4   Compute the weight α_t by (14.11);
5   Update the distribution D_{t+1} by (14.10);
6   **end**
7 **end**

Boosting a Strong Ranker

Suppose N weak rankers $\{w_i\}_{i=1}^{N}$ have been learned from a group of overlapped subsets of training data. To determine an overall stronger ranker, a boosting learning process (Algorithm 14.1) is designed as follows:

1. At step t, a boosting process first selects the best weak ranker w_{k_t} that minimises the following cost function:

$$k_t = \arg\min_i \sum_{s=1}^{|P|} D_t^s \mathbf{I}_{w_i^\top (\hat{x}_s^- - \hat{x}_s^+) \geq 0} \tag{14.9}$$

where D_t^s is the weight of pair-wise difference vectors at step t, $\sum_{s=1}^{|P|} D_t^s = 1$, and $\mathbf{I}$ is a Boolean function.

2. D_t^s is updated as follows:

$$D_{t+1}^s = F^{-1} D_t^s \exp\{\alpha_t (w_{k_t}^\top (\hat{x}_s^- - \hat{x}_s^+))\} \tag{14.10}$$

where F is a normaliser such that $\sum_{s=1}^{|P|} D_{t+1}^s = 1$, with an initialisation of $D_1^s = \frac{1}{|P|}$. The weight α_t is then determined by

$$\alpha_t = 0.5 \log \frac{1+r}{1-r}, \quad r = \sum_{s=1}^{|P|} D_t^s (w_{k_t}^\top (\hat{x}_s^+ - \hat{x}_s^-)) \tag{14.11}$$

To ensure that boosting learning both converges and updates the above weight iteratively, the input weak rankers w_i are normalised by $2 \max_{i,s} |w_i^\top (\hat{x}_s^- - \hat{x}_s^+)|$, so that $w_i^\top (\hat{x}_s^+ - \hat{x}_s^-) \in [-1, +1]$, as suggested by Freund et al. (2003).

Compared to a batch mode RankSVM, the advantages of an ensemble RankSVM are two-folds:

1. It is not required to select the best parameter C for each weak ranker using cross-validation. The ensemble learning process automatically selects the optimal value of C by assigning different weights to weak rankers with different C parameter values.

2. More importantly, the spatial complexity of this ensemble learning process is significantly smaller than batch RankSVM, making it scalable for learning a large person re-identification problem. Each weak ranker is learned on a small set of data and the boosting process is based on the data projection values of each weak ranker. To learn each weak ranker, the spatial complexity is

$$O\left(d\left(\frac{1}{n^2}\left(\frac{1}{L} - \frac{n}{L^2}\right)m^3 + \frac{1}{n}\left(\frac{1}{L} - \frac{1}{n}\right)m^2\right)\right) \qquad (14.12)$$

where d is the dimension of each image feature vector and n is the number of subsets. After learning each weak learner, the space complexity is

$$O\left(N\left(\left(\frac{1}{L} - \frac{1}{L^2}\right)m^3 + \left(\frac{1}{L} - 1\right)m^2\right)\right) \qquad (14.13)$$

where N is the total number of weak rankers. The number of features d required for re-identification is typically in the order of $d > 10^3$, and $N \ll d$. Therefore, the overall space complexity of an ensemble RankSVM is approximately only $(\frac{1}{n^2})$ of that of a standard batch RankSVM.

Prosser et al. (2010) and Zheng et al. (2011) show that improvement on person re-identification matching accuracy can be obtained by casting the problem as a ranking problem solved by either support vector ranking or relative distance learning. Their results also suggest that an ensemble RankSVM can yield comparable re-identification accuracy as a batch mode RankSVM but with a large reduction in computational complexity, gaining greater efficiency in model learning.

14.2 Context-Aware Search

The problem of matching people across significantly different camera views using visual appearance information *alone* is inherently ill-posed and extremely difficult. Even with a carefully selected set of features, for example a histogram of multiple colour and image gradient features, and with learning of ranking based relative matching score using RankSVM, the correct matching rate given a gallery set of moderate size can be very low.[4] In order to further improve person re-identification accuracy, contextual information must be utilised. In the following, we consider how to exploit multi-camera behaviour correlational context to facilitate person re-identification in a distributed multi-camera network.

---

[4]There were around 10% correct matches obtained from a gallery size of 532 people in the experiments carried out by Prosser et al. (2010).

In principle, expectation on object movement can be utilised to narrow down the possibility in selecting matching candidates when people's visual appearance exhibits a significant degree of ambiguity and uncertainty. To obtain a very accurate estimation on population inter-camera transition time distributions requires extensive manual labelling of a large number of people moving across camera views. These people must be observed at different times with different speeds and under different degrees of crowdedness. This is implausible in practice. Of course, if a reasonably good multi-camera person re-identification model is in place, the model can be deployed to collect automatically more statistics on population movement patterns across camera views, and to build over time accurate inter-camera people movement transition time distributions. This knowledge can then be iteratively used to improve the model over time. Without a good enough person re-identification model in place to bootstrap this iterative model learning and improvement process, alternative automatic object tracking by spatio-temporal correspondence or association by detection models does not work well in a crowded public space. This is especially true when camera views are not overlapped, and if the spatial and temporal imagery resolutions are also low. We have a Catch-22 problem.

A less accurate but more feasible approach to 'seeding' a solution towards solving this problem is considered as follows. Learning behaviour correlational context to estimate both inter-camera spatial topology and temporal topology is made possible by the methods discussed in Sect. 13.3. This approach does not place any reliance on object tracking or object association by detection between camera views. Given such a framework, model estimation on both correlations between regional activity patterns across different views and their inter-camera time delays can be employed to reduce the search space in the gallery set during a person re-identification matching process.

More precisely, given the bounding boxes of two people a and b observed in different camera views, either a similarity score or a ranking score between the two is computed and denoted as $S^{a,b}$. To incorporate the learned prior on activity correlation and inter-camera time delay into this score, the regions occupied by person a and b are determined (see Sect. 13.1) together with the associated inter-region correlation and time delay values (see Sect. 13.2). If a person's bounding box overlaps with N_r regions in the scene, the occupancy fractions of individual regions within the bounding box are computed and represented by a set of weights:

$$\boldsymbol{\mu} = \{\mu_i | i = 1, \ldots, N_r\} \tag{14.14}$$

where $\sum_{i=1}^{N_r} \mu_i = 1$. The weights are used to calculate the correlation between regions occupied by person a and b as follows:

$$\hat{\rho}^{a,b} = \sum_{i=1}^{N_r^a} \mu_i^a \left(\sum_{j=1}^{N_r^b} \mu_j^b \hat{\rho}_{\mathbf{x}_i, \mathbf{x}_j} \right) \tag{14.15}$$

where $\hat{\rho}_{\mathbf{x}_i,\mathbf{x}_j}$ is the maximum cross canonical correlation computed by (13.13). The corresponding time delay is given as

$$\hat{\tau}^{a,b} = \sum_{i=1}^{N_r^a} \mu_i^a \left(\sum_{j=1}^{N_r^b} \mu_j^b \, \hat{\tau}_{\mathbf{x}_i,\mathbf{x}_j} \right) \tag{14.16}$$

where $\hat{\tau}_{\mathbf{x}_i,\mathbf{x}_j}$ is computed by (13.12). The overall joint expectation and matching score is computed as

$$S_{\text{overall}}^{a,b} = \begin{cases} \overline{S}^{a,b} \hat{\rho}^{a,b}, & \text{if } 0 < t_{\text{gap}}^{a,b} < \alpha \hat{\tau}^{a,b} \\ 0, & \text{otherwise} \end{cases} \tag{14.17}$$

where $t_{\text{gap}}^{a,b}$ is the time gap of observing the two people in the two camera views, whilst α is a limiting factor that determines the maximum allowable transition time between cameras during person matching. In essence, the regional activity correlation score is used as a prior to bias the matching score for yielding a joint likelihood for re-identification. The time delay estimate is used as a weighted band filter to ignore the unlikely matches even if the matching score may be high.

Figure 14.2 show examples from person re-identification in an underground station scene. It is evident that despite poor image quality and drastic imagery feature variations across camera views, good matching can be obtained by combining spatial and temporal contextual information with appearance information. In contrast to using visual appearance features alone, activity correlation and time delay estimate avoid exhaustive matching and reduce spurious matches significantly. Passengers in an underground station commonly wear similar clothes. It is difficult to visually match the same person over a large camera network. On the contrary, with the inferred time delayed activity correlations employed as contextual information, the person re-identification problem is reduced to a context-aware search problem, resulting in greater efficiency and accuracy in finding the correct match.

14.3 Discussion

We considered addressing the problem of people re-identification across disjoint camera views. When only appearance information is available, a person re-identification problem can be casted into a ranking problem. A support vector ranking model is exploited to improve the reliability and accuracy in person re-identification under challenging viewing conditions.

RankSVM is a kernel learning based global ranking model. However, it suffers from a scalability limitation when applied to the person re-identification problem. A person re-identification model typically needs be trained by a large training data set in order to capture the significant appearance variations of different people. This makes RankSVM very expensive to compute and learn due to the large number of inequality constraints that must be satisfied in model learning. A primal-based

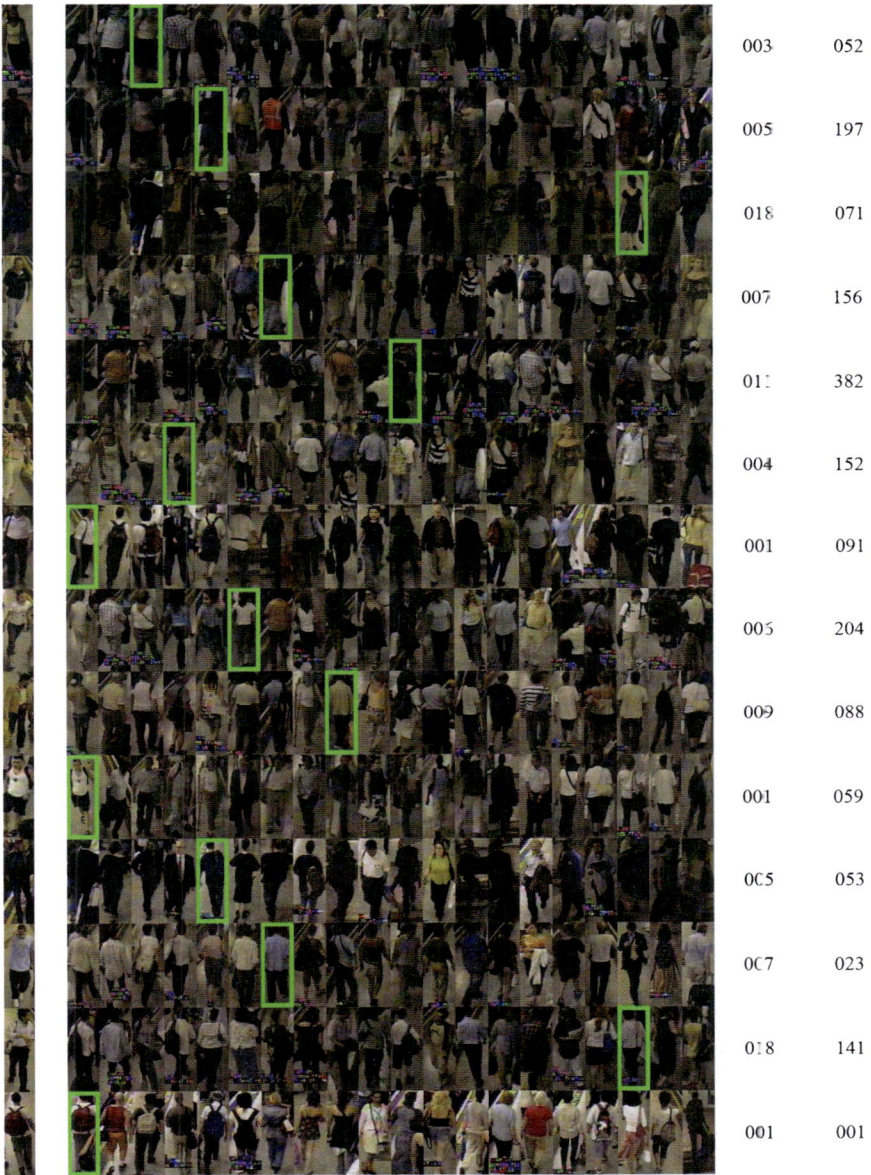

	003	052
	005	197
	018	071
	007	156
	011	382
	004	152
	001	091
	005	204
	009	088
	001	059
	0C5	053
	0C7	023
	018	141
	001	001

Fig. 14.2 Person re-identification in an underground station. The *first image* on the *left* of *each row* is a probe image. It is followed by the top 20 matched results, sorted from *left to right* using a context-aware search model, with the correct match highlighted by a *green bounding box*. The ranks returned by this model are compared to those from matching visual appearance features alone, shown at the rightmost two columns (smaller is better, with 001 denoting a Rank-1 match). It is evident that using visual appearance alone often cannot find the correct match in the top 20 ranks. Notice the significant visual ambiguity between images of the same person in different camera views. This is due to variations of pose, colour, lighting, as well as poor image quality

RankSVM can speed up the learning process of a RankSVM, but without addressing the spatial complexity problem. This can incur very high memory consumption in model learning, and can easily become unmanageable and unscalable. An ensemble learning mechanism is available for limiting this problem and making kernel based ranking computationally accessible for solving the person re-identification problem.

Matching people across significantly different camera views by visual appearance alone is inherently ill-posed and extremely difficult. Even with a carefully selected visual features and with learning of ranking based matching, the matching accuracy of a model remains being low. Spatial and temporal activity correlation based contextual information can be exploited to reduce the search space for possible candidate matching, and provide a context-aware search model for person re-identification. This context based approach to person re-identification brings about greater efficiency, accuracy and robustness in finding the correct match.

Learning a multi-camera behaviour correlation model and constructing a context-aware people search model across camera views are first steps towards a more comprehensive treatment of multi-camera behaviour analysis and interpretation. In the next, and final chapter of this book, we consider and illustrate plausible ways to build a global behaviour model, aiming for 'connecting the dots' and establishing a sense of global situational awareness. Such a model can be deployed to detect a global abnormal situation inaccessible to localised individual points of views. The problem of incrementally learning and adapting a global behaviour model is also addressed, necessary for coping with situational changes in a wider context beyond visual observations.

References

Chapelle, O., Keerthi, S.: Efficient algorithms for ranking with SVMs. Inf. Retr. **13**(3), 201–215 (2010)

Farenzena, M., Bazzani, L., Perina, A., Murino, V., Cristani, M.: Person re-identification by symmetry-driven accumulation of local features. In: IEEE Conference on Computer Vision and Pattern Recognition, San Francisco, USA, June 2010, pp. 2360–2367 (2010)

Freund, Y., Iyer, R., Schapire. R.E., Singer, Y.: An efficient boosting algorithm for combining preferences. J. Mach. Learn. Res. **4**, 933–969 (2003)

Gheissari, N., Sebastian, T.B., Rittscher, J., Hartley, R.: Person reidentification using spatiotemporal appearance. In: IEEE Conference on Computer Vision and Pattern Recognition, New York, USA, June 2006, pp. 1528–1535 (2006)

Gilbert, A., Bowden, R.: Tracking objects across cameras by incrementally learning inter-camera colour calibration and patterns of activity. In: European Conference on Computer Vision, Graz, Austria, pp. 125–136 (2006)

Gray, D., Tao, H.: Viewpoint invariant pedestrian recognition with an ensemble of localized features. In: European Conference on Computer Vision, Marseille, France, October 2008, pp. 262–275 (2008)

Hahnel, M., Klunder, D., Kraiss, K.F.: Color and texture features for person recognition. In: IEEE International Joint Conference on Neural Networks, pp. 647–652 (2004)

Javed, O., Shafique, K., Shah, M.: Appearance modeling for tracking in multiple non-overlapping cameras. In: IEEE Conference on Computer Vision and Pattern Recognition, pp. 26–33 (2005)

Joachims, T.: Optimizing search engines using clickthrough data. In: ACM SIGKDD International Conference on Knowledge Discovery and Data Mining, pp. 133–142 (2002)

Loy, C.C., Xiang, T., Gong, S.: Time-delayed correlation analysis for multi-camera activity understanding. Int. J. Comput. Vis. **90**(1), 106–129 (2010)

Madden, C., Cheng, E.D., Piccardi, M.: Tracking people across disjoint camera views by an illumination-tolerant appearance representation. Mach. Vis. Appl. **18**(3), 233–247 (2007)

Prosser, B., Gong, S., Xiang, T.: Multi-camera matching using bi-directional cumulative brightness transfer functions. In: British Machine Vision Conference, Leeds, UK, September 2008a

Prosser, B., Gong, S., Xiang, T.: Multi-camera matching under illumination change over time. In: ECCV Workshop on Multi-camera and Multi-modal Sensor Fusion Algorithms and Applications, Marseille, France, October 2008b

Prosser, B., Zheng, W., Gong, S., Xiang, T.: Person re-identification by support vector ranking. In: British Machine Vision Conference, Aberystwyth, UK, September 2010

Schapire, R.E., Singer, Y.: Improved boosting algorithms using confidence-rated predictions. Mach. Learn. **37**(3), 297–336 (1999)

Wang, H., Suter, D., Schindler, K.: Effective appearance model and similarity measure for particle filtering and visual tracking. In: European Conference on Computer Vision, Graz, Austria, pp. 606–618 (2006)

Wang, X., Doretto, G., Sebastian, T., Rittscher, J., Tu, P.: Shape and appearance context modeling. In: IEEE International Conference on Computer Vision, Rio de Janeiro, Brazil, October 2007, pp. 1–8 (2007)

Zheng, W., Gong, S., Xiang, T.: Associating groups of people. In: British Machine Vision Conference, London, UK, September 2009

Zheng, W., Gong, S., Xiang, T.: Person re-identification by probabilistic relative distance comparison. In: IEEE Conference on Computer Vision and Pattern Recognition, Colorado Springs, USA, June 2011

Chapter 15
Connecting the Dots

An eventual goal for automated visual analysis of distributed object behaviour is to bring about a coherent understanding of partially observed uncertain sensory data from the past and present, and to 'connect the dots' in finding a big picture of global situational awareness for explaining away anomalies and discovering hidden patterns of significance. To that end, we consider the computational task and plausible models for modelling global behaviours and detecting global abnormal activities across distributed and disjoint multiple cameras.

For constructing a global behaviour model for detecting holistic anomalies, besides model sensitivity and robustness (Chap. 8), model tractability and scalability are of a greater concern. A typical distributed camera network may consist of dozens to hundreds of cameras, many of which cover a wide-area scene of different distinctive activity semantic regions (Sects. 10.1 and 13.1). To build a model for discovering and describing global behaviour patterns emerging from a distributed network of local activity regions, the uncertainty and complexity in temporal delays between activities observed in different camera views need be addressed. Care must also be taken to ensure that the model computational cost and memory consumption can be controlled at a manageable level. For these reasons, the models described in Parts II and III cannot be applied directly as in the case of a single camera view. To address the problem, we revisit and further develop three models studied early for modelling behaviours in a single camera view.

15.1 Global Behaviour Segmentation

Let us first consider modelling global behaviours using a dynamic Bayesian network, specifically a hidden Markov model (HMM). Assume that a set of disjoint camera views have been decomposed into regions (Sect. 13.1), and the correlation strength and temporal delay between regional activities have been learned (Sect. 13.2) and represented using a regional activity affinity matrix $\mathbf{P}$ (13.14). A large camera network observing a distributed visual environment can capture a

S. Gong, T. Xiang, *Visual Analysis of Behaviour*,
DOI 10.1007/978-0-85729-670-2_15, © Springer-Verlag London Limited 2011

Fig. 15.1 A hidden Markov model with two time slices unrolled. Observation nodes are shown as *shaded circles* and hidden nodes are shown as *clear squares*

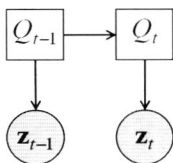

diverse range of activities occurring simultaneously. Not all the observed activities are correlated over space and time. We consider that uncorrelated activities do not contribute towards meaningful global behaviours. They shall be excluded from modelling. An emerging global behaviour comprising correlated local activities is discovered and modelled by taking the following steps.

1. A spectral clustering algorithm (Sect. 13.1) is employed to group regional activities through a regional activity affinity matrix **P** (13.14).
2. The resultant regional activity clusters are examined for discovering global behaviours. Those clusters that consist of cross-camera activity regions with the highest mean correlation strength are selected. Each selected cluster is considered to form a global behaviour.
3. For each discovered global behaviour consisting of χ regions, activity patterns in any one of the χ regions can be set as a reference point to align temporally activity patterns in other regions according to their temporal offsets $\hat{\tau}_{\mathbf{x}_i,\mathbf{x}_j}$, computed by (13.12). This temporal alignment process is to facilitate the construction of a time index *synchronised* global behaviour representation as follows.
4. Each temporally aligned activity region is represented by its average regional activity pattern (13.7) as a two-dimensional time series $(\hat{\mathbf{u}}, \hat{\mathbf{v}})$. All aligned activity regions of a global behaviour are used to form a time index synchronised global behaviour time series $\mathbf{z}_t$ of $\chi \times 2$ dimensions. At each time instance t, $\mathbf{z}_t$ is given by $\mathbf{z}_t = [\hat{u}'_{1,t}, \hat{v}'_{1,t}, \ldots, \hat{u}'_{\chi,t}, \hat{v}'_{\chi,t}]$, with the prime symbol indicating an aligned time series removing any temporal offset between camera views in this synchronised global behaviour. Note that this camera view alignment process does not remove any temporal ordering between activities in different camera views. It merely removes any arbitrary temporal gaps between disjoint camera views when activities are not visible. The global behaviour time series $\mathbf{z}_t$ is used as the input to train an HMM for capturing the temporal dynamics of this global behaviour over time.
5. The HMM is an ergodic model, that is, fully connected (Fig. 15.1), with Q_t being a discrete random variable and $Q_t \in \{q^i | i = 1, \ldots, K\}$. It is assumed that the model is first-order Markov so that $p(Q_t|Q_{1:t-1}) = p(Q_t|Q_{t-1})$, and the observations are conditionally first-order Markov, that is, $p(\mathbf{z}_t|Q_t)$. Automatic model order selection is performed on $\mathbf{z}_t$ using a Bayesian information criterion (BIC) score, so to find the optimal value of K (see Sect. 7.2.2). For model parameter estimation, the time aligned regional activity patterns at different time instances are clustered into K groups using a k-means clustering algorithm. The parameters of an HMM with K hidden states are initialised by the means and covariance matrices of the K groups, then estimated using the Baum–Welch algorithm (Baum et al. 1970).

One HMM is learned for each global behaviour, with multiple HMMs modelling different global behaviours. They can be used for real-time global behaviour based temporal segmentation, similar to what was considered for a single camera view in Sect. 7.2. However, the aim here is to segment automatically unseen multi-camera video streams into synchronised global activity phases based on *what is happening* not in a single camera view, but in a global behavioural space by connecting distributed local regional activities as 'dots' in a big picture. These global activity phases are obtained by inferring the hidden states Q_t at each time instance using an on-line filtering technique (Murphy 2002).

More precisely, given $\mathbf{z}_t$ observed as a continuous data stream, the probability of a specific hidden state $p(Q_t|\mathbf{z}_{1:t})$ is computed as a function of current input $\mathbf{z}_t$ and prior belief state $p(Q_{t-1}|\mathbf{z}_{1:t-1})$:

$$p(Q_t|\mathbf{z}_{1:t}) \propto p(\mathbf{z}_t|Q_t, \mathbf{z}_{1:t-1}) p(Q_t|\mathbf{z}_{1:t-1})$$

$$= p(\mathbf{z}_t|Q_t)\left(\sum_{Q_{t-1}} p(Q_t|Q_{t-1}) p(Q_{t-1}|\mathbf{z}_{1:t-1})\right) \qquad (15.1)$$

Based on the Markovian assumption, $p(\mathbf{z}_t|Q_t, \mathbf{z}_{1:t-1})$ is replaced by $p(\mathbf{z}_t|Q_t)$. Similarly, $p(Q_t|\mathbf{z}_{1:t-1})$ can be computed from the prior belief state under the Markovian assumption. To infer the activity phase Q_t^*, the probabilities $p(Q_t = q^i|\mathbf{z}_{1:t})$ are computed by (15.1). The most likely hidden state is then determined by choosing the hidden state that yields the highest probability:

$$Q_t^* = \arg\max_{q^i} p\left(Q_t = q^i|\mathbf{z}_{1:t}\right) \qquad (15.2)$$

We use a distributed underground station scene[1] monitored by a network of cameras to illustrate the effectiveness of using an HMM model for global behaviour based video temporal segmentation. A number of global behaviour patterns are discovered from the scene, including the platform activities monitored by Cam-6, Cam-7, Cam-8, and Cam-9, as well as escalator activities captured by Cam-4 and Cam-5. Figure 15.3 shows an example of the discovered escalator activities from this environment, where activity Region-46 in Cam-4 and Region-51 in Cam-5 are detected being highly correlated. Model inferred global behaviour phases and example video frames are shown in Figs. 15.2 and 15.3, respectively. It is evident that compared to the ground truth, the model gives very accurate global behaviour temporal segmentation. Loy et al. (2009) show quantitatively that compared to single view behaviour analysis, a global behaviour model is able to reduce visual ambiguities and produce accurate multi-camera view global behaviour segmentation over time.

---

[1]The scene layout, camera topology and example camera views of this environment are shown in Fig. 13.3. The computed local activity regions in different camera views can be seen in Fig. 13.4.

Fig. 15.2 An example of global behaviour temporal phases being inferred by an HMM. The activity corresponds to an escalator scene from an underground station. The Y-axis represents global behaviour phases and the X-axis represents the time index

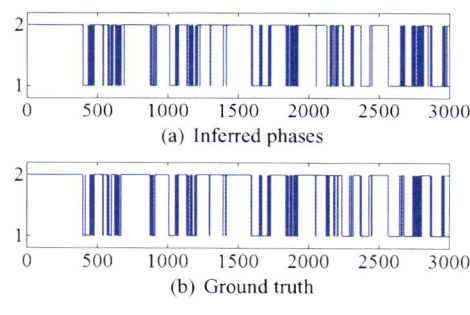

(a) Inferred phases

(b) Ground truth

Fig. 15.3 Example video frames from the model estimated global behaviour phases at the escalator scene from an underground station (Fig. 15.2(b)). *Phase 1*: passengers on an escalator approaching an exit; *Phase 2*: passengers moving clear of the escalator exit area

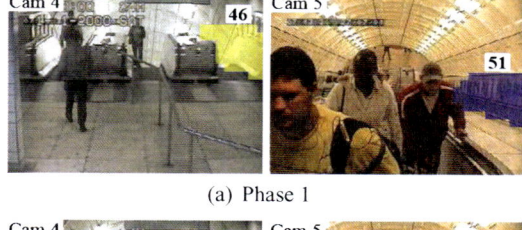

(a) Phase 1

(b) Phase 2

15.2 Bayesian Behaviour Graphs

To monitor all activities captured by a camera network and detect abnormal behaviours either local to individual camera views or global across camera views, it is desirable to build a single model for a camera network. A dynamic Bayesian network, as considered in the preceding section, is less suitable for this purpose due to limitations on its tractability and scalability. A static Bayesian graph[2] as a time-delayed probabilistic graphical model is a plausible alternative method.

A time-delayed probabilistic graphical model is capable of modelling global dependencies between activities based on learning their pair-wise correlations (Sect. 13.2). When learning pair-wise correlations, not all learned correlations are optimised. This is due to that they are modelled independently, resulting in likely redundant correlations being introduced. This problem can be addressed by learning the topology of a global behaviour Bayesian graph that aims to capture all activity correlations using a global optimisation. To that end, Loy et al. (2010) suggest a framework for learning globally optimised time-delayed correlations between

---

[2]In this book, we use the term 'Bayesian graph' and 'static Bayesian network' interchangeably.

distributed local activities. The technique is based on a time-delayed probabilistic graphical model. In this single graph model, a node represents activities in a semantic region of a camera view. Directed links between nodes encode the causal activity relationships between regions. The time-delayed dependencies among regions across camera views are globally optimised by a two-stage model structure learning process. An overview of this approach is shown in Fig. 15.4.

15.2.1 A Time-Delayed Probabilistic Graphical Model

A time-delayed probabilistic graphical model is defined as $B = \langle G, \Theta \rangle$, consisting of a directed acyclic graph G. The nodes in this graph represent a set of discrete random variables $\mathbf{X} = \{X_i | i = 1, 2, \ldots, n\}$, where n is the total number of regions across views; X_i is a discrete variable, representing the average regional activity pattern observed in the ith region (Sect. 13.1). Specific values taken by a variable X_i are denoted as x_i, whilst a stream of values x_i of variable X_i is denoted as $\mathbf{x}_i(t) = (x_{i,1}, \ldots, x_{i,t}, \ldots)$.

The model is defined by a set of parameters denoted as Θ, specifying the conditional probability distribution $p(X_i | \mathbf{Pa}(X_i))$. All the observations in the model are also discrete variables. The conditional probability distribution between a child node X_i and its parents $\mathbf{Pa}(X_i)$ in G is represented by a multinomial probability distribution. Hence, Θ consists of a set of parameters

$$\theta_{x_i | \mathbf{pa}(X_i)} = p\left(x_i | \mathbf{pa}(X_i)\right)$$

for each possible discrete value x_i of X_i, and $\mathbf{pa}(X_i)$ of $\mathbf{Pa}(X_i)$; $\mathbf{Pa}(X_i)$ denotes a set of parents of X_i, where $\mathbf{pa}(X_i)$ is an instantiation of $\mathbf{Pa}(X_i)$.

Conditional independence is assumed. That is, X_i is independent from its non-descendants given its parents. These dependencies are represented through a set of directed edges $\mathbf{E}$. Each edge points to a node from its parents on which the distribution is conditioned. Given any two variables X_i and X_j, a directed edge from X_i to X_j is represented by $X_i \rightarrow X_j$, where $(X_i, X_j) \in \mathbf{E}$ and $(X_j, X_i) \notin \mathbf{E}$. It should be pointed out that $p(X_i | \mathbf{Pa}(X_i))$ are not quantified by a common time index, but with relative time delays that are discovered using time-delayed mutual information analysis introduced in Sect. 13.2. Other notations used in the model are:

- The number of states of X_i is r_i.
- The number of possible configurations of $\mathbf{Pa}(X_i)$ is q_i.
- A set of discrete value x_i across all variables is given as $\mathbf{x} = \{x_i | i = 1, 2, \ldots, n\}$.
- A collection of m cases of $\mathbf{x}$ is denoted as $\mathcal{X} = \{\mathbf{x}_1, \ldots, \mathbf{x}_m\}$.
- The number of cases of $(x_i, \mathbf{pa}(X_i))$ in $\mathcal{X}$ is represented as $N_{x_i | \mathbf{pa}(X_i)}$, specifically $N_{ijk} = N_{x_i = k | \mathbf{pa}(X_i) = j}$.

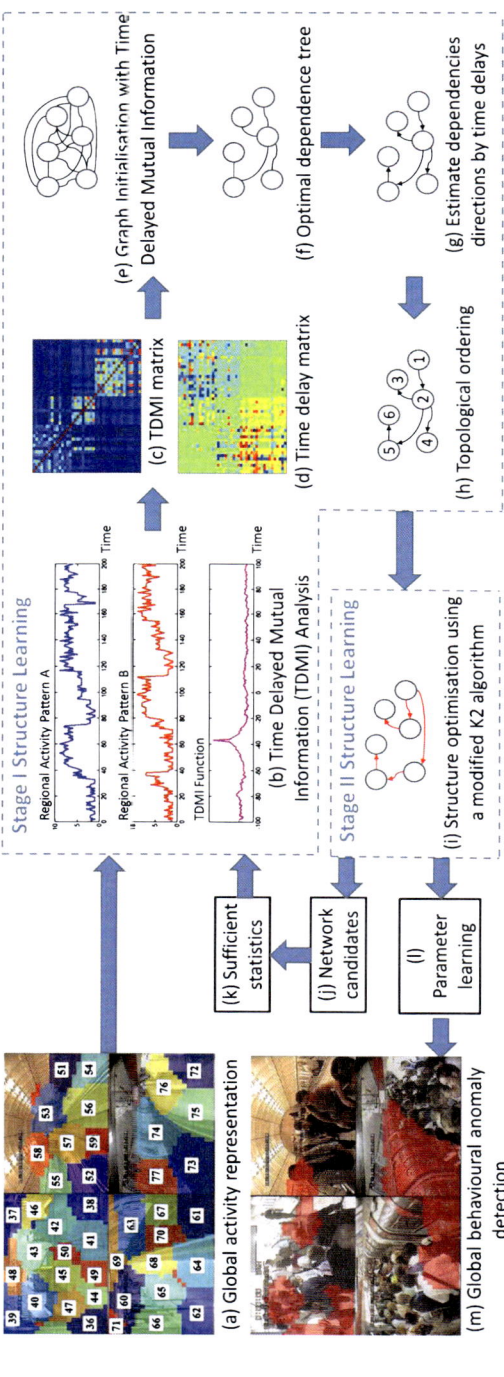

Fig. 15.4 Learning a global behaviour graph to model dependencies between local activities observed in disjoint multi-cameras using a time-delayed probabilistic graphical model

15.2.2 Bayesian Graph Structure Learning

A learning process is required to discover and quantify the optimal model structure of a time-delayed probabilistic graphical model B that encodes the time-delayed dependencies between behaviour graph nodes. This learning process can be carried out either in a batch mode, assuming a model learned from previous examples are valid for explaining away future observations; or incrementally in order to adapt the model to changes in visual context and behaviour relevance. In the following, we first describe a batch mode learning strategy. This shall facilitate the introduction of an incremental learning algorithm in Sect. 15.2.5 for model adaptation to context change.

In essence, the batch mode learning strategy is a two-stage process. In the first stage (Fig. 15.4(b)–(h)), the learning process aims to estimate a prior structure of a graphical model from pair-wise correlations estimated by time-delayed mutual information. This is to apply an ordering constraint on model structure. In the second stage (Fig. 15.4(i)), this ordering constraint is propagated in a scored-searching based learning in order to prune the graph structure by eliminating any candidate structure inconsistent with the time-delayed mutual information constraint. This two-stage learning process ensures accurate and tractable learning of globally optimised activity dependencies.

Constraint-Based Learning

The first-stage mutual information constrained learning has three steps:

1. Step-1, Figs. 15.4(b)–(d): Time-delayed mutual information analysis (Fraser and Swinney 1986) is explored to learn initial time-delayed association between each pair of regional activity patterns, as introduced in Sect. 13.2.2.
2. Step-2, Figs. 15.4(e)–(f): The aim is to generate an optimal dependence tree $\mathcal{T}$, for instance a Chow–Liu tree (Chow and Liu 1968), that best approximates the graph joint probability $p(\mathbf{X})$ by a product of second-order conditional and marginal distributions, given as

$$p(\mathbf{X}) = p(X_1, \ldots, X_n) = \prod_{i=1}^{n} p(X_i | X_{\psi(i)}) \qquad (15.3)$$

where a mapping function $\psi(i)$ with $0 \leq \psi(i) < n$ defines a tree graph $\mathcal{T}$ so that

$$X_{\psi(i)} = \begin{cases} \mathbf{Pa}(X_i), & \text{if } \psi(i) > 0 \\ \mathbf{Pa}(X_i) = \emptyset, & \text{if } \psi(i) = 0 \end{cases} \qquad (15.4)$$

The optimal dependence tree $\mathcal{T}$ is obtained based on the estimated time-delayed mutual information matrix $\mathbf{I}$. In particular, elements of $\mathbf{I}$ are assigned as weights to each possible edges of a graph with node set $\mathbf{X}$ that encodes no assertion of conditional independence. The Prim's algorithm (Prim 1957) is then applied to

---

Algorithm 15.1: Finding a prior graph structure and topological ordering

---

Input: An undirected weighted graph with a node set
    $\mathbf{X} = \{X_i | i = 1, 2, \ldots, n\}$, edge set $\mathbf{E}$, time-delayed mutual information
    matrix $\mathbf{I}$, and time delay matrix $\mathbf{D}$.
Output: Prior network structure G^{p} defined by $\mathbf{X}^{\mathrm{p}}$ and $\mathbf{E}^{\mathrm{p}}$, topological
    ordering $\prec$.

1   $\mathbf{X}^{\mathrm{p}} = X$, where X is an arbitrary node from $\mathbf{X}$;
2   $\mathbf{E}^{\mathrm{p}} = \emptyset$;
3   **while** $\mathbf{X}^{\mathrm{p}} \neq \mathbf{X}$ **do**
4   Choose an edge $(X_i, X_j) \in \mathbf{E}$ with maximum weight $\hat{I}_{ij}$, where $X_i \in \mathbf{X}^{\mathrm{p}}$
    and $X_j \notin \mathbf{X}^{\mathrm{p}}$;
5      $\mathbf{X}^{\mathrm{p}} = \mathbf{X}^{\mathrm{p}} \cup \{X_j\}$;
6      $\mathbf{E}^{\mathrm{p}} = \mathbf{E}^{\mathrm{p}} \cup \{(X_i, X_j)\}$;
7   **if** $\hat{\tau}_{ij} > 0$ **then**
8        $\mid$ $X_i \to X_i$;
9   **else**
10      $\mid$ $X_i \leftarrow X_j$;
11   **end**
12 **end**
13   $\prec$ = topological_sort $(\mathbf{E}^{\mathrm{p}})$;

---

find a subset of the edges that forms a tree structure that includes every node, in which the total weight of the tree is maximised.

3. Step-3, Figs. 15.4(g)–(h): Edge orientation is estimated. The undirected tree $\mathcal{T}$ is transformed to a directed prior network structure G^{p} by assigning orientations to the edges. Typically, one can assign edge orientations by either selecting a random node as a root node, or by performing conditional independence test (Chen et al. 2008) and scoring function optimisation over the graph (Chen et al. 2006). These methods are either inaccurate or require exhaustive search on all possible edge orientations. They are thus computationally costly. To overcome these problems, an effective and simple approximation can be used as follows.

a. One can orient the edges by tracing the time delays for a pair of nodes in the tree structure using $\mathbf{D}$ learned by the time-delayed mutual information analysis. In particular, if the activity patterns observed in X_i are lagging the patterns observed in X_j with a time delay τ, it is reasonable to assume that the distribution of X_i is conditionally dependent on X_j. The edge is therefore pointed from X_j to X_i.

b. With G^{p} defined by the edges, one can derive the ordering of variables $\prec$ by performing topological sorting (Cormen et al. 2009). In particular, the ordering $\prec$ specifies that a variable X_j can only be the parent of X_i if, and only if, X_j precedes X_i in $\prec$, i.e. $X_j \in \mathbf{Pa}(X_i)$ iff $X_j \prec X_i$.

The process for computing G^{p} and $\prec$ is summarised in Algorithm 15.1.

Time-Delayed Scored Searching

The second stage of the model structure learning process aims to optimise the initial model structure obtained from stage one. This is illustrated in Fig. 15.4(i). This is based on a re-formulation of a heuristic search technique, known as the K2 algorithm (Cooper and Herskovits 1992). The re-formulation is required to generate an optimised time-delayed dependency structure based on $\prec$ derived from the first-stage learning.[3] More precisely, the K2 algorithm iterates over each node X_i that has an empty parent set $\mathbf{Pa}(X_i)$ initially. Candidate parents are then selected in accordance with the node sequence specified by $\prec$. They are added incrementally to $\mathbf{Pa}(X_n)$ for which the addition will increase the score $S(G|\mathcal{X})$ of structure G given data set $\mathcal{X}$.

For the scoring function, we consider the Bayesian information criterion (BIC) (Schwarz 1978), which is both score equivalent and decomposable (Heckerman et al. 1995). Specifically, the BIC score for a given Bayesian graph structure is computed as

$$S_{\mathrm{BIC}}(G|\mathcal{X}) = \sum_{i=1}^{n} S_{\mathrm{BIC}}\big(X_i|\mathbf{Pa}(X_i)\big)$$

$$= \sum_{i=1}^{n}\sum_{t=1}^{m} \log p\big(x_{i,t}|\mathbf{pa}(X_i), \boldsymbol{\theta}_{x_{i,t}|\mathbf{pa}(X_i)}\big) - \log m \sum_{i=1}^{n} \frac{b_i}{2} \qquad (15.5)$$

where $b_i = q_i(r_i - 1)$ is the number of parameters needed to describe $p(X_i|\mathbf{Pa}(X_i))$.

The above formulation differs from the original K2 algorithm in that any addition of candidate parent is required not only to increase the graph score, but also to satisfy the constraint imposed by the time delays discovered from the first-stage learning. In addition, the score computation (15.5) is carried out by shifting parent's activity patterns with a relative delay to child node's activity patterns based on **D**. This is shown in line 6 of Algorithm 15.2.

Computational Complexity

The computational complexity of the first stage of model structure learning (Algorithm 15.1) is as follows:

1. The total number of possible activity region pairs to be considered for obtaining pair-wise time-delayed mutual information function (13.16) is in the order of $O(n^2)$, where n is the number of regions.

---

[3]Without the first-stage learning, $\prec$ can be set randomly. However, a randomly set $\prec$ does not guarantee the most probable model structure. Alternatively, one can apply the K2 algorithm exhaustively on all possible orderings to find a structure that maximises the score. This direct approach is computationally intractable even for a moderate number of nodes, since the space of ordering is $n!$ for a n-node Bayesian graph model.

Algorithm 15.2: A modified K2 algorithm that satisfies a time delay constraint

---

Input: A graph with a node set $\mathbf{X} = \{X_i | i = 1, 2, \ldots, n\}$, an ordering of nodes
$\prec$, an upper bound φ on the number the parents a node may have, time
delay matrix $\mathbf{D}$.

Output: Final graph structure G defined by $\{(X_i, \mathbf{Pa}(X_i)) \mid i = 1, 2, \ldots, n\}$.

1 **for** $i = 1$ *to* n **do**

2 $\quad$ $\mathbf{Pa}(X_i) = \emptyset$;

3 $\quad$ score$_{\text{old}} = S(X_i | \mathbf{Pa}(X_i))$;

4 $\quad$ OKToProceed = **true**;

5 $\quad$ **while** OKToProceed *and* $|\mathbf{Pa}(X_i)| < \varphi$ **do**

6 $\quad\quad$ Let $X_j \prec X_i$, $X_j \notin \mathbf{Pa}(X_i)$, with activity patterns $x_j(t + \tau)$,
$\quad\quad$ $\tau = \mathbf{D}(X_i, X_j) \leq 0$, which maximises $S(X_i | \mathbf{Pa}(X_i) \cup \{X_j\})$;

7 $\quad\quad$ score$_{\text{new}} = S(X_i | \mathbf{Pa}(X_i) \cup \{X_j\})$;

8 $\quad\quad$ **if** score$_{\text{new}}$ > score$_{\text{old}}$ **then**

9 $\quad\quad\quad$ score$_{\text{old}} = $ score$_{\text{new}}$;

10 $\quad\quad\quad$ $\mathbf{Pa}(X_i) = \mathbf{Pa}(X_i) \cup \{X_j\}$;

11 $\quad\quad$ **else**

12 $\quad\quad\quad$ OKToProceed = **false**;

13 $\quad\quad$ **end**

14 $\quad$ **end**

15 **end**

---

2. If the maximum time delay is limited to τ_{max}, the number of time-delayed mutual information calculations (13.18) is $\tau_{\text{max}} - 1$.
3. The overall complexity of time-delayed mutual information analysis (Step-1) is $O(n^2 \tau_{\text{max}})$.
4. The run time complexity of the optimal dependence tree approximation (Step-2) is $O(e \log n)$.
5. Topological sorting (Step-3) takes $O(n + e)$ time (Cormen et al. 2009), where e is the number of edges.

The complexity of the second stage of model structure learning (Algorithm 15.2) is as follows:

1. The **for** statement loops $O(n)$ times.
2. The **while** statement loops at most $O(\varphi)$ times once it is entered, where φ denotes the maximum number of parents a node may have.
3. Inside the **while** loop, line 11 in Algorithm 15.2 is executed for at most $n - 1$ times since there are at most $n - 1$ candidate parents consistent with $\prec$ for X_i. Hence, line 11 in Algorithm 15.2 takes $O(sn)$ time if one assumes each score evaluation takes $O(s)$ time.
4. Other statements in the **while** loop take $O(1)$ time.
5. Therefore, the overall complexity of the second stage of model structure learning is $O(sn)O(\varphi)O(n) = O(sn^2\varphi)$.

6. In the worst case scenario where one does not apply an upper bound to the number of parents a node may have, the time complexity becomes $O(sn^3)$ since $\varphi = n$.

Further Considerations

1. *What is the advantage from using a two-stage learning process?* Both stages of the model structure learning process are important for having a tractable solution in discovering and learning the time-delayed dependencies among regional activities. Specifically, without the first-stage structure learning, time delay information would not be available for constraining the search space in finding the optimal model structure. The computation would not be tractable. On the other hand, a poor model structure may be obtained if the tree structure is without the second-stage optimisation. This is because the tree structure can only approximate an optimal set of $n - 1$ first-order dependence relationships among the n variables. It does not estimate the target distribution, which may include more complex dependencies. Furthermore, studies show that the constraint-based learning can be sensitive to failures in independence tests (Friedman et al. 1999). Therefore, a second-stage scored-searching based optimisation is necessary to discover additional dependencies and correct potential error from the first-stage learning.

2. *Why is an exact learning method not used?* In this two-stage approach, a heuristic search algorithm is chosen for the second-stage structure learning instead of an exact learning algorithm. In general, an exact structure learning algorithm is computationally intractable for a large graph, since there are $2^{O(n^2 \log n)}$ directed acyclic graphs for a n-node network (Friedman and Koller 2000). A search using a typical exact algorithm would take exponential time on the number of variables n. For instance, it is $O(n2^n)$ for a dynamic programming based technique (de Campos et al. 2009). Such a high computational complexity prohibits its use from learning a typical camera network that consists of hundreds of local activity regions.

3. *Are there alternative heuristic search techniques?* Among various heuristic search algorithms, the K2 algorithm (Cooper and Herskovits 1992), upon which the described algorithm is based, is well suited for learning the dependency structure of a large camera network due to its superior computational efficiency given the ordering constraint. Specifically, with the ordering constraint, the search space of the re-formulated K2 algorithm is much smaller than that of a conventional search method without the ordering constraint (Auliac et al. 2008). In addition, the ordering constraint also helps in avoiding the costly acyclicity checks since the topological order already ensures acyclicity of structure. Furthermore, the K2 algorithm is more efficient than alternative methods such as the Markov chain Monte Carlo based structure learning (Murphy 2001), which requires a sufficiently long burn-in time to obtain a converged approximation for a large graph (Gilks et al. 1995).

15.2.3 Bayesian Graph Parameter Learning

After learning the model structure of a time-delayed probabilistic graphical model, the model parameters need be estimated (Fig. 15.4(l)). There are two methods for estimating the parameters of a probabilistic graphical model given complete observation. That is, values are assigned to *all* the variables of interest. The first method is maximum likelihood estimation (MLE). MLE aims to find a set of parameters that maximise the likelihood function of the model given observation. The second available method is Bayesian learning. This method begins with a prior probability over the model parameters and computes a posterior probability when new instances are observed. The posterior can be treated as the current belief of the model, and be used as a prior for the next learning iteration. This method has the attractive property that the belief state at time t is the same as the posterior after observing $\mathbf{x}_1, \ldots, \mathbf{x}_t$ from an initial prior belief state. In other words, one can update the posterior distribution sequentially and efficiently using a closed-form formula with conjugate prior.[4]

Bayesian learning is preferred for learning a global behaviour model because it offers a more tractable solution especially when the posterior distribution can be estimated sequentially using a closed-form formula with conjugate prior. With Bayesian learning, one can adopt the BDeu prior (Buntine 1991), a conjugate of a multinomial distribution over model parameters.

Figure 15.5 shows an example of a learned global Bayesian behaviour graph structure using a time-delayed probabilistic graphical model. This model aims to learn emerging global behaviours captured by a camera network in an underground station.[5] It is evident that most of the discovered activity dependencies are between regions from the same camera views which have short time delays. This is expected. However, the model also discovers a number of interesting, and not so obvious, inter-camera correlations, with correct activity directions on edge dependencies (Loy et al. 2010).

15.2.4 Cumulative Anomaly Score

A method for detecting global abnormal behaviour is to examine the log-likelihood $\log p(\mathbf{x}_t|\Theta)$ of the observations given a global behaviour model (Zhou and Kimber 2006). For measuring log-likelihood using a time-delayed probabilistic graphical model, an unseen global activity pattern can be considered as being abnormal if

$$\log p(\mathbf{x}_t|\Theta) = \sum_{i=1}^{n} \log p\left(x_{i,t}|\mathbf{pa}(X_i), \boldsymbol{\theta}_{x_i|\mathbf{pa}(X_i)}\right) < \text{Th} \qquad (15.6)$$

---

[4]Conjugate prior is a prior that has the same functional form as the posterior distribution (Bishop 2006).

[5]The global space layout of the station and computed local activity regions in different camera views are shown in Figs. 13.3 and 13.4, respectively.

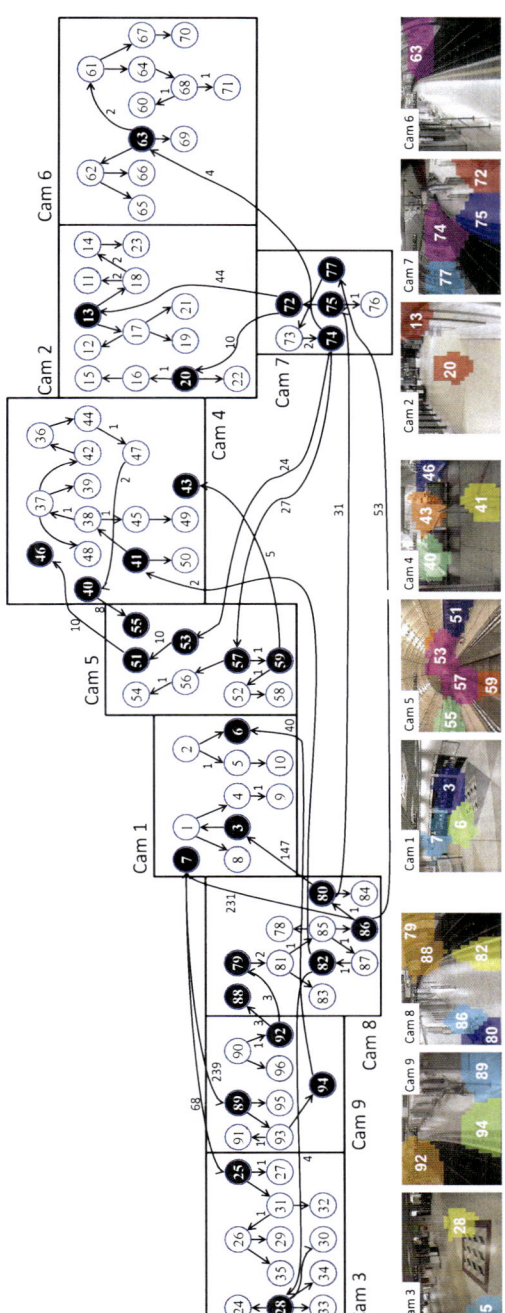

Fig. 15.5 An example of a global Bayesian behaviour graph learned using a two-stage structure learning method. *Edges* are labelled with the associated time delays discovered using time-delayed mutual information analysis. Those activity regions with corresponding graph nodes that are discovered as having inter-camera dependencies are *highlighted in black*

where Th is a pre-defined threshold, and $\mathbf{x}_t = \{x_{i,t} | i = 1, 2, \ldots, n\}$ are observations at time t for all n regions. However, given a crowded public scene captured in videos with low image resolution both spatially and temporally, observations $\mathbf{x}_t$ are inevitably noisy. A log-likelihood based anomaly detection model is likely to confuse a true behavioural anomaly with a noisy observation. This is because both can contribute to a low value in $\log p(\mathbf{x}_t | \Theta)$. They cannot be distinguished solely by the value of $\log p(\mathbf{x}_t | \Theta)$.

This problem is addressed by introducing a cumulative abnormality score. This measurement of abnormality alleviates the effect of noise by accumulating the temporal history of the likelihood of behavioural anomaly occurrences in each region over time. The rational for this measurement is based on the assumption that noise would not persist over sustained period of time and therefore can be filtered out when visual evidence is accumulated over time. Specifically, an abnormality score is set to zero at $t = 0$ and is computed on-the-fly for each graph node of a time-delayed probabilistic graphical model. This is used to monitor the likelihood of anomaly in each activity region. The log-likelihood of a given observation $x_{i,t}$ for the ith region at time t is computed as

$$\log p\left(x_{i,t} | \mathbf{pa}(X_i), \theta_{x_{i,t} | \mathbf{pa}(X_i)}\right) = \log \frac{N_{x_{i,t} | \mathbf{pa}(X_i)} + \frac{\eta}{r_i q_i}}{\sum_{k=1}^{r_i}(N_{x_{i,t}=k | \mathbf{pa}(X_i)} + \frac{\eta}{r_i q_i})} \tag{15.7}$$

If the log-likelihood is below a threshold Th_i, the abnormality score for $x_{i,t}$, denoted as $c_{i,t}$, is increased as

$$c_{i,t} = c_{i,t-1} + \left| \log p\left(x_{i,t} | \mathbf{pa}(X_i), \theta_{x_i | \mathbf{pa}(X_i)}\right) - \text{Th}_i \right| \tag{15.8}$$

Otherwise it is decreased from the previous abnormality score:

$$c_{i,t} = c_{i,t-1} - \delta\left(\left| \log p\left(x_{i,t} | \mathbf{pa}(X_i), \theta_{x_i | \mathbf{pa}(X_i)}\right) - \text{Th}_i \right|\right) \tag{15.9}$$

where δ is a decay factor controlling the rate of the decrease. The score value of $c_{i,t}$ is set to 0 whenever it becomes a negative number after a decrease, so that $c_{i,t} \geq 0, \forall\{i, t\}$ and a larger value indicating higher likelihood of being abnormal. For computing the log-likelihood in (15.7), the activity patterns in the region of a parent node are also taken into account based on the relative delay between the parent node and the child node.

A global behavioural anomaly is then detected at each time frame when the total of cumulative abnormality score across all the regions is greater than a threshold Th, that is

$$C_t = \sum_{i=1}^{n} c_{i,t} > \text{Th} \tag{15.10}$$

Overall, there are two thresholds to be set for global behavioural anomaly detection. Threshold Th_i is set automatically to the same value for all the nodes as

$$\text{Th}_i = \overline{LL} - \sigma_{LL}^2 \tag{15.11}$$

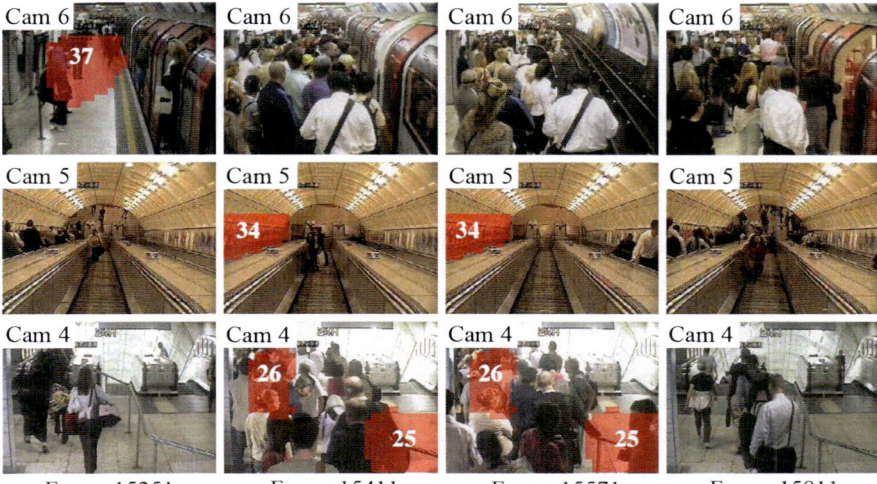

Fig. 15.6 An example of global behavioural anomaly detection in an underground station. This anomaly was caused by a faulty train, and first occurred in Cam-6 and Cam-7, and later propagated to Cam-5 and Cam-4 (see Fig. 13.3 for the scene layout). Regions that contributed to the anomaly are *highlighted* in *red*

where $\overline{LL}$ and σ^2_{LL} are the mean and variance of the log-likelihood computed over all the nodes for all the frames from a validation data set. The other threshold Th is set by the detection rate and false alarm rate requirements from specific applications.

Once a global behavioural anomaly is detected, the contributing local activities at individual regions can be located by examining $c_{i,t}$. In particular, $c_{i,t}$ for all regions are ranked in a descending order. The top ranked regions are considered to contribute towards a global behavioural anomaly where they count for a given fraction $P = [0, 1]$ of C_t.

An example of global behavioural anomaly detection is shown in Fig. 15.6. This anomaly occurred in the underground station scene shown in Fig. 13.3. The global anomaly signals an unexpected *combination* of local regional activity patterns:

> A faulty train was observed in Cam-6 and Cam-7 that led to overcrowding on a train platform. To prevent further congestion on the platform, passengers were disallowed to enter the platform by the escalator (Region-55 in Cam-5). This in turn caused congestion in front of the escalator entry zone in Cam-4.

It is evident that the model is successful in both detecting the anomaly across Cam-4, Cam-5 and Cam-6, and locating the contributing local regions of activity in each camera view.

15.2.5 Incremental Model Structure Learning

Up until now, we have only considered learning a global Bayesian behaviour graph model using an off-line batch mode. A batch mode learning process performs a single round model learning using a full training data set. In contrast, an incremental learning process updates a model at each time instance based on an on-line stream of continuous observations. Given a new observation $\mathbf{x}_t$ at time t, an incremental learning method for a Bayesian graph network produces a model B_t with a refined structure G_t and updated parameters Θ_t.

There are a number of reasons for model incremental learning. Visual interpretation of behaviour is inherently context dependent. Behavioural context is dynamic and may change over time. Changes in behavioural context, either gradual or abrupt, pose difficulties to global behaviour modelling. In particular, a gradual context change may involve gradual behaviour drift over time. For instance, different volumes of crowd flow occur at different time periods. The change may also be associated with changes in the definition and relevance of normality versus abnormality. For instance, when an activity pattern is observed but has never been seen before, it would be considered as being 'abnormal'. If similar patterns of activity were to appear repeatedly thereafter, it may now be considered as being normal. Intuitively, a gradual context change will have a gradual impact on both the model structure and parameters. On the other hand, an abrupt context change indicates a more drastic change in viewing conditions such as camera angle adjustments, removals or additions of cameras from or to a camera network. These changes can render both the current model structure and parameters obsolete. It is thus necessary to incrementally update a model based on new observations. In particular, gradual context change can be addressed by updating the parameters of a model incrementally. Abrupt change in visual context needs be handled by model structure adaptation.

Two different incremental learning methods can be considered. For incremental structure learning, one can employ a naïve method, in which all the observations seen so far, $\mathbf{x}_1, \ldots, \mathbf{x}_t$ are used to estimate the model graph structure G_t. This approach is expected to yield the most accurate model structure since all the observed information is used for the estimation. However, it also requires to store all previously seen instances or to keep a count of the number of times each distinct instantiation to all variables $\mathbf{X}$ is observed. Memory requirement grows linearly with the number of instances collected. It is infeasible to store all the data for model re-build if a model is required to operate for a long period of time. The second option requires potentially an enormous memory space given a large number of model variables. For instance, given a 50 variable graph network with binary states in each variable, the number of distinct instantiations is 2^{50}. It is infeasible to store the counts for all possible instantiations during incremental learning.

Alternatively, one can approximate a maximum a posteriori probability (MAP) network (Friedman and Goldszmidt 1997; Lam and Bacchus 1994), a network candidate being considered most probable given the data seen so far. All the past observations can be summarised using the network, which is then exploited as a prior in the next learning iteration for posterior approximation. This method is memory efficient because it only needs to store new instances observed since the last maximum a posteriori probability update. However, it may lead to poor model learning since subsequent model structures can be biased towards the initial model (Friedman and Goldszmidt 1997).

To overcome the limitations of both the naïve method and the maximum a posteriori probability method, we consider an incremental model structure learning strategy that takes constant time regardless the number of instances observed so far, and is tractable in terms of memory usage without sacrificing the accuracy of the structure learned. This method is derived following the general principle introduced by Friedman and Goldszmidt (1997). Specifically, an obsolete model structure is replaced by searching from a set of most probable structure candidates at the current time, which are stored in a frontier $\mathcal{F}$. The associated sufficient statistics $\boldsymbol{\xi}$ of $\mathcal{F}$ are kept to allow a tractable update of network parameters by Bayesian learning.

More precisely, there are four steps in the incremental model structure learning strategy:

1. Step-1, finding a topological order $\prec$: Similar to the batch mode learning scheme, there are two stages in the incremental model structure learning process. It commences with the estimation of the ordering of variables $\prec$ in the first-stage learning (Algorithm 15.3, lines 5–13). In particular, up-to-date cumulative time-delayed mutual information functions $\mathcal{I}^{\mathrm{acc}}(\tau)$ for each pair of regional activity patterns are estimated by accumulating past time-delayed mutual information functions, as follows.

   a. Let $\mathcal{I}_{ij}^{\mathrm{acc}}(\tau)$ between the ith region and the jth region be computed as

$$\mathcal{I}_{ij}^{\mathrm{acc}}(\tau) = \beta \overline{\mathcal{I}}_{ij}^{\mathrm{acc}}(\tau) + (1-\beta)\mathcal{I}_{ij}(\tau) \qquad (15.12)$$

   where β denotes an update coefficient that controls the updating rate, $\overline{\mathcal{I}}_{ij}^{\mathrm{acc}}(\tau)$ represents the cumulative time-delayed mutual information function found in previous learning iteration, and $\mathcal{I}_{ij}(\tau)$ denotes a time-delayed mutual information function computed using $\mathbf{x}_{t-h+1:t}$.

   b. Given $\mathcal{I}_{ij}^{\mathrm{acc}}(\tau)$, $\mathbf{I}$, $\mathbf{D}$, and $\prec$ are updated using the same procedures as for the batch mode learning process (Sect. 15.2.2).

2. Step-2, building a frontier $\mathcal{F}$: After computing G^{p} and $\prec$, the learning process proceeds to the second stage. We first construct a frontier $\mathcal{F}$ based on $\prec$ and a model structure estimated from the previous iteration G_{t-1} (Algorithm 15.3, line 14). Formally, $\mathcal{F}$ is defined by a set of families composed of X_i and its parent set $\mathbf{Pa}(X_i)$:

$$\mathcal{F} = \left\{ \left(X_i, \mathbf{Pa}_j(X_i)\right) \mid 1 \le i \le n, 1 \le j \le \Omega \right\} \qquad (15.13)$$

   where Ω denotes the total number of different parent sets $\mathbf{Pa}_j(X_i)$ associated with X_i. A node may be associated with multiple different parent sets.[6] To prevent proliferation of parent set combinations and to constrain the search space to a set of most promising model structures, less probable structure candidates can be pruned from joining the final scoring process. In particular,

---

[6]This is because we construct $\mathcal{F}$ by including existing families in G_{t-1} as well as using different combinations of candidate parents of X_i consistent with $\prec$. With this strategy, $\mathcal{F}$ is enriched by a diverse set of model structure candidates that could be simpler, or more complex, than G_{t-1} through combining different families in $\mathcal{F}$.

Algorithm 15.3: Two-stage incremental Bayesian graph model structure learning

---

Input: Data stream $(\mathbf{x}_1, \ldots, \mathbf{x}_t, \ldots)$; an upper bound, φ, on the number the parents a node may have; number of past instances to keep, h; an initial model structure, G_0; a set of sufficient statistics, $\boldsymbol{\xi}_0 = \xi(G_0)$; update coefficient β.

Output: G_t and $\boldsymbol{\xi}_t$.

1 **for** t *from* $1, 2, \ldots$ **do**
2 $\quad$ $G_t = G_{t-1}, \boldsymbol{\xi}_t = \boldsymbol{\xi}_{t-1}$;
3 $\quad$ Receive $\mathbf{x}_t$. Compute C_t [(15.10)];

4 $\quad$ **if** $t \bmod h = 0$ *and* $|\{C_i | t - h + 1 \leq i \leq t, C_i < \mathrm{Th_{CAS}}\}| \geq \frac{h}{2}$ **then**
$\quad\quad$ // Stage Cne
5 $\quad\quad$ Compute $\mathcal{I}(\tau)$ using $\mathbf{x}_{t-h+1:t}$ [(13.18) and (13.16)];
6 $\quad\quad$ **if** $t = 1$ **then**
7 $\quad\quad\quad$ Set $\mathcal{I}^{\mathrm{acc}}(\tau) = \mathcal{I}(\tau)$;
8 $\quad\quad$ **else**
9 $\quad\quad\quad$ Update $\mathcal{I}^{\mathrm{acc}}(\tau)$ using $\mathcal{I}(\tau)$ with updating rate β [(15.12)];
10 $\quad\quad$ **end**
11 $\quad\quad$ $\overline{\mathcal{I}}^{\mathrm{acc}}(\tau) = \mathcal{I}^{\mathrm{acc}}(\tau)$;
12 $\quad\quad$ Compute $\mathbf{D}$ and $\mathbf{I}$ using $\mathcal{I}^{\mathrm{acc}}(\tau)$ [see Sect. 13.2.2];
13 $\quad\quad$ Find the ordering of variables $\prec$ [Algorithm 15.1];

$\quad\quad$ // Stage Two
14 $\quad\quad$ Create $\mathcal{F}$ based on $\prec$ and G_{t-1};
15 $\quad\quad$ Obtain $\boldsymbol{\xi}_t$ by updating each record in $\boldsymbol{\xi}_{t-1}$ using $\mathbf{x}_{t-h+1:t}$;
16 $\quad\quad$ Search for highest scored G_t from $\mathcal{F}$ [Algorithm 15.2];
17 $\quad$ **end**
18 **end**

---

a. Combinations of parent set for X_i are formed by selecting only a set of most probable parents $\mathbf{mpp}_i$ consistent with $\prec$, with $|\mathbf{mpp}_i| \leq \varphi < n$. Parameter φ denotes the maximum number of parents a node may have, $\mathbf{mpp}_i$ contains parents that return the highest time-delayed mutual information among other candidate parents. For instance, given $\varphi = 2$ and the $\mathbf{mpp}$ of X_1 are X_2 and X_3, possible parents sets of X_1 would be $\{(X_2), (X_3), (X_2, X_3), (\emptyset)\}$.

b. In general, the maximum number of parent combinations a node may have is

$$\Omega = 1 + \sum_{k=1}^{\varphi} \binom{\varphi}{k} \tag{15.14}$$

3. Step-3, updating sufficient statistics $\boldsymbol{\xi}$: Since $\mathcal{F}$ at time t may be different from that in $t - 1$, one needs to update the associated sufficient statistics of each family in $\mathcal{F}$ (Algorithm 15.3, line 15). In a network characterised by multinomial

distributions, the sufficient statistics that quantify its conditional probability distribution are given as

$$\xi(G) = \{ \mathbf{N}_{X_i | \mathbf{Pa}(X_i)} \mid 1 \leq i \leq n \} \tag{15.15}$$

where $\mathbf{N}_{X_i | \mathbf{Pa}(X_i)} = \{ N_{x_i | \mathbf{pa}(X_i)} \}$ are extracted from $\mathbf{x}_{t-h+1 \,:\, t}$, and a set of such sufficient statistics at time t is denoted as $\boldsymbol{\xi}_t$. Given $\mathcal{F}$, $\boldsymbol{\xi}_t$ is updated based on $\mathbf{x}_{t-h+1:t}$ as follows:

$$\boldsymbol{\xi}_t = \boldsymbol{\xi}_{t-1} \cup \left\{ \mathbf{N}_{X_i | \mathbf{Pa}(X_i)} \mid \left(X_i, \mathbf{Pa}(X_i) \right) \in \mathcal{F} \right\} \tag{15.16}$$

The updated sufficient statistics $\boldsymbol{\xi}_t$ will be used in the next step for network scoring. After the incremental two-stage model structure learning, $\boldsymbol{\xi}_t$ will also be used for parameter update by Bayesian learning.

4. Step-4, scoring a network: In this step, one wishes to search for the optimal structure G_t within $\mathcal{F}$ to replace G_{t-1} (Algorithm 15.3, line 16). This is achieved through comparing the scores returned by a set of candidate structures that can be evaluated using the records in $\boldsymbol{\xi}_t$. That is:

$$G_t = \underset{\{ G' \mid \xi(G') \subset \xi_t \}}{\arg \max} \; S^*(G' | \boldsymbol{\xi}_t) \tag{15.17}$$

where $S^*(\cdot)$ denotes a modified version of the original score $S(\cdot)$ defined in (15.5). The score needs be modified because one may start collecting new sufficient statistics or may remove redundant one at different times, due to addition or removal of families from $\mathcal{F}$ during incremental model structure learning. Different number of instances $\mathbf{N}_{X_i | \mathbf{Pa}(X_i)}$ recorded in a family's sufficient statistics would affect the final score value. For instance, a lower score may be assigned to a family that observes more instances. To avoid unfair comparison of different candidate structures, it is necessary to average the score yielded by each family with the total instances recorded in its sufficient statistics. In particular, one can follow the method proposed by Friedman and Goldszmidt (1997) to modify S_{BIC} as

$$S^*_{\mathrm{BIC}}\left(X_i | \mathbf{Pa}(X_i) \right) = \frac{S_{\mathrm{BIC}}(X_i | \mathbf{Pa}(X_i))}{\sum_{(x_i | \mathbf{pa}(X_i))} N_{x_i | \mathbf{pa}(X_i)}} \tag{15.18}$$

As this incremental learning strategy includes the previous network structure G_{t-1} in $\mathcal{F}$ and its sufficient statistics in every learning iteration, the process shall improve *monotonically* as it must return a structure G_t that scores at least as well as G_{t-1}, i.e. $S^*(G_t | \boldsymbol{\xi}) \geq S^*(G_{t-1} | \boldsymbol{\xi})$.

In practice, the model structure learning process is invoked after receiving h instances, $\mathbf{x}_{t-h+1 \,:\, t}$, to ensure sufficient information for learning the time-delayed mutual information functions. In addition, there must be at least half of the h instances scoring below a pre-defined filtering threshold $\mathrm{Th}_{\mathrm{CAS}}$ during behavioural anomaly detection.

Similar to Th_i (15.11), the threshold $\mathrm{Th}_{\mathrm{CAS}}$ is obtained from a validation set. Specifically, after Th_i is obtained, we compute C_t for every frame and set $\mathrm{Th}_{\mathrm{CAS}}$ as

$\frac{\sum_{t=1}^{l} C_t}{2l}$, where l is the total number of frames of the validation data set. The filtering step is formulated to prevent excessive number of outliers from being inadvertently incorporated into the model updating process.

15.3 Global Awareness

So far, we have considered two global behaviour models based on a dynamic Bayesian network (DBN) and a static Bayesian graph, respectively. Both models have strengths and weaknesses. A DBN model is able to capture temporal order information about behaviour patterns explicitly, therefore being sensitive in detecting behavioural anomalies caused by temporal order violation. However, it is also sensitive to noise and has a high computational cost. In contrast, a static Bayesian graph is computationally more efficient and suitable for on-line processing, including on-the-fly incremental model adaptation. However, it is intrinsically limited in modelling the dynamic characteristics of a global behaviour. In the following, we consider an alternative model which can be seen as a compromise of the two aforementioned models. It has the advantage of a static Bayesian graph on tractability and scalability. By embedding the temporal order information into a topic model, it also has the strength of a DBN on representing explicitly dynamic information.

15.3.1 Time-Ordered Latent Dirichlet Allocation

In a latent Dirichlet allocation (LDA) model (Blei et al. 2003), each video word is placed into a document. The temporal order information about occurrences of video words is therefore lost. To overcome this problem, Li et al. (2009) consider a time-ordered-LDA model for global behaviour correlation modelling by capturing explicitly the temporal order information in a topic model.

In a time-ordered-LDA model, a document $\mathbf{w}$ corresponds to a video clip of T continuous sliding-windows. Each video frame is composed of C camera views and K_R semantic regions. Assume that K_R semantic regions have been segmented into C camera views. These regions are now referred globally as $\{R_r\}$ where $1 \leq r \leq K_R$. Global behaviours are then represented by a codebook of K_R video words obtained by clustering the regional activity representation described in Sect. 13.1. Each of the video words represents the regional behaviour associated with the rth region.

To enable the model time-ordered, each video word is indexed by both the region label r and the sliding-window index t. This increases the size of codebook V to $T \times K_R$, and V can be re-written as $\{v_r^t\}$ where v_r^t is the word extracted from the rth region in the tth sliding-window, $1 \leq r \leq K_R$ and $1 \leq t \leq T$. With the introduction of sliding-window index t, a document is now composed of a set of successive sliding-windows: $\mathbf{w} = \{\mathbf{w}_1, \ldots, \mathbf{w}_t, \ldots, \mathbf{w}_T\}$, where $\mathbf{w}_t$ is the tth sliding-window in the video clip. Each sliding-window now contains different types of words as the sliding-window index t is different.

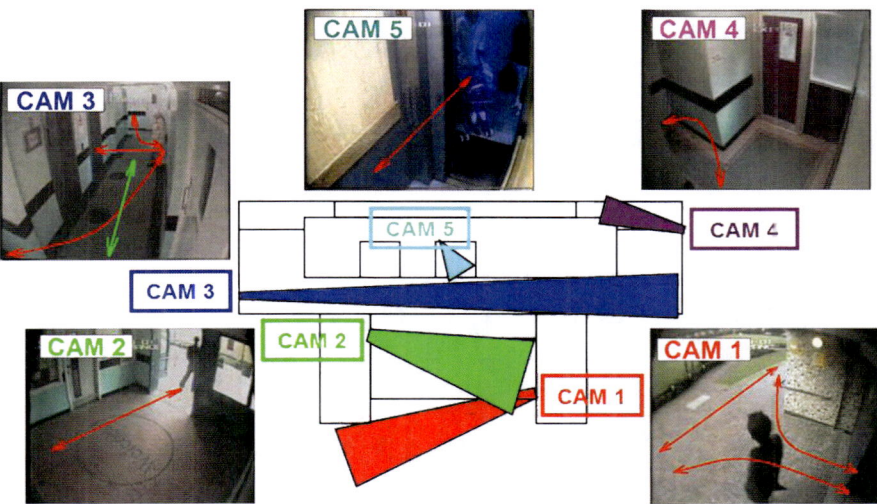

Fig. 15.7 The layout of and views from a multi-camera system in a residential housing environment

Compared to a standard LDA, a time-ordered-LDA differs in how a document is represented by video words. The parameter learning and inference methods are identical to that of LDA (Blei et al. 2003). Yet, this simple extension of LDA brings about a benefit of capturing explicitly the temporal characteristics of visual behaviours. Furthermore, on-line global behaviour prediction and anomaly detection is now made possible.

To illustrate the effectiveness of a time-ordered-LDA on global behaviour modelling, we consider a visual environment where a camera network comprising five cameras was installed to monitor both the inside and outside areas of a residential building. Figure 15.7 shows examples of the views with paths of typical and diverse activities of people in each view, and the topology of the camera network. In particular, Cam-1, Cam-4 and Cam-5 monitored the front entrance, the back exit and lift of the building, respectively. These cameras were connected through Cam-2 and Cam-3 monitoring the lobby and lift lobby, respectively. Typical global behaviours in this environment are people walking through Cam-1 and Cam-2 and either waiting for the lift in Cam-3 or using the staircase through Cam-4. Although the paths of regular behaviours in Fig. 15.7 seem to suggest that these behaviours are relatively simple and predictable, the actual behaviours can be rather complex and uncertain. For example, people can either wait for the lift in Cam-3 or use the stairs in Cam-4 depending on which floor the lift was and which floor they wanted to go. Both of these cannot be detected visually in the five camera views. When waiting in the lift lobby, some people preferred to walk around, shown by the green arrow in Cam-3, whilst others stood still, depending on personal preference. As shown in Fig. 15.7, most camera views are non-overlapping and with low resolution. The whole building is poorly lit with unstable lighting, especially in Cam-1 and Cam-2.

Fig. 15.8 Five camera views in a residential housing environment are decomposed into local activity regions

(a) An example topic learned by a time-ordered-LDA

(b) An example topic learned by a standard LDA

Fig. 15.9 Example topics learned by different topic models

The viewing conditions make this visual environment challenging for modelling global behaviours. The five camera views were decomposed into local regions using the method described in Sect. 13.1. These local regions are illustrated in Fig. 15.8.

The topics learned using a tim-ordered-LDA model correspond to typical global behaviours with local activity patterns in different camera views occurring in certain temporal orders. An example of the learned topics is shown in Fig. 15.9(a), with each topic highlighted by the top 2 local behaviours as words. Note that each word is associated with a sliding-window index t. It can be seen clearly in Fig. 15.9(a) that this topic corresponds to the global behaviour of people reaching the ground floor by the staircase (Cam-4, Regions-29 and Region-30), walking pass the lift lobby (Cam-3, Region-21 and Region-23) and the front lobby (Cam-3, Region-14 and Region-16), and appearing outside the building (Cam-1, Region-5 and Region-10). For comparison, a topic learned using a standard LDA is depicted in Fig. 15.9(b), which corresponds to the same global behaviour as in Fig. 15.9(a). However, this topic only indicates that those local regional activities are expected to take place in the same video clip, as a document. It reveals no information about by what temporal order they take place.

15.3.2 On-line Prediction and Anomaly Detection

Prediction

Consider that a document $\mathbf{w}$ is a video clip of T successive sliding-windows: $\mathbf{w} = \{\mathbf{w}_1, \ldots, \mathbf{w}_t, \ldots, \mathbf{w}_T\}$. A cumulative temporal document at index t, denoted as $\mathbf{w}_{1:t} = \{\mathbf{w}_1, \ldots, \mathbf{w}_t\}$, is computed by all the sliding-windows up to t. Clearly, two successive cumulative temporal documents $\mathbf{w}_{1:t}$ and $\mathbf{w}_{1:t+1}$ have the following relationship: $\mathbf{w}_{1:t+1} = \{\mathbf{w}_{1:t}, \mathbf{w}_{t+1}\}$, and $\mathbf{w}_{1:T} = \mathbf{w}$.

Given a cumulative temporal document $\mathbf{w}_{1:t}$, a model can make prediction for the next sliding-window $\mathbf{w}_{t+1}$ by evaluating how likely a regional activity pattern will be observed in each of the K_R regions cross all camera views. This is expressed as $P(\widehat{w}_{t+1}^r | \mathbf{w}_{1:t})$, where $\widehat{w}_{t+1}^r$ is the video word corresponding to the occurrence of a local activity in the rth region in frame $t + 1$. Its value is computed as

$$P\left(\widehat{w}_{t+1}^r | \mathbf{w}_{1:t}\right) = \frac{P(\widehat{w}_{t+1}^r \mathbf{w}_{1:t})}{P(\mathbf{w}_{1:t})} \tag{15.19}$$

where $P(\widehat{w}_{t+1}^r, \mathbf{w}_{1:t})$ is the joint probability of $\widehat{w}_{t+1}^r$ and $\mathbf{w}_{1:t}$. The profile of K topics inferred from $\mathbf{w}_{1:t}$, denoted as a K-component vector $\gamma_{1:t}$, is used to compute $P(\widehat{w}_{t+1}^r, \mathbf{w}_{1:t})$. More precisely, (15.19) is re-written as

$$P\left(\widehat{w}_{t+1}^r | \mathbf{w}_{1:t}\right) = \frac{P(\widehat{w}_{t+1}^r, \mathbf{w}_{1:t} | \alpha, \beta, \gamma_{1:t})}{P(\mathbf{w}_{1:t} | \alpha, \beta, \gamma_{1:t})} \tag{15.20}$$

This results in an approximation of the log-likelihood of $P(\widehat{w}_{t+1}^r | \mathbf{w}_{1:t})$ as

$$\log P\left(\widehat{w}_{t+1}^r | \mathbf{w}_{1:t}\right) \approx L\left(\gamma_{1:t}, \phi\left(\widehat{w}_{t+1}^r, \mathbf{w}_{1:t}\right); \alpha, \beta\right) - L\left(\gamma_{1:t}, \phi(\mathbf{w}_{1:t}); \alpha, \beta\right) \tag{15.21}$$

where $L(*)$ represents the lower bound of $\log P(*)$. To compute $L(\gamma_{1t}, \phi(\mathbf{w}_{1:t}); \alpha, \beta)$, a standard procedure of variational inference can be adopted (Blei et al. 2003), in which $\gamma_{1:t}$ and $\phi(\mathbf{w}_{1:t})$ are inferred through iterative updates. For computing $L(\gamma_{1:t}, \phi(\widehat{w}_{t+1}^r, \mathbf{w}_{1:t}); \alpha, \beta)$, $\gamma_{1:t}$ is set to a constant and only ϕ is updated using $\widehat{w}_{t+1}^r$ and $\mathbf{w}_{1:t}$. Following the same procedure, the likelihoods of the occurrences of local activities in all regions in the next sliding-window can be computed.

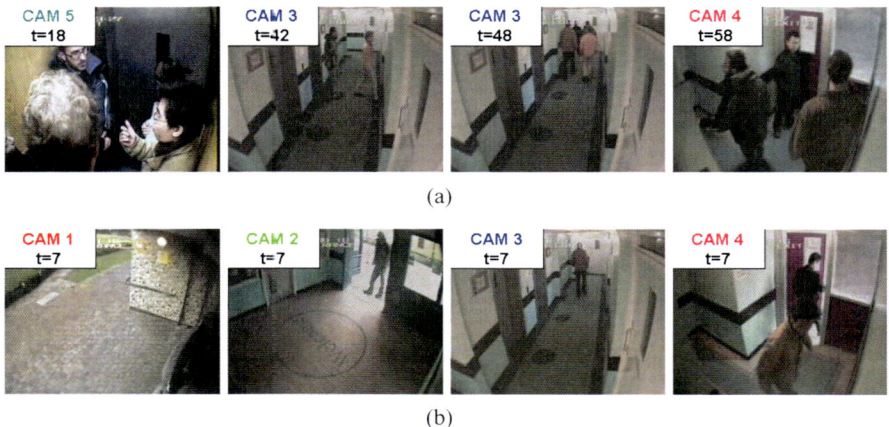

(a)

(b)

Fig. 15.10 Examples of abnormal global behaviours detected in a residential housing environment

Anomaly Detection

Using the on-line prediction procedure described above, global behavioural anomalies are detected on-line as follows:

1. At sliding-window t, using the time-ordered-LDA model parameters α, β to infer the topic profile $\gamma_{1:t}$.
2. Compute the likelihoods of local behaviour occurrences for all K_R regions in the next sliding-window $t + 1$ using $\log P(\widehat{w}_{t+1}^r | \mathbf{w}_{1:t})$ (15.21).
3. Given observations at time $t + 1$, $\mathbf{w}_{t+1}$, compute an anomaly score $A_s = \sum_r \log P(\widehat{w}_{t+1}^r | \mathbf{w}_{1:t})$ for the regions where local activities have been observed. This sliding-window as a document is deemed as being abnormal if $A_s < \mathrm{Th}_f$, where Th_f is a global behaviour anomaly threshold.
4. If a document at $t + 1$ is deemed as being abnormal, locate the contributing local activity regions by examining all new observations in $\mathbf{w}_{t+1}$ with corresponding $\log P(\widehat{w}_{t+1}^r | \mathbf{w}_{1:t})$. If $\log P(\widehat{w}_{t+1}^r | \mathbf{w}_{1:t}) < \mathrm{Th}_b$, the corresponding local regional activity is identified as being one of the causes of the global anomaly; Th_b is a local anomaly identification threshold.

Figure 15.10 shows some examples of global behavioural anomalies detected in the residential housing environment. In Fig. 15.10(a), a group of people moved out of a lift, but went to the back exit in Cam-4 instead of using the front entrance in Cam-1 as expected. This results in abnormal co-occurrences of local activities over the whole video clip. Figure 15.10(b) shows an anomaly where people loitered, and caused unexpected temporal order violations between local activities across different camera views.

Li et al. (2009) show that a time-ordered-LDA model is able to give better anomaly detection compared to a standard LDA model, especially when a global behavioural anomaly occurs very quickly, for example, the anomaly detected in

Fig. 15.10(b). For detecting these anomalies, the local activity patterns only exhibit abnormal temporal orders during a short duration, which has a negligible effect on the co-occurrence relationship between local activities captured by a standard LDA model. Without modelling the temporal order relationships explicitly between local activity patterns, a standard LDA model is unable to detect these subtle global anomalies.

15.4 Discussion

We considered three different models for distributed global behaviour modelling. All three models aim to discover and quantify the correlations between local activities with unknown time delays within and across non-overlapping multi-camera views. The first model is based on a dynamic Bayesian network (DBN), which models explicitly the dynamics of behaviour so it can be sensitive to temporal anomalies. However, a DBN is also sensitive to noise and has poor tractability and scalability. It is more suitable for temporal segmentation of behaviour rather than anomaly detection. To address the limitation on its tractability and scalability, instead of building a single model for an entire camera network, separate models are considered for subsets of correlated behaviours discovered automatically.

The second model is a time-delayed probabilistic graphical model. This model is essentially a static Bayesian graph in which different graph nodes represent activities in different decomposed regions from different views. Directed graph links between nodes in the model encode regional time-delayed dependencies. As a static Bayesian graph, it has better tractability and scalability, particularly given an efficient learning algorithm. However, its improved tractability and scalability is obtained at the cost of losing sensitivity to temporal anomalies. To overcome this problem, a third model is considered based on modifications to a probabilistic topic model (PTM), so that it can capture temporal order information ignored by a standard PTM. The model aims to strike a balance between sensitivity in anomaly detection and robustness against noise. Once learned, this time-ordered-LDA model can be used for detecting and localising anomalies either within a camera view or across different camera views.

The models and learning strategies considered for multi-camera behaviour analysis during the course of Part IV share some common ground with those discussed in Part III for group behaviour modelling in a single view. More specifically, in order to cope with the complexity of multi-object behaviours in multi-camera views and the uncertainty due to incomplete visual information, these methods exploit the ideas of (1) decomposing a complex behaviour into behaviour constituents and focusing on the modelling of the correlations among them; (2) discovering and learning behavioural context for context-aware behaviour anomaly detection; and (3) taking a flexible learning strategy according to the availability of data annotation and human feedback. The key difference is that with the introduction of uncertain temporal gaps between distributed camera views in a camera network, special care needs be taken to cope with the temporal delays between local behaviours observed at individual camera views in order to construct an effective global model for discovering emerging holistic behavioural patterns of interest.

References

Auliac, C., Frouin, V., Gidrol, X., d'Alche-Buc, F.: Evolutionary approaches for the reverse-engineering of gene regulatory networks: a study on a biologically realistic dataset. BMC Bioinform. **9**(91), 1–14 (2008)

Baum, L.E., Petrie, T., Soules, G., Weiss, N.: A maximization technique occurring in the statistical analysis of probabilistic functions of Markov chains. Ann. Math. Stat. **41**(1), 164–171 (1970)

Bishop, C.M.: Pattern Recognition and Machine Learning. Springer, Berlin (2006)

Blei, D.M., Ng, A.Y., Jordan, M.I.: Latent Dirichlet allocation. J. Mach. Learn. Res. **3**, 993–1022 (2003)

Buntine, W.: Theory refinement on Bayesian networks. In: Uncertainty in Artificial Intelligence, Los Angeles, USA, pp. 52–60 (1991)

Chen, X., Anantha, G., Wang, X.: An effective structure learning method for constructing gene networks. Bioinformatics **22**(11), 1367–1374 (2006)

Chen, X., Anantha, G., Lin, X.: Improving Bayesian network structure learning with mutual information-based node ordering in the K2 algorithm. IEEE Trans. Knowl. Data Eng. **20**(5), 628–640 (2008)

Chow, C., Liu, C.: Approximating discrete probability distributions with dependence trees. IEEE Trans. Inf. Theory **14**(3), 462–467 (1968)

Cooper, G.F., Herskovits, E.: A Bayesian method for the induction of probabilistic networks from data. Mach. Learn. **9**(4), 309–347 (1992)

Cormen, T.H., Leiserson, C.E., Rivest, R.L., Stein, C.: Introduction to Algorithms. MIT Press, Cambridge (2009)

de Campos, C.P., Zeng, Z., Ji, Q.: Structure learning of Bayesian networks using constraints. In: International Conference on Machine Learning, pp. 113–120 (2009)

Fraser, A.M., Swinney, H.L.: Independent coordinates for strange attractors from mutual information. Phys. Rev. **33**(2), 1134–1140 (1986)

Friedman, N., Goldszmidt, M.: Sequential update of Bayesian network structure. In: Uncertainty in Artificial Intelligence, pp. 165–174 (1997)

Friedman, N., Koller, D.: Being Bayesian about network structure. In: Uncertainty in Artificial Intelligence, pp. 201–210 (2000)

Friedman, N., Nachman, I., Peér, D.: Learning Bayesian network structure from massive datasets: the "Sparse Candidate" algorithm. In: Uncertainty in Artificial Intelligence, pp. 206–215 (1999)

Gilks, W.R., Richardson, S., Spiegelhalter, D. (eds.): Markov Chain Monte Carlo in Practice. Chapman & Hall, London (1995)

Heckerman, D., Geiger, D., Chickering, D.M.: Learning Bayesian networks: the combination of knowledge and statistical data. Mach. Learn. **20**(3), 197–243 (1995)

Lam, W., Bacchus, F.: Using new data to refine a Bayesian network. In: Uncertainty in Artificial Intelligence, pp. 383–390 (1994)

Li, J., Gong, S., Xiang, T.: Discovering multi-camera behaviour correlations for on-the-fly global activity prediction and anomaly detection. In: IEEE International Workshop on Visual Surveillance, Kyoto, Japan, October 2009

Loy, C.C., Xiang, T., Gong, S.: Multi-camera activity correlation analysis. In: IEEE Conference on Computer Vision and Pattern Recognition, Miami, USA, June 2009, pp. 1988–1995 (2009)

Loy, C.C., Xiang, T., Gong, S.: Time-delayed correlation analysis for multi-camera activity understanding. Int. J. Comput. Vis. **90**(1), 106–129 (2010)

Murphy, K.P.: Active learning of causal Bayes net structure. Technical report, University of California, Berkeley (2001)

Murphy, K.P.: Dynamic Bayesian networks: representation, inference and learning. PhD thesis, University of California at Berkeley (2002)

Prim, R.C.: Shortest connection networks and some generalizations. Bell Syst. Tech. J. **36**, 1389–1401 (1957)

Schwarz, G.: Estimating the dimension of a model. Ann. Math. Stat. **6**(2), 461–464 (1978)

Zhou, H., Kimber, D.: Unusual event detection via multi-camera video mining. In: IEEE International Conference on Pattern Recognition, pp. 1161–1166 (2006)

Epilogue

We set out in this book to address the problem of visual analysis of behaviours of objects, in particular of people. Understanding and interpreting behaviour is central to social interaction and communication between humans. To study computational models of behaviour by constructing automatic recognition systems and devices may help us with better understanding of how the human visual system bridges sensory mechanisms and semantic understanding. Automated visual analysis of behaviour also provides the key building blocks for an artificial intelligent vision system. Computational modelling and analysis of object behaviour through visual observation offers the potential for a wide range of applications including visual surveillance for crime prevention and detection, asset and facility management, video indexing and search, robotics and personalised healthcare, human computer interaction, animation and computer games.

Visual analysis of behaviour is a challenging problem. It requires to solve a host of difficult problems in computer vision including object detection, segmentation, tracking, motion trajectory analysis, pattern classification, and time series data modelling and prediction. An automated system for behaviour understanding and interpretation must address two challenging problems of computational complexity and uncertainty. Object behaviours in general exhibit complex spatio-temporal dynamics in a highly dynamical and uncertain environment. Visual analysis of behaviour is also subject to noise, incompleteness and uncertainty in sensory data. To further compound the problem, semantic meaning of a behaviour is highly context dependent, and behavioural context may not be measurable given visual information alone. To address these difficulties, statistical machine learning and the exploitation of human knowledge have become critical.

In this book, we study plausible computational models and tractable algorithms that are capable of automatic visual analysis of behaviour in complex and uncertain visual environments, ranging from well-controlled private spaces to highly crowded public scenes. In particular, we consider algorithms and methodologies capable of representing, learning, recognising, interpreting and predicting behaviours. The study of automatic visual analysis of behaviour is concerned with different types of behaviour ranging from facial expression and hand gesture to group behaviour observed from a network of distributed cameras. The book aims to reflect the current

S. Gong, T. Xiang, *Visual Analysis of Behaviour*,
DOI 10.1007/978-0-85729-670-2, © Springer-Verlag London Limited 2011

trends, progress and challenges on visual analysis of behaviour and highlight some of the open questions.

Despite the best efforts of computer vision researchers in the past two decades, deploying automated visual analysis of behaviour to a realistic environment is still in its infancy. We hope that in the not too distance future, those techniques considered in this book will be matured enough to give birth to a host of practical computer vision applications that will have a profound impact on human life.

Index

Printed by Books on Demand, Germany